Quantum Information in Gravitational Fields

Quantum Information in Gravitational Fields

Marco Lanzagorta

US Naval Research Laboratory

Morgan & Claypool Publishers

Rights & Permissions
To obtain permission to re-use copyrighted material from Morgan & Claypool Publishers, please contact info@morganclaypool.com.

ISBN   978-1-627-05330-3 (ebook)
ISBN   978-1-627-05329-7 (print)

DOI   10.1088/978-1-627-05330-3

Version: 20140401

IOP Concise Physics
ISSN 2053-2571 (online)
ISSN 2054-7307 (print)

A Morgan & Claypool publication as part of IOP Concise Physics

Morgan & Claypool Publishers, 40 Oak Drive, San Rafael, CA, 94903, USA

This book is dedicated to the memory of my uncle Jorge and my furry kids: Chanel, Oliver, Oscar, and the cat in the clear plastic house.

Contents

Preface

One of the major scientific thrusts in recent years has been to try to harness quantum phenomena to increase dramatically the performance of a wide variety of classical information processing devices. In particular, it is generally accepted that quantum computers and communication systems promise to revolutionize our information infrastructure.

With the prospect of satellite-based quantum communications, it is necessary to understand the basic dynamics of qubits orbiting Earth. It is equally important to know the way gravitation affects the performance of quantum computers. Finally, it is worth considering if these gravitational effects can somehow be harnessed in such a way as to develop new quantum sensing devices to measure minute variations in the local gravitational field.

This book offers a concise discussion of quantum information in the presence of classical gravitational fields described by Einstein's general theory of relativity. Besides a basic description of the fundamental physical principles of quantum fields in curved spacetimes, we also offer some new results on steganographic quantum communications in inertial frames, qubits in Schwarzschild spacetime, the spin–curvature coupling in Schwarzschild spacetime, qubits in Kerr spacetimes and the performance of quantum technologies operating in gravitational fields. A significant part of our discussion is restricted to qubits represented by spin-$\frac{1}{2}$ particles (e.g. electrons). In a subsequent book we intend to cover the case of spin-1 particles (e.g. photons).

Our approach makes strong emphasis on the smooth interplay between the physics and mathematics of general relativity and relativistic quantum mechanics. To this end, the book takes a unifying group-theoretic approach to these topics. Thus, our description of relativistic spin-$\frac{1}{2}$ particles is based on the representations of the Poincare group, while gravitational effects are introduced through Lorentz transformations in local inertial frames properly defined by tetrad fields. In particular, we analyse how spin-$\frac{1}{2}$ qubits are affected by a stationary spherically symmetric gravitational field (Schwarzschild metric) and by a stationary axisymmetric gravitational field (Kerr metric). In addition, we discuss an example that shows the coupling between a qubit's spin and spacetime curvature. Finally, we discuss gravitational effects in the context of quantum communications, entanglement, Einstein–Podolsky–Rosen (EPR) experiments, quantum computation and sensing.

The structure of this book is as follows. In chapter 2 we give a brief overview of Einstein's general theory of relativity. We discuss a few advanced topics, such as tetrad fields and the non-geodetic motion due to spin–curvature coupling, which are not commonly covered in introductory textbooks. Chapter 3 offers a review of relativistic quantum field theory. We place a strong emphasis on the group theoretic structure of quantum field theory. This theory will be used in chapter 4 to discuss quantum information within the context of special relativity. As characteristic examples, we present the analysis of a simple steganographic communication protocol and quantum teleportation as observed in inertial frames of reference.

The structure of quantum field theory in the presence of gravitational fields is discussed in chapter 5. We place special emphasis on the description of Dirac spinors in curved spacetimes. As specific applications of the general equations obtained in chapter 5, we analyse the case of spinors in Schwarzschild and Kerr spacetimes in chapters 6 and 8, respectively. Chapter 7 presents an example of the coupling between spin and curvature for a Dirac spinor in Schwarzschild spacetime. And in chapter 9, the results from the previous three chapters are analysed within the context of quantum information systems. In particular, we discuss the case of quantum communication, computation and sensing devices in the presence of a gravitational field. Finally, we offer our conclusions in chapter 10, where we further discuss the intricate relationship between quantum information and gravity.

This book assumes the reader is familiar with the basic principles of non-relativistic quantum mechanics, special relativity and quantum information theory. Our discussion of general relativity and relativistic quantum fields is brief, but all the relevant equations are presented in the text. In addition, the reader is not required to have any specialized knowledge of quantum field theory in curved spacetimes.

Acknowledgements

I am grateful to Keye Martin for his advice and encouragement. I am also thankful to Johnny Feng, Tanner Crowder, Joe Czika, Jeffrey Uhlmann and Marcelo Salgado for reviewing this manuscript and for providing valuable corrections, suggestions and insights. And, of course, I deeply appreciate the encouragement and perseverance provided by my editors Mike Morgan and Joel Claypool. Finally, I kindly acknowledge the support received by the US Naval Research Laboratory.

Biography

Dr Marco Lanzagorta

Dr Marco Lanzagorta is a Research Physicist at the US Naval Research Laboratory in Washington, DC. In addition, Dr Lanzagorta is Affiliate Associate Professor and Member of the Graduate Faculty at George Mason University, and co-editor of the Quantum Computing series of graduate lectures published by Morgan and Claypool. He is a recognized authority on the research and development of advanced information technologies and their application to combat and scientific systems. He has over 100 publications in the areas of physics and computer science, and he has authored the books *Quantum Radar* (2011) and *Underwater Communications* (2012). He received a doctorate degree in theoretical physics from Oxford University in the United Kingdom. In the past, Dr Lanzagorta was Technical Fellow and Director of the Quantum Technologies Group of ITT Exelis, and worked at the European Organization for Nuclear Research (CERN) in Switzerland, and at the International Centre for Theoretical Physics (ICTP) in Italy.

Chapter 1

Introduction

One of the major scientific thrusts in recent years has been to try to harness quantum phenomena to dramatically increase the performance of a wide variety of classical devices [1–5]. These advances in quantum information science have had a considerable impact on the development of quantum information processing devices such as communication, computation and sensor systems.

1.1 Quantum information

The *bit* is the unit of classical information. In particular, a classical bit is a scalar variable which has a single value of either 0 or 1. Clearly, the bit has a value that is unique, deterministic and unambiguous.

On the other hand, the unit of quantum information is the *qubit*, which has different properties than its classical counterpart [3–5]. A qubit is more general in the sense that it represents a state defined by a pair of complex numbers, $\{\alpha, \beta\}$, which together express the probability that a reading of the value of the qubit will give a value of 0 or 1. It is well known that quantum information obeys very unique properties:

- **Superposition:** A qubit can be in the state of 0, 1, or some mixture – referred to as a superposition – of the 0 and 1 states.
- **Destructive measurements:** Upon reading the value of a qubit we obtain a bit of classical information with certain probability, and any other bit states that were in the superposition along the measured state are lost.
- **No-cloning:** In general, quantum information cannot be copied.
- **Entanglement:** The measurement of a qubit may affect the state of other qubits. Alternatively, knowledge of the individual states does not yield a complete picture of the whole system.

Furthermore, substantial research efforts have shown that these properties can be exploited to enable the design of novel communication, computation and sensing devices.

1.2 Quantum communications

It is generally accepted that quantum communication systems promise to revolutionize our information infrastructure. More specifically, research over the past decade has shown that quantum information seems to offer the possibility of perfectly secure communications, something that is impractical with classical information.

In the context of modern cryptography, *perfect security* means that the *ciphertext* (the encrypted message) reveals no information about the *plaintext* (the original unencrypted message) [6, 7]. In other words, no matter how much ciphertext the adversary is able to intercept, there is not enough information to obtain the plaintext.

It is important to remark that perfect security does not depend on what is computationally feasible, as it places no bounds on the computational resources available to the adversary. That is, a cryptosystem is perfectly secure if it cannot be broken, even with *infinite* computational power. This is the reason why perfect security is often referred to as *unconditional security*.

A less restrictive class of ciphers are those that are *computationally secure* [6, 7]. These are ciphers that are difficult to break, but not impossible. In this case the adversary requires a certain amount of computational resources to be available to break the code in a timely manner.

In general, computationally secure ciphers are implemented using a mathematical problem that is known to be difficult to solve. In such a case, the security proof of the cipher is mapped to the solution of a hard mathematical problem and the cipher is said to be *provably secure*. Note that the security proof is relative to some other mathematical problem and it does not provide an absolute proof of the computational difficulty of the problem [6, 7]. For instance, the efficient prime factorization of a large coprime integer (an integer that can be expressed as the product of two prime numbers) is believed to be a difficult mathematical problem. Breaking any cipher that involves prime factorization is equivalent to solving the hard mathematical problem of finding an efficient factorization algorithm.

Ciphers can also be classified according to their cryptographic keys in symmetric and asymmetric ciphers [6, 7]. *Symmetric ciphers* use the same (secret) key for encryption and decryption. It can be shown that these ciphers can be designed to be perfectly secure. The best example of a perfectly secure symmetric cipher is the *one-time pad* (also known as the Vernam pad in honor of its inventor, Gilbert Vernam) [8]. However, the big challenge is to distribute the secret key among the authorized users.

On the other hand, *asymmetric ciphers* use a public key for encryption and a private key for decryption. Because the encryption key is made public, it is very easy to distribute the key among the users. However, these ciphers cannot be made perfectly secure, but they are computationally secure.

As an example of a computationally secure asymmetric cipher, let us consider the case of RSA (Rivest–Shamir–Adleman), which probably is the most widely used cipher. Indeed, most online shopping and banking transactions are protected by RSA and its variants. Without going into details, the security of RSA is based on the computationally difficult problem of finding the prime factorization of a large coprime integer [6, 7].

That being said, let us note that prime factorization is *believed* to be a very difficult computational problem. That is, the hardness of the factorization problem is only an unproved mathematical assumption, as there is no formal proof about the hardness of this problem. Therefore, the security of RSA is based on an unproved mathematical assumption. As a consequence, at any time in the future someone could come up with an efficient classical algorithm able to efficiently factorize large integers.

On the other hand, the security of quantum key distribution (QKD) protocols *seems* to be based on the laws of physics, rather than on unproved mathematical assumptions [9–13]. Indeed, the destructive nature of the measurement procedure means that eavesdroppers can be detected. Also, the no-cloning property means that quantum information cannot be forged. It is crucial to remark that, although this is true in theory, physical implementations of QKD introduce imperfections into the protocol. The extent of these imperfections and their impact on the security of QKD is a topic that continues to generate controversy.

Recent years have seen much theoretical and experimental research towards the development of practical QKD systems in optical fibres [14], free-space between two ground stations [15], free-space between a satellite and a ground station [16–19] and for communications with underwater vehicles [20]. Furthermore, today QKD is a commercially available technology that was successfully deployed during the 2008 Swiss Federal elections and the 2010 FIFA World Cup competition [21].

1.3 Quantum computing

Arguably, the principal motivation of most quantum information science efforts has been the creation of a quantum computer able to run *Shor's algorithm* for cryptoanalysis applications (i.e. secret code breaking) [3, 4]. Perhaps the most important result so far in the area of quantum computing, Shor's algorithm provides a protocol to break RSA which is exponentially faster than the best-known classical method [22]. Here, the problem of prime factorization, which is the backbone of RSA, is reduced to a periodicity problem using some basic results from number theory (Euclid's theorem, Euler's theorem and Euclid's algorithm). For a large coprime integer N, this process takes $\mathcal{O}(\log(N))$ computational steps. Then, the algorithm uses a quantum Fourier transform in $\mathcal{O}(\log^3(N))$ steps to find the period of the function. Therefore, the overall complexity of Shor's algorithm is $\mathcal{O}(\log^3(N))$, which is exponentially faster than the best known classical factorization technique, the *general number field sieve* running with complexity $\mathcal{O}(poly(N))$.

Grover's algorithm is another well-known application of quantum computation [3, 4]. Let us consider the computational problem of finding a specific element in a completely unordered and unstructured database. The best classical solution is the exhaustive search of the N elements that make the database with complexity $\mathcal{O}(N)$. On the other hand, Grover's algorithm achieves a quadratic improvement and is able to give the right answer with high probability in $\mathcal{O}(\sqrt{N})$ computational steps [23].

Even though a full-blown quantum computer able to factorize a large number is still far into the future, many theoretical and experimental results have shown that it may be possible to harness quantum phenomena to solve a variety of smaller-scale computational problems (e.g. simulation of quantum systems) [24–27].

1.4 Quantum sensors

Advances in quantum information science have had a considerable impact on the development of quantum sensors: sensing devices that exploit quantum phenomena to increase their sensitivity [28–34]. Examples of quantum sensors recently proposed by the scientific community include radar [35, 36], lidar [37], raman spectrometers [38], magnetometers [39] and gravitometers [28, 40].

Furthermore, the interaction between quantum information science and quantum sensing has been substantial; for example, quantum sensors can be described mathematically as noisy quantum channels. In addition, quantum control techniques developed in the context of quantum computation are useful to harness quantum sensing hardware.

It is important to remark that although quantum sensing is not as mature as quantum computation, it offers simpler engineering challenges. For instance, a quantum computer requires a large number of qubits on arbitrary superpositions with coherence times long enough to perform complex computations using a large variety of gates. On the other hand, quantum sensors require a relatively small number of qubits on specific entangled states and only a handful of quantum operations.

As a consequence, it seems that the development of quantum sensors offers an enticing near-term option for the practical application of the quantum information technologies required for the construction of a hypothetical quantum computer.

1.5 Relativistic quantum information

A considerable amount of work has been done to understand how entanglement and quantum information are affected by Lorentz transformations in the context of special relativity [41–45]. Some of this work has been generalized to study quantum information in accelerated frames [46–49]. And some other efforts have studied the effects of gravitation and spacetime curvature on quantum information [50–57].

As we will discuss in the following chapters, quantum information is an intrinsically relativistic theory with a direct coupling to gravitational fields described by Einstein's general theory of relativity. In contrast to the much more abstract concept of a classical bit of information, *quantum information invariably gravitates*. Therefore, *gravitation* could be added to the list of unique features that characterize quantum information given in section 1.1. Indeed, although it is very difficult to think about the gravitational interaction of the bit '0', it is possible to give a formal mathematical description of an arbitrary qubit interacting with a gravitational field.

These observations have a clear impact on the philosophical underpinnings of information theory. In addition, the gravitation of quantum information implies that gravity has a non-trivial effect on the performance of quantum technologies. Indeed, classical gravitational fields affect the performance of quantum communication and computation systems. For instance, a 1 MHz quantum computer running Grover's algorithm on Earth's surface is limited to a computational space of $N \approx 10^{36}$ database elements before the quantum complexity advantage disappears owing to gravitational effects (assuming that all running time constants on the quantum computer are exactly equal to one).

On the other hand, quantum gravimeters can perform significantly better than their classical counterparts. In particular, it seems that it is possible to design gravimeters able to detect small variations in the local gravitational field (i.e. significantly better than $\delta g/g \approx 10^{-10}$).

1.6 Summary

Quantum information is a new paradigm that promises to revolutionize our information technology infrastructure. To achieve this goal, it is important to understand clearly the unique properties of qubits. In particular, gravitational interactions affect the theoretical complexity of communication and computation devices. On the other hand, the coupling between quantum information and gravity can be exploited to design novel gravimetry sensors.

Bibliography

[1] Bruß D and Leuchs G (eds) 2007 *Lectures on Quantum Information* (New York: Wiley)
[2] Kok P and Lovett B W 2010 *Introduction to Optical Quantum Information Processing* (Cambridge: Cambridge University Press)
[3] Lanzagorta M and Uhlmann J 2008 *Quantum Computer Science* (San Rafael, CA: Morgan & Claypool)
[4] Nielsen M A and Chuang I L 2000 *Quantum Computation and Quantum Information* (Cambridge: Cambridge University Press)
[5] Vedral V 2006 *Introduction to Quantum Information Science* (Cambridge: Cambridge University Press)
[6] Menezes A, van Oorschot P and Vanstone S (ed) 1996 *Handbook of Applied Cryptography* (Boca Raton, FL: CRC Press)
[7] Stinson D R 2002 *Cryptography: Theory and Practice* (London: Chapman & Hall/CRC)
[8] Shannon C 1949 Communication theory of secrecy systems *Bell Syst. Tech. J.* **28** 656715
[9] Bouwmeester D, Ekert A and Zelinger A (ed) 2000 *The Physics of Quantum Information* (Berlin: Springer)
[10] Gisin N *et al* 2001 Quantum cryptography *Rev. Mod. Phys.* **74** 175
[11] Kollmitzer C and Pivk M (ed) 2010 *Applied Quantum Cryptography* (Berlin: Springer)
[12] Sergienko A V (ed) 2006 *Quantum Communications and Cryptography* (London: Taylor & Francis)
[13] Van Assche G 2006 *Quantum Cryptography and Secret Key Distillation* (Cambridge: Cambridge University Press)
[14] Hiskett P A *et al* 2006 Long-distance quantum key distribution in optical fibre *New J. Phys.* **8** 193
[15] Schmitt-Manderbach T *et al* 2007 Experimental demonstration of free-space decoy-state quantum key distribution over 144 km *Phys. Rev. Lett.* **98** 010504
[16] Hughes R J *et al* 2000 Quantum cryptography for secure satellite communications *Proceedings of the IEEE Aerospace Conference* 2000 (Piscataway, NJ: IEEE) vol 1 1803

[17] Rarity J G *et al* 2002 Ground to satellite secure key exchange using quantum cryptography *New J. Phys.* **4** 82

[18] Villoresi P *et al* 2004 Space-to-ground quantum-communication using an optical ground station: A feasibility study *Proc. SPIE* **5551** 113

[19] Yin J *et al* 2013 Experimental single-photon transmission from satellite to earth arXiv:1306.0672v1 [quant-ph]

[20] Lanzagorta M 2012 *Underwater Communications* (San Rafael, CA: Morgan & Claypool)

[21] IdQuantique 2013 http://www.idquantique.com, last retrieved on August 12.

[22] Shor P W 1997 Polynomial-time algorithms for prime factorization and discrete logarithms on a quantum computer *SIAM J. Comput.* **26** 1484–509

[23] Grover L K 1996 A fast quantum mechanical algorithm for database search *Proceedings, 28th Annual ACM Symposium on the Theory of Computing* (May 1996) 212

[24] Aspuru-Guzik A and Walther P 2012 Photonic quantum simulators *Nat. Phys.* **8** 285–91

[25] Jones N C *et al* 2012 Faster quantum chemistry simulation on fault-tolerant quantum computers *New J. Phys.* **14** 115023

[26] Lanyon B P *et al* 2010 Towards quantum chemistry on a quantum computer *Nature Chemistry* **2** 106–11

[27] Whitfield J D, Love P J and Aspuru-Guzik A 2013 Computational complexity in electronic structure *Phys. Chem. Chem. Phys.* **15** 397–411

[28] Didomenico L D, Lee H, Kok P and Dowling J P 2004 Quantum interferometric sensors *Proceedings of SPIE Quantum Sensing and Nanophotonic Devices*

[29] Dowling J P 2008 Quantum optical metrology – the lowdown on high-NOON states, *Contemp. Phys.* **49** 125–43

[30] Giovannetti V, Lloyd S and Maccone L 2001 Quantum enhanced positioning and clock synchronization *Nature* **412** 417

[31] Giovannetti V, Lloyd S and Maccone L 2004 Quantum-enhanced measurements: beating the standard quantum limit *Science* **306** 1330–6

[32] Giovannetti V, Lloyd S and Maccone L 2006 Quantum metrology *Phys. Rev. Lett.* **96** 010401

[33] Huver S D, Wildfeuer C F and Dowling J P 2008 Entangled Fock states for robust quantum optical metrology, imaging, and sensing *Phys. Rev.* A **78** 063828

[34] Lloyd S 2008 Enhanced sensitivity of photodetection via quantum illumination *Science* **321** 1463–5

[35] Lanzagorta M 2013 Amplification of radar and lidar signatures using quantum sensors *Proc. SPIE 8734, Active and Passive Signatures* IV, 87340C

[36] Lanzagorta M 2011 *Quantum Radar* (San Rafael, CA: Morgan & Claypool)

[37] Shapiro J H 2007 Quantum pulse compression laser radar *Proc. SPIE* **6603** 660306

[38] Lanzagorta M 2012 Entangled-photons Raman spectroscopy *Proc. SPIE* 8382, *Active and Passive Signatures III*, 838207

[39] Lanzagorta M 2011 *Nanomagnet-based magnetic anomaly detector* US Patent No. 8,054070 B1, Nov. 8

[40] von Borzeszkowski H and Mensky M B 2001 Gravitational effects on entangled states and interferometer with entangled atoms *Phys. Lett.* A **286** 102–6

[41] Alsing P M and Milburn G J 2002 On entanglement and Lorentz invariance *Quantum Inf. Comput.* **2** 487

[42] Bergou A J, Gingrich R M and Adami C 2003 Entangled light in moving frames *Phys. Rev.* A **68** 042102

[43] Gingrich R M and Adami C 2002 Quantum entanglement of moving bodies *Phys. Rev. Lett* **89** 270402

[44] Terno D R 2006 Quantum information in loop quantum gravity *J. Phys.: Conf. Ser.* **33** 469–74

[45] Soo C and Lin C C Y 2004 Wigner rotations, Bell States, and Lorentz invariance of entanglement and von Neumann entropy *Int. J. Quantum Info.* **2** 183–200

[46] Alsing P M, McMahon D and Milburn G J 2006 Teleportation in a non-inertial frame arXiv: quant-ph/0311096

[47] Bradler K, Hayden P and Panangaden P 2012 Quantum communication in Rindler spacetime *Comm. Math. Phys.* **312** 361–98

[48] Bruschi D E, Fuentes I and Louko J 2012 Voyage to alpha centauri: entanglement degradation of cavity modes due to motion *Phys. Rev.* D **85** 061701(R)

[49] McMahon D, Alsing P M and Embid P 2006 The Dirac Equation in Rindler Space: A Pedagogical Introduction arXiv:gr-qc/0601010

[50] Alsing P M, Fuentes-Schuller I, Mann R B and Tessier T E 2006 Entanglement of Dirac fields in non-inertial frames *Phys. Rev.* A **74** 032326

[51] Alsing P M, Stephenson G J and Kilian P 2009 Spin-induced non-geodesic motion, gyroscopic precession, Wigner rotation and EPR correlations of massive spin-$\frac{1}{2}$ particles in a gravitational field arXiv:0902.1396v1 [quant-ph]

[52] Alsing P M and Stephenson G J 2009 The Wigner rotation for photons in an arbitrary gravitational field arXiv:0902.1399v1 [quant-ph]

[53] Ball J L, Fuentes-Schuller I and Schuller F P 2006 Entanglement in an expanding spacetime *Phys. Lett.* A **359** 550–4

[54] Fuentes-Schuller I and Mann R B 2005 Alice falls into a black hole: entanglement in non-inertial frames *Phys. Rev. Lett.* **95** 120404

[55] Lanzagorta M and Uhlmann J 2013 Quantum computation in gravitational fields forthcoming, available upon request

[56] Lanzagorta M 2012 Effect of gravitational frame dragging on orbiting qubits arXiv:1212.2200 [quant-ph]

[57] Terashima H and Ueda M 2004 Einstein–Rosen correlation in gravitational field *Phys. Rev.* A **69** 032113

Chapter 2

Special and general relativity

The best-known description of classical gravitational fields is given by *Einstein's general theory of relativity*. A detailed discussion of this theory is outside the scope of this book, and the reader is encouraged to peruse the rich literature available on the topic [1–7].

In this chapter we will limit our discussion to a brief overview of the basic concepts, equations and notation that emerge from the special and general theories of relativity. Our emphasis is to present those equations that will be required to understand the behaviour of quantum particles in the presence of gravitational fields. In particular, we will give a description of the tetrad formalism, which is probably the simplest way to describe the interaction of spin-$\frac{1}{2}$ particles with gravitational fields. Furthermore, we will discuss the classical coupling between spin and curvature, a non-trivial topic which is usually ignored in introductory books to the general theory of relativity.

2.1 Special relativity

The special theory of relativity requires the line element:

$$dS^2 = -dt^2 + dx^2 + dy^2 + dz^2 \tag{2.1}$$

to be an invariant for all inertial observers in flat spacetime (i.e. in the absence of gravitational fields). Notice that we are using *natural units* ($c = 1$). We can explicitly express the invariance of dS^2 in tensor notation as:

$$dS^2 = \eta_{\mu\nu} dx^\mu dx^\nu \tag{2.2}$$

where:

$$x^\mu = (t, x, y, z) \tag{2.3}$$

is the *contravariant four-position* and $\eta_{\mu\nu}$ is the *Minkowski tensor* with $(-, +, +, +)$ signature.

doi:10.1088/978-1-627-05330-3ch2

In particular, with this choice of signature in the Minkowski tensor, we have:

$$dS^2 = (-1 + v^2)\, dt^2 \quad \Rightarrow \quad dS^2 \leqslant 0 \tag{2.4}$$

where:

$$v \equiv \sqrt{\left(\frac{dx}{dt}\right)^2 + \left(\frac{dy}{dt}\right)^2 + \left(\frac{dz}{dt}\right)^2} \tag{2.5}$$

and:

$$v \leqslant 1 \tag{2.6}$$

because the speed of light in vacuum ($c = 1$) is considered to be maximal.

Under a change of coordinates:

$$x^\mu \to \tilde{x}^\mu \quad \Rightarrow \quad d\tilde{x}^\mu = \frac{\partial \tilde{x}^\mu}{\partial x^\alpha} dx^\alpha \tag{2.7}$$

the invariance of the line element requires that:

$$d\tilde{S}^2 = dS^2 \quad \Rightarrow \quad \frac{\partial \tilde{x}^\mu}{\partial x^\alpha} \frac{\partial \tilde{x}^\nu}{\partial x^\beta} \eta_{\mu\nu} = \eta_{\alpha\beta} \tag{2.8}$$

which is called the *Lorentz condition*. After some tensor algebra it can be shown that the Lorentz condition implies that:

$$\frac{\partial^2 \tilde{x}^\mu}{\partial x^\alpha \partial x^\beta} = 0 \tag{2.9}$$

which means that the coordinate transformation has to be linear in its variables:

$$\tilde{x}^\alpha = \Lambda^\alpha{}_\beta x^\beta + a^\alpha \tag{2.10}$$

where:

$$\Lambda^\alpha{}_\beta = \frac{\partial \tilde{x}^\alpha}{\partial x^\beta} \tag{2.11}$$

is a constant tensor that does not depend on the coordinates [7]. In addition, the Lorentz condition also implies that:

$$\Lambda^\mu{}_\alpha \Lambda^\nu{}_\beta \eta_{\mu\nu} = \eta_{\alpha\beta} \tag{2.12}$$

Let us note that Λ is a rank-2 tensor with 16 components. However, equation (2.12) only has ten linearly independent equations. As we will see later, this implies that the coordinate transformation tensor Λ restricted by the Lorentz condition only involves six free parameters. Notice, however, that the Lorentz condition does not impose any conditions on the four components of a^α.

On the other hand, the *covariant four-position* is defined by:

$$x_\alpha = \eta_{\alpha\beta} x^\beta = (-t, x, y, z) \tag{2.13}$$

and transforms as:

$$x_\alpha \to \tilde{x}_\alpha = \Lambda_\alpha^{\ \beta} x_\beta + a_\alpha \tag{2.14}$$

where:

$$\Lambda_\alpha^{\ \beta} = \frac{\partial x^\beta}{\partial \tilde{x}^\alpha} \tag{2.15}$$

which is also independent of the coordinates. Furthermore, these expressions make clear that $\eta_{\mu\nu}$ and $\eta^{\mu\nu}$ are used to lower and rise indices, respectively.

Furthermore, the transformation tensors for covariant and contravariant four-vectors are related by:

$$\Lambda_\alpha^{\ \beta} = \eta_{\alpha\mu} \eta^{\beta\nu} \Lambda^\mu_{\ \nu} \tag{2.16}$$

and as a consequence:

$$\Lambda_\alpha^{\ \beta} \Lambda^\alpha_{\ \gamma} = \eta_{\alpha\mu} \eta^{\beta\nu} \Lambda^\mu_{\ \nu} \Lambda^\alpha_{\ \gamma} = \delta^\beta_\gamma \tag{2.17}$$

which is equivalent to saying that the inverse of the transformation tensor is given by:

$$(\Lambda_\alpha^{\ \beta})^{-1} = \Lambda^\beta_{\ \alpha} \tag{2.18}$$

Let us assume an infinitesimal transformation given by:

$$\Lambda^\alpha_{\ \beta} = \delta^\alpha_{\ \beta} + \xi^\alpha_{\ \beta} \qquad a^\alpha = \epsilon^\alpha \tag{2.19}$$

where:

$$|\xi^\alpha_{\ \beta}| \ll 1 \tag{2.20}$$

for all possible values of the indices α and β. Then, the Lorentz condition implies that:

$$\begin{aligned}
\eta_{\alpha\beta} &= \eta_{\mu\nu} \Lambda^\mu_{\ \alpha} \Lambda^\nu_{\ \beta} \\
&= \eta_{\mu\nu} (\delta^\mu_{\ \alpha} + \xi^\mu_{\ \alpha})(\delta^\nu_{\ \beta} + \xi^\nu_{\ \beta}) \\
&= \eta_{\mu\nu} \delta^\mu_{\ \alpha} \delta^\nu_{\ \beta} + \eta_{\mu\nu} \delta^\mu_{\ \alpha} \xi^\nu_{\ \beta} + \eta_{\mu\nu} \xi^\mu_{\ \alpha} \delta^\nu_{\ \beta} + \mathcal{O}(\xi^2) \\
&= \eta_{\alpha\beta} + \xi_{\alpha\beta} + \xi_{\beta\alpha} + \mathcal{O}(\xi^2) \tag{2.21}
\end{aligned}$$

And as a consequence, at $\mathcal{O}(\xi)$ we have:

$$\xi_{\alpha\beta} = -\xi_{\beta\alpha} \tag{2.22}$$

In other words, the most general transformation has a total of ten free parameters (six from the antisymmetric rank-2 tensor ξ and four from the four-vector ϵ).

2.2 Lorentz transformations

The *Poincare group* is the group of coordinate transformations that satisfy the Lorentz condition [8–10]. More formally, the Poincare group is denoted by IO(3, 1) and corresponds to the isometry group of Minkowski spacetime (with three spatial and one temporal dimensions). In the context of special relativity, the isometries correspond to affine transformations in $\mathbb{R}^4$ that preserve the *Minkowski metric* (e.g. those that satisfy the Lorentz condition)[1]. From equation (2.19) we can observe that the Poincare group consists of four translations over the four space-time dimensions (represented by the four ϵ^α parameters), as well as rotations and boosts over the three spatial dimensions (represented by the six independent parameters in $\xi^\alpha{}_\beta$).

The *Lorentz Group* is a subgroup of the Poincare group and consists of the six rotations and boosts (but does not include the translations) [8–10]. That is, Lorentz transformations are exclusively composed of rotations and boosts. As such, the Lorentz group is often called the *homogeneous Lorentz group*, whereas the Poincare group is referred as the *inhomogeneous Lorentz group*. The Lorentz group is denoted by O(3, 1). Also, the Poincare group can be written as the semidirect product:

$$IO(3, 1) = \mathbb{R}^{(3+1)} \rtimes O(3, 1) \qquad (2.23)$$

of the spacetime translations and the Lorentz groups[2].

However, it is often the case where we are just interested in the orthochronous subgroup of Lorentz transformations $O^+(3, 1)$ that preserve the direction of time and the proper subgroup of Lorentz transformations SO(3, 1) that preserve orientation. Then, the *proper, orthochronous Lorentz group* is denoted by $SO^+(3, 1)$.

Furthermore, it can be shown that the Lorentz group is a *Lie group*. As a consequence, O(3, 1) has an associated smooth manifold. This manifold has four connected components, which means that an arbitrary Lorentz transformation in O(3, 1) can be written as the semidirect product of a $SO^+(3, 1)$ transformation and an element of the discrete group:

$$\{1, P, T, PT\} \qquad (2.24)$$

where P and T are the *space inversion* and *time reversal* operators, respectively.

---

[1] Let us recall that an *affine transformation* is a transformation that preserves straight lines. That is, parallel lines remain parallel after the effect of an affine transformation.

[2] A semidirect product is constructed from a subgroup and a *normal subgroup* (i.e. a subgroup that is invariant under conjugation by members of the group) [10]. For instance, it is easy to check that translations form a normal subgroup of the Euclidean group $E(n)$. Indeed, let us consider as an example a rotation R and a translation T. Then, we have: $RTR^{-1} = T$, and therefore $RT = TR$. In general, for any transformation $g \in E(n)$, we have $gT = Tg$. In the case of special relativity, it is important that the Poincare group is obtained from a semi-direct product of $\mathbb{R}^{(3+1)}$ and O(3, 1), which means that every element of IO(3, 1) can be written in a *unique* way as a product of an element of $\mathbb{R}^{(3+1)}$ and an element of O(3, 1).

In a similar manner, the *proper, orthochronous Poincare group* is denoted by $ISO^+(3, 1)$. It is important to remark that, as first observed by Wigner, the fundamental group that describes elementary quantum particles is *not* the Lorentz group, but the Poincare group [11].

Let us now consider, for example, a Lorentz boost in the x^1 direction with speed $\mathbf{v} = v\hat{\mathbf{i}}$ given by:

$$
\begin{aligned}
\tilde{x}^0 &= \gamma(x^0 + \beta x^1) \\
\tilde{x}^1 &= \gamma(\beta x^0 + x^1) \\
\tilde{x}^2 &= x^2 \\
\tilde{x}^3 &= x^3
\end{aligned}
\tag{2.25}
$$

where:

$$
\gamma \equiv \frac{1}{\sqrt{1 - \beta^2}} \qquad \beta = v
\tag{2.26}
$$

Because:

$$
\gamma^2 - \beta^2 \gamma^2 = 1
\tag{2.27}
$$

we can parametrize the Lorentz boost transformation using hyperbolic trigonometric functions over an angular variable φ in the following way:

$$
\gamma = \cosh\varphi \qquad \gamma\beta = \sinh\varphi \qquad v = \tanh\varphi
\tag{2.28}
$$

This parametrization leads to the following matrix representation of the coordinates transformation:

$$
\begin{pmatrix} \tilde{x}^0 \\ \tilde{x}^1 \\ \tilde{x}^2 \\ \tilde{x}^3 \end{pmatrix} = \begin{pmatrix} \cosh\varphi & \sinh\varphi & 0 & 0 \\ \sinh\varphi & \cosh\varphi & 0 & 0 \\ 0 & 0 & 1 & 0 \\ 0 & 0 & 0 & 1 \end{pmatrix} \begin{pmatrix} x^0 \\ x^1 \\ x^2 \\ x^3 \end{pmatrix}
\tag{2.29}
$$

which clearly corresponds to a hyperbolic rotation by an angle φ in four-dimensional space.

A Lorentz boost over the x^1 direction by an infinitesimal angle $|\delta\varphi| \ll 1$ is given by:

$$
B_x(\delta\varphi) \approx 1 + iK_x\delta\varphi
\tag{2.30}
$$

and the generator of the infinitesimal Lorentz transformation is:

$$
K_x = \frac{1}{i}\frac{\partial B_x}{\partial \varphi}\bigg|_{\varphi=0} = -i\begin{pmatrix} 0 & 1 & 0 & 0 \\ 1 & 0 & 0 & 0 \\ 0 & 0 & 0 & 0 \\ 0 & 0 & 0 & 0 \end{pmatrix}
\tag{2.31}
$$

where B_x is the matrix shown in equation (2.29) [12]. We can obtain similar expressions for the the infinitesimal boost generators in the other two directions:

$$K_y = -i \begin{pmatrix} 0 & 0 & 1 & 0 \\ 0 & 0 & 0 & 0 \\ 1 & 0 & 0 & 0 \\ 0 & 0 & 0 & 0 \end{pmatrix}$$

$$K_z = -i \begin{pmatrix} 0 & 0 & 0 & 1 \\ 0 & 0 & 0 & 0 \\ 0 & 0 & 0 & 0 \\ 1 & 0 & 0 & 0 \end{pmatrix} \tag{2.32}$$

And the three generators for infinitesimal rotations are given by:

$$J_x = -i \begin{pmatrix} 0 & 0 & 0 & 0 \\ 0 & 0 & 0 & 0 \\ 0 & 0 & 0 & 1 \\ 0 & 0 & -1 & 0 \end{pmatrix}$$

$$J_y = -i \begin{pmatrix} 0 & 0 & 0 & 0 \\ 0 & 0 & 0 & -1 \\ 0 & 0 & 0 & 0 \\ 0 & 1 & 0 & 0 \end{pmatrix} \tag{2.33}$$

$$J_z = -i \begin{pmatrix} 0 & 0 & 0 & 0 \\ 0 & 0 & 1 & 0 \\ 0 & -1 & 0 & 0 \\ 0 & 0 & 0 & 0 \end{pmatrix}$$

These infinitesimal generators can be used to construct more general finite transformations [12]. For example, let us consider a finite angle θ obtained from an infinitesimal angle $\delta\theta$ as follows:

$$\theta = \lim_{N \to \infty} N\delta\theta \tag{2.34}$$

then, a finite rotation over the 'z' axis by θ is given by:

$$R_z(\theta) = \lim_{N \to \infty} (R_z(\delta\theta))^N$$

$$= \lim_{N \to \infty} (1 + iJ_z\delta\theta)^N$$

$$= \lim_{N \to \infty} \left(1 + iJ_z\frac{\theta}{N}\right)^N$$

$$= e^{iJ_z\theta} \tag{2.35}$$

Therefore, in general:

$$R_{\mathbf{n}}(\theta) = e^{i\mathbf{J}\cdot\boldsymbol{\theta}} = e^{i\mathbf{J}\cdot\mathbf{n}\theta}$$
$$B_{\mathbf{m}}(\varphi) = e^{i\mathbf{K}\cdot\boldsymbol{\varphi}} = e^{i\mathbf{K}\cdot\mathbf{m}\varphi} \tag{2.36}$$

where:

$$\mathbf{J} = (J_x, J_y, J_z) \tag{2.37}$$
$$\mathbf{K} = (K_x, K_y, K_z) \tag{2.38}$$

and:

$$\boldsymbol{\theta} = \mathbf{n}\theta$$
$$\boldsymbol{\varphi} = \mathbf{m}\varphi \tag{2.39}$$

represent a rotation angle θ over the direction $\mathbf{n}$ and a boost angle φ over the direction $\mathbf{m}$.

2.3 Lagrangian dynamics

As we know from classical mechanics, the *principle of least action* is used to study the dynamics of material particles [13, 14]. That is, the functional integral known as the *action* (denoted by S), is such that its variation δS is zero for the actual motion of the system. In a strict sense, the action is a minimum only for infinitesimal lengths in the path of integration, whereas for arbitrary lengths it takes an extremal value, which is not necessarily a minimum [14].

To satisfy the *principle of special relativity*, which states the invariance of the laws of physics under Lorentz transformations, the action of a relativistic particle has to be a scalar and an invariant under Lorentz transformations. Furthermore, the integrand must be a differential of first order. For a free particle, the only possible scalar that can be constructed under these restrictions is the line element dS [4].

To avoid notational confusion, in what follows we will use the *proper time* dτ instead of the line element dS:

$$d\tau^2 \equiv -dS^2 \tag{2.40}$$

and as a consequence:

$$d\tau^2 = (1 - v^2)\, dt^2 \geqslant 0 \Rightarrow d\tau = \sqrt{1 - v^2}\, dt \tag{2.41}$$

Then, the action S for a particle that moves from point a to point b can be written as:

$$S = -m \int_a^b d\tau \tag{2.42}$$

where m is the mass of the particle and the integral is taken along the world line of a particle between two world points a and b. Furthermore, it is easy to show that the total proper time has a maximal value for a straight world line connecting two events. However, it is possible to make the value of the integral arbitrarily small, by performing the integral over curved lines connecting both events. Therefore, the expression for the action requires the minus sign, otherwise the action would not have a minimum.

We can rewrite the action as follows:

$$S = -m \int_a^b \frac{d\tau^2}{d\tau}$$

$$= m \int_a^b \frac{dx^\mu dx_\mu}{d\tau}$$

$$= \int_a^b p_\mu dx^\mu \tag{2.43}$$

where the *four-momentum* of the particle is defined as:

$$p^\mu \equiv m \frac{dx^\mu}{d\tau} \tag{2.44}$$

where $x^\mu(\tau)$ represents the path of the particle.

Alternatively, if we notice that:

$$\frac{d\tau^2}{dt^2} = -\frac{dx^\mu dx_\mu}{dt^2} = 1 - \mathbf{v} \cdot \mathbf{v} = 1 - v^2 \tag{2.45}$$

then:

$$\frac{dx^0}{d\tau} = \frac{dt}{d\tau} = \gamma \tag{2.46}$$

and we can write the action as:

$$S = -m \int_a^b d\tau$$

$$= -m \int_a^b \frac{dt}{\gamma}$$

$$= -m \int_a^b \sqrt{1 - v^2} \, dt \tag{2.47}$$

where γ is given by equation (2.26).

Furthermore, recalling that in the *configuration space* of the system the action is the integral over time of the Lagrangian L, we get:

$$S = \int_a^b L \, dt = -m \int_a^b \sqrt{1 - v^2} \, dt \tag{2.48}$$

and therefore, the Lagrangian results in:

$$L = -m\sqrt{1 - v^2} \qquad (2.49)$$

for a free relativistic particle of mass m.

In configuration space in Cartesian coordinates, the Lagrangian for a free relativistic particle has the formal expression:

$$L(q_i, \dot{q}_i, t) = -m\sqrt{1 - \sum_{i=1}^{3} \dot{q}_i^2} \qquad (2.50)$$

where:

$$\dot{q}_i = v_i \quad \Rightarrow \quad \sum_{i=1}^{3} \dot{q}_i^2 = v^2 \qquad (2.51)$$

where q_i represents the *canonical coordinates* of the mechanical system. Furthermore, the *canonical momentum* for a free relativistic particle with three degrees of freedom can also be obtained through the equation:

$$p_i = \frac{\partial L}{\partial \dot{q}_i} = m\gamma\dot{q}_i \qquad i = 1, 2, 3 \qquad (2.52)$$

while the *Hamilton equations* are written as:

$$\dot{q}_i = \frac{\partial H}{\partial p_i} \qquad \dot{p}_i = -\frac{\partial H}{\partial q_i} \qquad i = 1, 2, 3 \qquad (2.53)$$

where H is the *Hamiltonian* of the system given by:

$$H(p_i, q_i, t) = \sum_{i=1}^{3} p_i\dot{q}_i - L \qquad (2.54)$$

and can be easily computed to be:

$$H = m\gamma \sum_{i=1}^{3} \dot{q}_i^2 + \frac{m}{\gamma}$$

$$= m\gamma v^2 + \frac{m}{\gamma}$$

$$= m\gamma \qquad (2.55)$$

or alternatively, we can write H as:

$$H = \sqrt{\sum_{i=1}^{3} p_i^2 + m^2} = \sqrt{p^2 + m^2} \qquad (2.56)$$

where:

$$p^2 = \sum_{i=1}^{3} p_i^2 \qquad (2.57)$$

These expressions make it evident that the Hamiltonian is the energy expressed in terms of the momentum.

Let us recall that the *principle of least action* states that:

$$\delta S = 0 \qquad (2.58)$$

where S is considered as a functional of the time, the canonical coordinates and their canonical time derivatives:

$$S = S[q(t), \dot{q}(t), t] \qquad (2.59)$$

and S reaches a minimum for the canonical variables that describe the actual motion of the particle. Indeed, the motion of the particle is such that the proper time interval $d\tau^2$ is maximal along a straight world line [4]. However, because $dS^2 = -d\tau^2 \leqslant 0$ for a material particle (with the equality reached for a massless particle traveling at the speed of light), then dS^2 is minimal in value, but maximal in magnitude, along a straight world line.

As usual, the dynamics of the actual motion of the particle is described by the *Euler–Lagrange equations*:

$$\frac{d}{dt} \frac{\partial L}{\partial \dot{q}_i} - \frac{\partial L}{\partial q_i} = 0 \qquad i = 1, 2, 3 \qquad (2.60)$$

which in the case of a free relativistic particle lead to:

$$\frac{d}{dt} \left(\frac{m \dot{q}_i}{\sqrt{1 - \sum_{i=1}^{3} \dot{q}_i^2}} \right) = 0 \quad \Rightarrow \quad \frac{dp_i}{dt} = 0 \qquad i = 1, 2, 3 \qquad (2.61)$$

and because L does not depend explicitly on time, then:

$$\frac{dH}{dt} = \frac{\partial H}{\partial t} + \sum_i \frac{\partial H}{\partial q_i} \dot{q}_i + \sum_i \frac{\partial H}{\partial p_i} \dot{p}_i = \frac{\partial H}{\partial t} = 0 \qquad (2.62)$$

as we would have expected for a massive particle ($m \neq 0$).

Notice that the above discussion considered the Lagrangian in the configuration space of the system defined by $(q_i(t), \dot{q}_i(t), t)$, which is not in an explicitly covariant notation. On the other hand, in a general four-dimensional pseudo-Riemannian manifold, the explicitly covariant Lagrangian $\mathcal{L}$ is written as:

$$\mathcal{L} = \mathcal{L}\left(x^\mu, \frac{dx^\mu}{d\tau} \right) = \mathcal{L}(x^\mu, \dot{x}^\mu) \qquad (2.63)$$

where we have defined:

$$\dot{x}^{\mu} \equiv \frac{dx^{\mu}}{d\tau} \tag{2.64}$$

Then, the action takes the form:

$$S = -m \int_{a}^{b} d\tau = \int_{a}^{b} \mathcal{L} \, d\tau \tag{2.65}$$

In this case, the Euler–Lagrange equations are written as:

$$\frac{d}{d\tau} \frac{\partial \mathcal{L}}{\partial \dot{x}^{\mu}} - \frac{\partial \mathcal{L}}{\partial x^{\mu}} = 0 \tag{2.66}$$

where:

$$\mathcal{L} = m \frac{dx^{\mu}}{d\tau} \frac{dx_{\mu}}{d\tau} \tag{2.67}$$

is an explicitly covariant Lagrangian for a free relativistic massive particle. Needless to say, both approaches can be generalized to more complex physical systems, and both of them reach exactly the same dynamical equations. Indeed, using the explicitly covariant Lagrangian for a free relativistic massive particle, the Euler–Lagrange equations lead to:

$$\frac{dp^{\mu}}{d\tau} = 0 \qquad p^{\mu} = (m\gamma, m\gamma\mathbf{v}) \tag{2.68}$$

which is the explicitly covariant generalization of equation (2.61).

2.4 The principle of equivalence

The *principle of equivalence* states that, at every point in an arbitrary gravitational field, it is possible to choose a locally inertial coordinate system such that, in a very small region of space around the point in question, the laws of nature take the same form as in an unaccelerated Cartesian coordinate system without gravitation [1, 3, 7, 15].

Some authors consider a distinction between a *strong principle of equivalence* and a *weak principle of equivalence*. Although the former refers to the invariance of *all* the laws of nature, the second simply entails the invariance of the laws of motion of freely falling particles [7].

2.4.1 Particle dynamics in a gravitational field

Let us first consider a free massive particle (zero net force applied and moving at a subluminal speed $v < 1$). Then, Newton's second law states that:

$$\frac{d^2 x^{\mu}}{d\tau^2} = 0 \tag{2.69}$$

where, in natural units, the *proper time* $d\tau$ is given by:

$$d\tau^2 = -dS^2 \quad \Rightarrow \quad \tau = \int_{\mathcal{P}} \sqrt{-dS^2} \tag{2.70}$$

integrated over a time-like path $\mathcal{P}$ with $dS^2 < 0$.

Now, let us assume that the particle is immersed in some arbitrary gravitational field. In this case, Newton's second law implies that:

$$\frac{d^2 x^\mu}{d\tau^2} \neq 0 \tag{2.71}$$

However, if we invoke the principle of equivalence, we can find a locally inertial coordinate system ξ^α in a small neighbourhood around the particle such that:

$$\frac{d^2 \xi^\alpha}{d\tau^2} = 0 \tag{2.72}$$

Now, the locally inertial coordinate system is only valid in a small region of spacetime, so, in general, it will depend on the spacetime coordinates:

$$\xi^\alpha = \xi^\alpha(x^\mu) \tag{2.73}$$

Therefore, the second derivative of the locally inertial coordinates ξ^α can be rewritten as:

$$\begin{aligned}
\frac{d^2 \xi^\alpha}{d\tau^2} &= \frac{d}{d\tau}\left(\frac{d\xi^\alpha}{d\tau}\right) \\[6pt]
&= \frac{d}{d\tau}\left(\frac{\partial \xi^\alpha}{\partial x^\mu}\frac{dx^\mu}{d\tau}\right) \\[6pt]
&= \frac{\partial \xi^\alpha}{\partial x^\mu}\frac{d^2 x^\mu}{d\tau^2} + \frac{\partial^2 \xi^\alpha}{\partial x^\mu \partial x^\nu}\frac{dx^\mu}{d\tau}\frac{dx^\nu}{d\tau}
\end{aligned} \tag{2.74}$$

and then, the dynamical equation expressed in the general coordinate system x^μ has the form:

$$\frac{d^2 x^\beta}{d\tau^2} + \frac{\partial x^\beta}{\partial \xi^\alpha}\frac{\partial^2 \xi^\alpha}{\partial x^\mu \partial x^\nu}\frac{dx^\mu}{d\tau}\frac{dx^\nu}{d\tau} = 0 \tag{2.75}$$

If we define the *affine connection* as:

$$\Gamma^\beta_{\mu\nu} = \frac{\partial x^\beta}{\partial \xi^\alpha}\frac{\partial^2 \xi^\alpha}{\partial x^\mu \partial x^\nu} \tag{2.76}$$

the dynamics of a particle in a gravitational field are described by:

$$\frac{d^2 x^\beta}{d\tau^2} + \Gamma^\beta_{\mu\nu}\frac{dx^\mu}{d\tau}\frac{dx^\nu}{d\tau} = 0 \tag{2.77}$$

That is, all the dynamical effects generated by the gravitational field are incorporated into the affine connection.

Furthermore, the proper time can be written as:

$$d\tau^2 = -\eta_{\alpha\beta}d\xi^\alpha d\xi^\beta$$

$$= -\eta_{\mu\nu}\left(\frac{\partial\xi^\mu}{\partial x^\alpha}dx^\alpha\right)\left(\frac{\partial\xi^\nu}{\partial x^\beta}dx^\beta\right)$$

$$= -\eta_{\mu\nu}\frac{\partial\xi^\mu}{\partial x^\alpha}\frac{\partial\xi^\nu}{\partial x^\beta}dx^\alpha dx^\beta$$

$$= -g_{\alpha\beta}dx^\alpha dx^\beta \tag{2.78}$$

where:

$$g_{\alpha\beta} \equiv \eta_{\mu\nu}\frac{\partial\xi^\mu}{\partial x^\alpha}\frac{\partial\xi^\nu}{\partial x^\beta} \tag{2.79}$$

is the *metric tensor*. Notice that, in general, the metric tensor will depend on the spacetime coordinates:

$$g_{\mu\nu} = g_{\mu\nu}(x^\alpha) \tag{2.80}$$

but for the sake of notational simplicity, we will omit the functional dependence in most equations. So, for example, the symmetry of the metric tensor is simply denoted as:

$$g_{\mu\nu} = g_{\nu\mu} \tag{2.81}$$

In the most general case of a test particle with mass m, we define the four-momentum as:

$$p^\mu \equiv mu^\mu \tag{2.82}$$

where:

$$u^\mu \equiv \frac{dx^\mu}{d\tau} \tag{2.83}$$

and the normalization condition is given by:

$$p^\mu p_\mu = -m^2 \tag{2.84}$$

which is a consequence of the invariance of the line element over general coordinate transformations:

$$dS^2 = g_{\mu\nu}dx^\mu dx^\nu \Rightarrow \frac{dS^2}{d\tau^2} = g_{\mu\nu}\frac{dx^\mu}{d\tau}\frac{dx^\nu}{d\tau} = \frac{p^\mu p_\mu}{m^2} \tag{2.85}$$

and therefore:

$$dS^2 = -d\tau^2 \Rightarrow p^\mu p_\mu = -m^2 \tag{2.86}$$

which reduces to the non-gravitational relativistic mass–energy relation:

$$E^2 - \mathbf{p} \cdot \mathbf{p} = m^2 \tag{2.87}$$

in the absence of gravitation [7].

2.4.2 Torsion

The affine connection is not necessarily symmetric upon exchange of its lower indices. Indeed, this tensor is symmetric only if we can exchange the derivatives of the locally inertial coordinate system:

$$\Gamma^\alpha_{\mu\nu} = \Gamma^\alpha_{\nu\mu} \quad \Leftrightarrow \quad \frac{\partial^2 \xi^\alpha}{\partial x^\mu \partial x^\nu} = \frac{\partial^2 \xi^\alpha}{\partial x^\nu \partial x^\mu} \tag{2.88}$$

If the affine connection is symmetric in the lower indices, then the underlying manifold is said to be *torsion free*. In a natural way, the *torsion tensor* is defined as:

$$T^\alpha_{\mu\nu} \equiv \Gamma^\alpha_{\mu\nu} - \Gamma^\alpha_{\nu\mu} \tag{2.89}$$

However, within the context of traditional applications of classical gravitation, it can be shown that if a (pseudo-) Riemannian manifold can be embedded in some (pseudo-) Euclidean space of higher dimension, then the manifold has to be torsion free [3]. Furthermore, because of symmetry arguments, general relativity demands an affine connection with vanishing torsion [4]. Thus, the most widely used metric tensors, such as those that correspond to Schwarzschild and Kerr spacetimes, correspond to torsion-free affine connections. Indeed, this will be the case in all the examples discussed in this book.

However, it is important to mention that spacetime torsion is caused by the intrinsic quantum mechanical angular momentum of elementary particles [16]. Furthermore, general relativity cannot accommodate the coupling between spin and angular momentum. Therefore, general relativity appears to be insufficient to properly describe quantum fields with spin in the presence of classical gravitational fields [16]. Some extensions to general relativity, such as the *Einstein–Cartan theories*, have been developed to describe spacetimes with some degree of torsion [16, 17]. In these theories, torsion is related to the density of intrinsic angular momentum. That is, just as the energy–momentum tensor generates spacetime curvature, the intrinsic angular momentum (spin) tensor generates torsion in the spacetime.

Notice that the torsion tensor transforms as a tensor under general coordinates transformations. Now, because of the principle of equivalence, we know that we can find a local inertial coordinate system in which the components of the affine connections are zero ($\Gamma^\alpha_{\mu\nu} = 0$) at any given point in spacetime[3]. Then, because the affine connection vanishes at this point, so does the torsion tensor $T^\alpha_{\mu\nu} = 0$. Furthermore, as we will see in the next section, tensor equations remain valid in all frames of

---

[3] However, the derivatives of the affine connection are not necessarily zero at this point in spacetime.

reference. That is, $T^{\alpha}_{\mu\nu} = 0$ for all points in spacetime[4]. As a consequence, any gravitation theory based on the principle of equivalence needs to have zero torsion.

The degree of torsion in a manifold is important to determine the effects of a gravitational field on falling particles. For instance, in the specific case of torsion-free manifolds, the affine connection is related to the metric tensor through:

$$\Gamma^{\alpha}_{\mu\nu} = \frac{1}{2}g^{\sigma\alpha}\left(\frac{\partial g_{\nu\sigma}}{\partial x^{\mu}} + \frac{\partial g_{\mu\sigma}}{\partial x^{\nu}} - \frac{\partial g_{\nu\mu}}{\partial x^{\sigma}}\right) \tag{2.90}$$

Otherwise, the above equation needs to be modified by adding to the right-hand side a term known as the *contortion* given by:

$$K^{\alpha}_{\mu\nu} \equiv \frac{1}{2}(T^{\alpha}_{\mu\nu} + T_{\mu\ \nu}^{\ \alpha} + T_{\nu\mu}^{\ \ \alpha}) \tag{2.91}$$

As a consequence of these expressions, the effects of a gravitational field on the dynamics of a particle are incorporated into the specific mathematical structure of the metric tensor. In other words, if we have knowledge of the value of the metric tensor at all points of spacetime, then it is possible to know the exact dynamic equations that describe the motion of particles under the influence of the gravitational field.

2.4.3 Geodesics and geodesic congruences

There is a simple geometric interpretation of the dynamical equations for a particle in a gravitational field. Indeed, it can be shown that the motion of particles in a gravitational field will follow a *geodesic*, and:

$$\frac{\mathrm{d}^2 x^{\beta}}{\mathrm{d}\tau^2} + \Gamma^{\beta}_{\mu\nu}\frac{\mathrm{d}x^{\mu}}{\mathrm{d}\tau}\frac{\mathrm{d}x^{\nu}}{\mathrm{d}\tau} = 0 \tag{2.92}$$

is often referred as the *geodesic equation* [1, 3, 5]. That is, a particle moving between two points in spacetime under the influence of a gravitational field will follow the path that *extremizes* the length between two points. In this case, length is measured with the proper time or the line element as follows:

$$\mathrm{d}\tau^2 = -\mathrm{d}S^2 \Rightarrow \tau = \int_{\mathcal{P}}\sqrt{-g_{\mu\nu}\mathrm{d}x^{\mu}\mathrm{d}x^{\nu}} \tag{2.93}$$

integrated over a time-like path $\mathcal{P}$ with $\mathrm{d}S^2 < 0$.

Notice that for a manifold with a Riemann metric, the extremal length $\mathrm{d}S^2$ takes the minimum distance. That is, geodesics are curves that take the least distance between two points. The situation is the opposite for manifolds with a Lorentz metric (often called *pseudo-Riemannian metrics*). Indeed, in this case the geodesics are curves of maximal length as measured by the absolute value of the Lorentz distance $\mathrm{d}S^2$ [6].

---

[4] In contrast, notice that $\Gamma^{\alpha}_{\mu\nu} = 0$ is not an equation valid everywhere, because $\Gamma^{\alpha}_{\mu\nu}$ does *not* transform as a tensor.

Because of Heisenberg's uncertainty principle, quantum states $\Psi(x)$ have to be considered as superpositions of states traversing different world lines. As we will see in chapter 5, consideration of such path superpositions in the case of geodetic motion of quantum particles leads to the concept of *geodesic congruences* [18]. These are sets of curves defined in an open region of spacetime [1]. Furthermore, every point in that region of spacetime belongs to one, and only one, curve.

Let us imagine a family of non-crossing geodesics given by $x^\mu(r,s)$, where r is the parameter that identifies the geodesic evolution, and s is the parameter that labels the geodesic path under consideration. Then, the vectors tangent to the geodesics are:

$$T^\mu = \frac{\partial x^\mu}{\partial r} \tag{2.94}$$

and the *deviation vectors* are:

$$S^\mu = \frac{\partial x^\mu}{\partial s} \tag{2.95}$$

Then, we can define the *relative velocity of the geodesics* as:

$$V^\mu \equiv T^\alpha \nabla_\alpha S^\mu \tag{2.96}$$

and the *relative acceleration of the geodesics* as:

$$A^\mu \equiv T^\alpha \nabla_\alpha V^\mu \tag{2.97}$$

Furthermore, it can be shown that the relative acceleration between two neighbouring geodesics is given by the *geodesic deviation equation*:

$$A^\mu = R^\mu{}_{\alpha\beta\gamma} T^\alpha T^\beta S^\gamma \tag{2.98}$$

which, as expected, is proportional to the spacetime curvature $R^\mu{}_{\alpha\beta\gamma}$ (described in section 2.7). This result can be interpreted as a manifestation of gravitational tidal forces [1][5].

2.5 The principle of general covariance

Our previous discussion considered the case of a free particle in a gravitational field. We will now consider a particle in a gravitational field which is also being accelerated by some arbitrary non-gravitational external forces.

The *principle of general covariance* states that a physical equation remains valid in the presence of a gravitational field if the following two conditions are met:

    1. The equation holds in the absence of a gravitational field when the metric tensor $g_{\mu\nu}$ reduces to the Minkowski tensor $\eta_{\mu\nu}$.

---

[5] In a sense, the relative velocity and relative acceleration of the geodesics describe the density variation of the congruence. Indeed, one could imagine a geodesic congruence as a flow of particles, and the actual shape of the flow is determined by A^μ and V^μ: in some areas the flow may be wide, whereas in others the flow may be very narrow.

2. The physical equation is *general covariant*, in the sense that it preserves its form under a general transformation of coordinates $x \to \tilde{x}$.

In what follows, we will briefly discuss some of the implications of the principle of general covariance.

2.5.1 Tensor analysis

Because of the principle of general covariance, it is useful to have equations written in an explicit covariant way. To do so we require the use of tensors that obey specific transformation rules under a general transformation of coordinates. In the context of general relativity, general transformations of coordinates are described by the *general linear group* GL(4,ℝ) made of all real, regular (invertible), 4 × 4 matrices [8].

Then, under a general transformation of coordinates:

$$x^\mu \to \tilde{x}^\mu \tag{2.99}$$

contravariant tensors transform as:

$$A^{\mu\nu\cdots\rho} \to \tilde{A}^{\mu\nu\cdots\rho} = A^{\alpha\beta\cdots\gamma}\frac{\partial\tilde{x}^\mu}{\partial x^\alpha}\frac{\partial\tilde{x}^\nu}{\partial x^\beta}\cdots\frac{\partial\tilde{x}^\rho}{\partial x^\gamma} \tag{2.100}$$

whereas *covariant tensors* transform as:

$$A_{\mu\nu\cdots\rho} \to \tilde{A}_{\mu\nu\cdots\rho} = A_{\alpha\beta\cdots\gamma}\frac{\partial x^\alpha}{\partial\tilde{x}^\mu}\frac{\partial x^\beta}{\partial\tilde{x}^\nu}\cdots\frac{\partial x^\gamma}{\partial\tilde{x}^\rho} \tag{2.101}$$

Covariant and contravariant tensors are related through the metric tensor:

$$A_\mu = g_{\mu\nu}A^\nu \qquad A^\mu = g^{\mu\nu}A_\nu \qquad g_{\mu\nu}g^{\nu\lambda} = \delta_\mu^\lambda \tag{2.102}$$

Thus, any equation expressed through a tensor identity remains invariant under an arbitrary coordinate system [7]. For example, let us consider the following tensor equation in the x coordinate system:

$$A^\mu = B^\mu \tag{2.103}$$

Then, under a transformation to the $\tilde{x}$ coordinate system we obtain:

$$A^\mu \to \tilde{A}^\mu = A^\nu\frac{\partial\tilde{x}^\mu}{\partial x^\nu} = B^\nu\frac{\partial\tilde{x}^\mu}{\partial x^\nu} = \tilde{B}^\mu \Rightarrow \tilde{A}^\mu = \tilde{B}^\mu \tag{2.104}$$

which has the exact same form as in the x coordinate system (even though A^μ and $\tilde{A}^\mu$ may have different component values).

The metric tensor transforms as:

$$g_{\mu\nu} \to \tilde{g}_{\mu\nu} = g_{\alpha\beta}\frac{\partial x^\alpha}{\partial\tilde{x}^\mu}\frac{\partial x^\beta}{\partial\tilde{x}^\nu} \tag{2.105}$$

Then, if we define g as the determinant of the metric tensor:

$$g \equiv \det(g_{\mu\nu}) \tag{2.106}$$

it will transform as:

$$g \to \tilde{g} = g \left| \frac{\partial x}{\partial \tilde{x}} \right|^2 = g \mathcal{J}^{-2} \tag{2.107}$$

where we have defined the *Jacobian* of the transformation $x \to \tilde{x}$ as:

$$\mathcal{J} \equiv \left| \frac{\partial \tilde{x}}{\partial x} \right| = \det\left(\frac{\partial \tilde{x}^\mu}{\partial x^\nu} \right) \tag{2.108}$$

Furthermore, the volume element transforms as:

$$d^4 x \to d^4 \tilde{x} = \mathcal{J} d^4 x \tag{2.109}$$

and as a consequence, the quantity:

$$\sqrt{-g}\, d^4 x \tag{2.110}$$

is known as the *invariant volume element* and remains invariant under general coordinate transformations:

$$\sqrt{-g}\, d^4 x \to \sqrt{-\tilde{g}}\, d^4 \tilde{x} = \sqrt{-g}\, d^4 x \tag{2.111}$$

The determinant of the metric tensor is also important to simplify the expression for the contraction of the affine connection:

$$\Gamma^\mu_{\mu\alpha} = \frac{1}{2} \partial_\alpha \ln(-g) = \frac{1}{\sqrt{-g}} \partial_\alpha \sqrt{-g} \tag{2.112}$$

where we have used the straightforward result:

$$\Gamma^\mu_{\mu\alpha} = \frac{1}{2} g^{\mu\nu} \partial_\alpha g_{\mu\nu} \tag{2.113}$$

and the fact that, for an arbitrary matrix M:

$$Tr\left(\frac{1}{M} \partial_\alpha M \right) = \partial_\alpha \ln \det M \tag{2.114}$$

2.5.2 Covariant derivatives

It is important to recall that not all quantities that have a Greek subindex behave as tensors. As a case in point, the standard derivative of a contravariant four-vector is not a tensor under general coordinate transformations:

$$\frac{\partial A^\mu}{\partial x^\lambda} \to \frac{\partial \tilde{x}^\mu}{\partial x^\nu} \frac{\partial x^\rho}{\partial \tilde{x}^\lambda} \frac{\partial A^\nu}{\partial x^\rho} + \frac{\partial^2 \tilde{x}^\mu}{\partial x^\nu \partial x^\rho} \frac{\partial x^\rho}{\partial \tilde{x}^\lambda} A^\nu \tag{2.115}$$

but the *covariant derivative* of a contravariant four-vector defined as:

$$\frac{DA^\mu}{Dx^\lambda} \equiv \frac{\partial A^\mu}{\partial x^\lambda} + \Gamma^\mu_{\lambda\rho} A^\rho \tag{2.116}$$

2-18

transforms as a tensor:

$$\frac{\mathcal{D}A^\mu}{\mathcal{D}x^\lambda} \rightarrow \frac{\partial \tilde{x}^\mu}{\partial x^\nu} \frac{\partial x^\rho}{\partial \tilde{x}^\lambda} \frac{\mathcal{D}A^\nu}{\mathcal{D}x^\rho} \tag{2.117}$$

Notice, however, that the affine connection $\Gamma^\mu_{\alpha\beta}$ does *not* transform as a tensor.

In a similar way, the covariant derivative for a covariant four-vector is given by:

$$\frac{\mathcal{D}A_\mu}{\mathcal{D}x^\lambda} \equiv \frac{\partial A_\mu}{\partial x^\lambda} - \Gamma^\rho_{\mu\lambda} A_\rho \tag{2.118}$$

whereas the covariant derivative of a scalar ϕ coincides with the standard derivative:

$$\frac{\mathcal{D}\phi}{\mathcal{D}x^\lambda} = \frac{\partial \phi}{\partial x^\lambda} \tag{2.119}$$

Similarly, the covariant derivative of a four-vector along a curve $x^\mu(\tau)$ is given by:

$$\frac{\mathcal{D}A^\mu}{\mathcal{D}\tau} = \frac{\mathrm{d}A^\mu}{\mathrm{d}\tau} + \Gamma^\mu_{\nu\lambda} \frac{\mathrm{d}x^\lambda}{\mathrm{d}\tau} A^\nu$$

$$\frac{\mathcal{D}B_\mu}{\mathcal{D}\tau} = \frac{\mathrm{d}B_\mu}{\mathrm{d}\tau} - \Gamma^\lambda_{\mu\nu} \frac{\mathrm{d}x^\nu}{\mathrm{d}\tau} B_\lambda \tag{2.120}$$

for contravariant and covariant vectors, respectively. If a vector A^μ is such that:

$$\frac{\mathcal{D}A^\mu}{\mathcal{D}\tau} = 0 \tag{2.121}$$

then it is said to satisfy the *parallel transport equation* [1, 3, 7]. This equation implies that a parallel field of vectors is defined on each point of the curve $x^\mu(\tau)$ by the parallel transport of the vector A^μ along the curve. That is, the vector A^μ is transported along the curve without any changes to its magnitude and orientation.

For notational simplicity it is convenient to introduce the shorthand notation for the standard and the covariant derivative operators:

$$\partial_\mu \equiv \frac{\partial}{\partial x^\mu} \qquad \mathcal{D}_\mu \equiv \frac{\mathcal{D}}{\mathcal{D}x^\mu} \tag{2.122}$$

Some useful properties of the covariant derivative include:

$$\mathcal{D}_\lambda \delta^\nu_\mu = 0$$

$$\mathcal{D}_\lambda g_{\mu\nu} = 0 \tag{2.123}$$

and as a consequence, the raising and lowering of indices commutes with the covariant derivative:

$$\mathcal{D}_\lambda A_\mu = \mathcal{D}_\lambda (g_{\mu\nu} A^\nu) = g_{\mu\nu} \mathcal{D}_\lambda A^\nu \tag{2.124}$$

Clearly, in the absence of gravity and in the standard Cartesian system of coordinates, the affine connection vanishes everywhere and the covariant derivative reduces to the standard derivative. Notice that this is not necessarily true if we are using some other system of coordinates. Let us consider, for example, a free material particle moving along a straight line that does not cross the origin of the standard Cartesian coordinate system. The coordinates of this particle can be written as:

$$x^{\mu}(\tau) = (\tau\sqrt{a^2+1}, a\tau, b, 0)$$

$$\Rightarrow \frac{dx^{\mu}}{d\tau} = (\sqrt{a^2+1}, a, 0, 0)$$

$$\Rightarrow \frac{d^2x^{\mu}}{d\tau^2} = 0$$

(2.125)

as expected for free particles in zero gravity. On the other hand, in spherical coordinates:

$$r(\tau) = \sqrt{a^2\tau^2 + b^2}$$

$$\theta(\tau) = \arcsin\left(\frac{b}{\sqrt{a^2\tau^2 + b^2}}\right)$$

(2.126)

where we have ignored the trivial $\varphi = 0$ expression. Then, the speed of the particle expressed in these coordinates is given by:

$$u^r = \frac{dr}{d\tau} = \frac{a^2\tau}{\sqrt{a^2\tau^2 + b^2}}$$

$$u^{\theta} = \frac{d\theta}{d\tau} = \frac{-ab}{a^2\tau^2 + b^2}$$

(2.127)

and the acceleration is calculated to be:

$$\frac{d^2r}{d\tau^2} = \frac{a^2b^2}{(a^2\tau^2 + b^2)^{3/2}}$$

$$\frac{d^2\theta}{d\tau^2} = \frac{2a^3b\tau}{(a^2\tau^2 + b^2)^2}$$

(2.128)

As such, with non-zero time derivatives, one could wrongly believe that this is an accelerated system. However, if we had used the covariant derivatives, we would have obtained:

$$\frac{\mathcal{D}u^r}{\mathcal{D}\tau} = 0$$

$$\frac{\mathcal{D}u^{\theta}}{\mathcal{D}\tau} = 0$$

(2.129)

which is the right description of a free particle in the absence of a gravitational field. Furthermore, we can see that even though there is no gravitation, the affine connection is different from zero as a consequence of using the spherical coordinates system.

Finally, let us note that, even though the principle of equivalence guarantees the existence of a local inertial frame where $\Gamma^{\mu}_{\alpha\beta} = 0$ for all points in spacetime, this is only true for a small region of space surrounding the point in question. The affine connection may vanish in a small region of this point, but because it is not a tensor field, we cannot expect that it will be zero everywhere else for arbitrary coordinate systems. Furthermore, the affine connection may be zero in this small region of space, but its derivatives may be non-vanishing.

2.5.3 The coordinate basis

The geometric interpretation of the vector notation used in this book is grounded on a *local coordinate basis of the tangent space* T_p given by the set of ordinary partial derivatives $\{\partial_\mu\}$. Indeed, it is not possible in general relativity to define a vector as an arrow that stretches from one point towards another point in a curved manifold [1].

A local coordinate basis of the tangent space formalizes the notion of vectors parallel to some determined coordinate axes [1, 6]. In this way a four-vector $\mathbf{V}$ is formally given by:

$$\mathbf{V} = V^\mu \partial_\mu \tag{2.130}$$

And under a general coordinate transformation:

$$\mathbf{V} \rightarrow \tilde{\mathbf{V}} = \tilde{V}^\mu \tilde{\partial}_\mu = \frac{\partial \tilde{x}^\mu}{\partial x^\nu} V^\nu \frac{\partial x^\alpha}{\partial \tilde{x}^\mu} \partial_\alpha = V^\mu \partial_\mu = \mathbf{V} \tag{2.131}$$

where we have used the transformation rules for the contravariant vector V^μ and the covariant local basis element ∂_μ. In this case, the vector $\mathbf{V}$ remains unchanged, but its components V^μ change under a general transformation of coordinates.

Notice that the elements of T_p are the contravariant tensors of the type $A^{\mu\nu\cdots}$. Similarly, we can define the cotangent space T_p^* which elements are the covariant vectors of the type $B_{\alpha\beta}\ldots$ [1, 6]. In this case, the coordinate basis is given by the set of *one forms* $\{dx^\mu\}$ and an arbitrary vector is written in terms of its covariant components as:

$$\mathbf{B} = B_\mu dx^\mu \tag{2.132}$$

For most of this book we will work using the tensor components $T^{\mu\nu\cdots}{}_{\alpha\beta\ldots}$ without making explicit reference to the coordinate bases $\{\partial_\mu\}$ and $\{dx^\mu\}$.

2.5.4 The minimal substitution rule

By invoking the principle of general covariance, it is possible to asses the effects of a gravitational field on a physical system by following these four steps (often referred as the *minimal substitution rule*):

1. Write down the equation that holds in the absence of gravitation using objects that behave as tensors under Lorentz transformations.
2. Replace the Minkowski tensor $\eta_{\mu\nu}$ with the metric tensor $g_{\mu\nu}$.

3. Replace standard derivatives with covariant derivatives.
4. Replace the non-invariant volume element d^4x with the invariant volume element $\sqrt{-g}\, d^4x$.

The resulting equation will be generally covariant and will remain valid in the presence of a gravitational field[6].

For example, the four-velocity of a massive particle is given by:

$$u^\mu = \frac{dx^\mu}{d\tau} \qquad (2.133)$$

with the normalization:

$$u^\mu u_\mu = -1 \qquad (2.134)$$

which is a result of imposing:

$$dS^2 = -d\tau^2 \qquad (2.135)$$

Then, Newton's Law in the absence of gravitation can be written as:

$$f^\mu = m\frac{du^\mu}{d\tau} \qquad (2.136)$$

where f^μ describes non-gravitational forces. Notice that the above equation is already expressed in terms of mathematical objects that behave as tensors under a Lorentz transformation. Then, the same equation in the presence of a gravitational field is given by:

$$f^\mu = m\frac{\mathcal{D}u^\mu}{\mathcal{D}\tau} = m\left(\frac{du^\mu}{d\tau} + \Gamma^\mu_{\alpha\beta}u^\alpha u^\beta\right) \qquad (2.137)$$

where f^μ represents all non-gravitational forces and the effects of gravitation are included by the term that involves the affine connection.

It is important to remark that the principle of general covariance holds as long as we consider a region of spacetime that is very small compared with the scale of the entire gravitational field [7]. Indeed, it is only in a small region of space that the principle of equivalence guarantees a coordinate system in which the effects of gravitation are absent. So, for instance, the size of the Moon ($\approx 1.74 \times 10^3$ km) is not much smaller than the Earth–Moon distance ($\approx 3.84 \times 10^5$ km). Therefore, the dynamical analysis of the effects of the Earth's gravitational field on the Moon's trajectory cannot be accurately calculated by invoking the principle of equivalence. In such a case, one can proceed to compute the effect of the Earth's gravitational field on infinitesimal elements of lunar mass, and then integrate over the entire volume. Alternatively, one could solve Einstein's field equations (described later in this chapter) for two bodies: the Earth and the Moon.

---

[6] As we will explicitly see in an example in section 5.5, the straightforward application of the minimal substitution rule may lead to ambiguities [6]. As such, it is necessary to have a good understanding of all the approximation and relationships used to obtain a given physical equation in flat spacetime, before we can apply the minimum substitution rule to obtain a generally covariant expression. Furthermore, section 6.7 discusses how the radiation emitted by an accelerated particle is not a generally covariant concept, and therefore the application of the minimal substitution rule to Larmor's equation leads to an incorrect result.

2.5.5 The energy–momentum tensor

Finally, let us recall that the *symmetric energy–momentum tensor* $T_{\mu\nu}$ describes all the energetic aspects of the matter field in consideration (i.e. energy density, pressure and stress) [7]. The energy–momentum tensor can be defined as the flux of momentum p^μ across a surface of constant x^ν. The mathematical expression of $T_{\mu\nu}$ can be obtained using variational calculus and Noether's theorem [19]. It can be shown that the energy–momentum tensor is given by:

$$T_{\mu\nu} \equiv 2 \frac{1}{\sqrt{-g}} \frac{\delta S_M}{\delta g^{\mu\nu}} \tag{2.138}$$

and equivalently:

$$T^{\mu\nu} \equiv -2 \frac{1}{\sqrt{-g}} \frac{\delta S_M}{\delta g_{\mu\nu}} \tag{2.139}$$

which satisfy:

$$\mathcal{D}_\mu T^{\mu\nu} = 0 \tag{2.140}$$

and:

$$T^{\mu\nu} = T^{\nu\mu} \tag{2.141}$$

where S_M is the action for the matter field[7].

2.5.6 The Euler–Lagrange equations

According to the principle of general covariance, the action for a particle in a gravitational field is given by:

$$S[x^\mu, \dot{x}^\mu] = -m \int_a^b \mathrm{d}\tau \tag{2.142}$$

where:

$$\dot{x}^\mu \equiv \frac{\mathrm{d}x^\mu}{\mathrm{d}\tau} \tag{2.143}$$

This expression makes explicit that the gravitational field is completely related to the change of the metric of the spacetime [4]. The proper time interval is related to the metric tensor through:

$$\mathrm{d}\tau^2 = -g_{\mu\nu}\mathrm{d}x^\mu \mathrm{d}x^\nu \tag{2.144}$$

---

[7] To derive these expressions we have used the fact that:

$$\delta(\delta_\nu^\mu) = 0 \Rightarrow \delta(g^{\mu\alpha}g_{\nu\alpha}) = \delta(g^{\mu\alpha})g_{\nu\alpha} + g^{\mu\alpha}\delta(g_{\nu\alpha}) = 0$$
$$\Rightarrow \delta(g^{\mu\alpha})g_{\nu\alpha} = -g^{\mu\alpha}\delta(g_{\nu\alpha})$$
$$\Rightarrow \delta(g^{\mu\alpha})\delta_\alpha^\beta = -g^{\nu\beta}g^{\mu\alpha}\delta(g_{\nu\alpha})$$
$$\Rightarrow \delta(g^{\mu\nu}) = -g^{\mu\alpha}g^{\nu\beta}\delta(g_{\alpha\beta})$$

and the integral is evaluated on some trajectories $x^\mu(\tau)$ and $\dot{x}^\mu(\tau)$, and the limits of integration are determined by the proper times τ_a and τ_b [4]. Similarly, for an arbitrary Lagrangian $\mathcal{L}$, the action is written as:

$$S[x^\mu, \dot{x}^\mu] = \int_a^b \mathcal{L} \, d\tau \qquad (2.145)$$

In both cases, the principle of least action states that the variation of the action has to be zero for the real trajectory $(x^\mu, \dot{x}^\mu)$:

$$\delta S = 0 \qquad (2.146)$$

In curved space, the variation of the action is evaluated with respect to the changes:

$$\begin{aligned} x^\mu &\rightarrow x^\mu + \delta x^\mu \\ g_{\mu\nu} &\rightarrow g_{\mu\nu} + (\partial_\alpha g_{\mu\nu})\delta x^\alpha \end{aligned} \qquad (2.147)$$

because the gravitational field is related to the change of the metric of the spacetime [1]. Notice that the variation of the metric tensor is obtained from a Taylor series expansion and it involves the standard derivative, not the covariant derivative. This is correct because we have to interpret the components of the metric tensor as functions of spacetime in some specific coordinate system [1].

Therefore, with the above variation on the coordinates and the metric tensor, the Euler–Lagrange equations for an arbitrary Lagrangian $\mathcal{L}$ are found to be:

$$\frac{d}{d\tau}\left(\frac{\partial \mathcal{L}}{\partial \dot{x}^\mu}\right) - \frac{\partial \mathcal{L}}{\partial x^\mu} = 0 \qquad (2.148)$$

where we notice that they involve the standard derivative, not the covariant derivative, for the reasons explained earlier [1]. Needless to say, we could use the Euler–Lagrange equation to derive the geodesic equation we found before.

2.6 The Hamilton–Jacobi equations

Let us recall that the action of a particle in a gravitational field is given by:

$$S[q, \dot{q}] = -m \int_a^b d\tau \qquad (2.149)$$

where the integral is evaluated on some trajectories q and $\dot{q}$, and the limits of integration are determined by the proper times τ_a and τ_b [4].

As we will discuss in chapter 5, the Hamilton–Jacobi equations are necessary to understand better the dynamics of quantum particles in curved spacetime. To this end, let us recall that the Hamilton–Jacobi equations consider the action S evaluated on a real trajectory $(q, \dot{q})$ such that $\delta S = 0$, and it is expressed as a function of the coordinates of the upper integration limit:

$$S[q(\tau_b), \dot{q}(\tau_b)] = -m \int_a^b d\tau \qquad (2.150)$$

where $q^\mu(\tau)$ satisfies the Euler–Lagrange equations. In other words, the action is considered as an indefinite integral [13, 14].

In such a case, the momentum, Lagrangian and Hamiltonian are related to the action through the following expressions:

$$p^\mu = \frac{\partial S}{\partial q_\mu}$$

$$\mathcal{L} = \frac{\mathrm{d}S}{\mathrm{d}\tau} \tag{2.151}$$

$$\mathcal{H} = -\frac{\partial S}{\partial \tau}$$

and the *Hamilton–Jacobi equation* is written as:

$$\frac{\partial S}{\partial \tau} + \mathcal{H}\left(q^\mu, \frac{\partial S}{\partial q^\mu}\right) = 0 \tag{2.152}$$

where $\mathcal{H}$ is the Hamiltonian of the system and we have used an explicitly covariant notation [4, 13, 14]. In other words, solving for S in the Hamilton–Jacobi equation is entirely equivalent to the solution found with the Euler–Lagrange equations.

Recalling that:

$$g_{\mu\nu}p^\mu p^\nu = -m^2 \tag{2.153}$$

the Hamilton–Jacobi equation for a particle in a gravitational field is given by:

$$g^{\mu\nu}\frac{\partial S}{\partial x^\mu}\frac{\partial S}{\partial x^\nu} + m^2 = 0 \tag{2.154}$$

which is equivalent to the equations of motion found in section 2.4 [4].

2.7 Einstein's field equations

As previously discussed, the effects of a gravitational field on the dynamics of a particle are fully determined by the metric tensor. The big challenge, of course, is to determine the metric tensor from an arbitrary distribution of mass and energy.

The metric tensor is related to the energy–momentum tensor $T_{\mu\nu}$ through the *Einstein field equations* given by:

$$R_{\mu\nu} - \frac{1}{2}g_{\mu\nu}R = -8\pi G T_{\mu\nu} \tag{2.155}$$

where:

$$R^\lambda{}_{\mu\nu\kappa} \equiv \frac{\partial \Gamma^\lambda_{\mu\nu}}{\partial x^\kappa} - \frac{\partial \Gamma^\lambda_{\mu\kappa}}{\partial x^\nu} + \Gamma^\eta_{\mu\nu}\Gamma^\lambda_{\kappa\eta} - \Gamma^\eta_{\mu\kappa}\Gamma^\lambda_{\nu\eta}$$

$$R_{\mu\kappa} \equiv R^\lambda{}_{\mu\lambda\kappa} \tag{2.156}$$

$$R \equiv g^{\mu\kappa}R_{\mu\kappa}$$

are the *Riemann tensor* (also known as the *curvature tensor*), the *Ricci tensor* and the *curvature scalar*, respectively, and G is the *universal gravitational constant* [1, 3, 5, 7].

Clearly, the Einstein field equations are a system of second-order partial differential equations on the elements of the metric tensor. By any means, the solutions to these equations are hard to find and only a few *exact solutions* exist for specific symmetries of the spacetime [20, 21]. In this text we will limit our discussion to a stationary isotropic spacetime produced by a spherically-symmetric object of mass M (the Schwarzschild solution) and a stationary axisymmetric spacetime produced by a rotating spherical object of mass M and angular momentum Ma (the Kerr solution). These solutions will be discussed in more detail in chapters 6 and 8, respectively.

The notion of curvature is quintessential in general relativity, as it describes the gravitational interaction. In other words, the physical description of a gravitational fields is entirely found in the curvature tensor. For instance, in a total absence of gravitational fields, the curvature tensor is identically zero everywhere in the manifold:

$$R^\lambda_{\ \mu\nu\kappa} = 0 \qquad (2.157)$$

This implies that the affine connection and its derivatives have to be zero everywhere. Furthermore, the above expression is valid everywhere for arbitrary coordinate systems because $R^\lambda_{\ \mu\nu\kappa}$ is a tensor field.

On the other hand, the principle of equivalence guarantees that, for every point in an arbitrary gravitational field, we can find a local inertial coordinate system such that the laws of physics take the same form as in an unaccelerated frame without gravitation. However, for this to happen, we need to be restricted to a very small region of space around the point in question. That is, around this specific point under consideration, the affine connection is zero, but not its derivatives. Also, because the affine connection is not a tensor, we cannot expect that the affine connection will be zero everywhere for arbitrary coordinate systems.

Therefore, the true condition to determine the absence of a gravitational field is to have a curvature tensor that takes the zero value *everywhere* for arbitrary coordinate systems. As such, it may be difficult to determine the presence of a gravitational field by simple visual inspection of the metric tensor. This is specifically true if the metric tensor is not expressed in traditional Cartesian coordinates.

At this point it is important to clarify the approximation that will be made throughout the book during the analysis of the dynamics of bodies moving through curved spacetime. First, we will assume that we can neglect the self-gravitational field of the body, so we will be considering *test* bodies immersed in an external gravitational field. As a consequence, the solutions to the Einstein field equations do not involve the energy–momentum tensor of the test body. Second, we will assume that the spatial extension of the body is much smaller than the characteristic spacetime curvature scale. That is, the test body is considered as a small *particle*. Thus, we will be considering *test particles* moving in an external gravitational field [22].

2.8 Principles of conservation

The equations that describe the motion of a test particle can be determined using variational principles [3]. For instance, we can use the general covariant Lagrangian for a free-falling particle of mass m:

$$\mathcal{L} = m\, \dot{x}^{\mu} \dot{x}^{\nu} g_{\mu\nu}(x) \tag{2.158}$$

and the associated Euler–Lagrange equation:

$$\frac{\mathrm{d}}{\mathrm{d}\tau}\left(\frac{\partial \mathcal{L}}{\partial \dot{x}^{\mu}}\right) - \frac{\partial \mathcal{L}}{\partial x^{\mu}} = 0 \tag{2.159}$$

At this point we want to explore the quantities that are conserved when a test particle interacts with a gravitational field.

To this end, let us write the Euler–Lagrange equation as:

$$\begin{aligned}
0 &= \frac{\mathrm{d}}{\mathrm{d}\tau}\left(\frac{\partial \mathcal{L}}{\partial \dot{x}^{\alpha}}\right) - \frac{\partial \mathcal{L}}{\partial x^{\alpha}} \\[1em]
&= \frac{\mathrm{d}}{\mathrm{d}\tau}\left(2\dot{x}^{\nu} g_{\nu\alpha}(x) + \dot{x}^{\mu}\dot{x}^{\nu}\frac{\partial}{\partial \dot{x}^{\alpha}} g_{\mu\nu}(x)\right) m \\[1em]
&\quad - m\dot{x}^{\mu}\dot{x}^{\nu}\frac{\partial}{\partial x^{\alpha}} g_{\mu\nu}(x) \\[1em]
&= 2m\frac{\mathrm{d}\dot{x}_{\alpha}}{\mathrm{d}\tau} - m\dot{x}^{\mu}\dot{x}^{\nu}\frac{\partial}{\partial x^{\alpha}} g_{\mu\nu}(x)
\end{aligned} \tag{2.160}$$

because:

$$\frac{\partial}{\partial \dot{x}^{\alpha}} g_{\mu\nu}(x) = 0 \tag{2.161}$$

and as a consequence:

$$\frac{\mathrm{d}\dot{x}_{\alpha}}{\mathrm{d}\tau} = \frac{\dot{x}^{\mu}\dot{x}^{\nu}}{2}\frac{\partial}{\partial x^{\alpha}} g_{\mu\nu}(x) \tag{2.162}$$

Therefore, if the metric tensor is independent of the four-coordinate x^{i}, then:

$$\frac{\mathrm{d}\dot{x}_{i}}{\mathrm{d}\tau} = 0 \tag{2.163}$$

and $\dot{x}_{i}$ is constant over the entire geodesic path of the particle [1, 3].

Let us recall that, in the most general case of a test particle with mass m, we define the four-momentum as:

$$p^{\mu} \equiv m u^{\mu} \tag{2.164}$$

Then, if the metric tensor is independent of the four-coordinate i, then, the ith component of the covariant momentum four-vector is conserved:

$$\frac{dp_i}{d\tau} = 0 \tag{2.165}$$

over the entire geodesic path of the particle.

Notice that the conservation equation involves the 'standard' derivative along the geodesic path instead of the covariant derivative. Indeed, the covariant derivative for a particle on a geodetic path is always zero:

$$\frac{\mathcal{D}p_\mu}{\mathcal{D}\tau} = \frac{p^\nu \mathcal{D}_\nu p_\mu}{m} = 0 \tag{2.166}$$

for all the covariant components of the four-momentum. However, this equation also implies:

$$0 = \frac{p^\nu \partial_\nu p_\mu - \Gamma^\sigma_{\alpha\mu} p^\alpha p_\sigma}{m}$$

$$= \frac{dp_\mu}{d\tau} - \frac{1}{2m} (\partial_\mu g_{\alpha\beta}) p^\alpha p^\beta \tag{2.167}$$

And therefore:

$$\frac{dp_\mu}{d\tau} = \frac{1}{2m} (\partial_\mu g_{\alpha\beta}) p^\alpha p^\beta \tag{2.168}$$

which clearly reduces to the invariance conclusion stated before if the metric tensor is independent of a given coordinate.

2.9 Killing vectors

Let us assume we have a metric that is *form invariant* under a general coordinate transformation. That is:

$$x \to \tilde{x} \Rightarrow g_{\mu\nu}(x) \to \tilde{g}_{\mu\nu}(\tilde{x}) : \tilde{g}_{\mu\nu}(y) = g_{\mu\nu}(y) \qquad \forall y \tag{2.169}$$

In other words, the functional relation between the metric and its argument remains invariant. Notice that, although the tensor transformation rule of the metric is given by:

$$g_{\mu\nu}(x) = \frac{\partial \tilde{x}^\alpha}{\partial x^\mu} \frac{\partial \tilde{x}^\beta}{\partial x^\nu} \tilde{g}_{\alpha\beta}(\tilde{x}) \tag{2.170}$$

the form invariant condition implies that:

$$g_{\mu\nu}(x) = \frac{\partial \tilde{x}^\alpha}{\partial x^\mu} \frac{\partial \tilde{x}^\beta}{\partial x^\nu} g_{\alpha\beta}(\tilde{x}) \tag{2.171}$$

Any coordinate transformation that leads to a form invariant metric is called an *isometry* [1, 7, 15].

Let us consider now an infinitesimal coordinate transformation of the form:

$$x^\mu \rightarrow \tilde{x}^\mu = x^\mu + \epsilon \xi^\mu \quad |\epsilon| \ll 1 \qquad (2.172)$$

Then, it can be shown that, if a metric tensor is form invariant under this transformation, the four-vectors ξ^μ need to obey:

$$\mathcal{D}_\mu \xi_\nu + \mathcal{D}_\nu \xi_\mu = 0 \qquad (2.173)$$

If this is the case, the four-vector field ξ^μ is called a *Killing vector* of the metric $g_{\mu\nu}$, and the expression above is known as the *Killing equation* [1, 5, 7]. Furthermore, it can be noted that any linear combination of Killing vectors is another Killing vector. Therefore, the vector space spanned by the Killing vectors determines the infinitesimal isometries of the metric tensor.

Let us consider the case of a metric tensor that is independent of the $\tilde{\sigma}$ coordinate. As discussed in the previous section, the $\tilde{\sigma}$ component of the covariant four-momentum is conserved along the geodesic:

$$\partial_{\tilde{\sigma}} g_{\mu\nu} = 0 \Rightarrow \frac{\mathrm{d}p_{\tilde{\sigma}}}{\mathrm{d}\tau} = 0 \qquad (2.174)$$

However, there is an alternative method for arriving at this same conclusion using Killing vectors. Indeed, the independence of $g_{\mu\nu}$ on the $\tilde{\sigma}$ component also implies that the metric tensor is form invariant under the transformation:

$$x^{\tilde{\sigma}} \rightarrow \tilde{x}^{\tilde{\sigma}} = x^{\tilde{\sigma}} + a^{\tilde{\sigma}} \qquad (2.175)$$

which suggests the existence of a Killing vector. Indeed, in component notation we can write the Killing vector as:

$$\xi^\mu = \delta^\mu_{\tilde{\sigma}} \qquad (2.176)$$

and therefore:

$$p_{\tilde{\sigma}} = \xi^\mu p_\mu \qquad (2.177)$$

It is important to remark that the previous two equations do not have unbalanced indices. The symbol $\tilde{\sigma}$ represents a fixed label, and should not be confused with a covariant index.

The derivative of $p_{\tilde{\sigma}}$ is given by:

$$\frac{\mathrm{d}p_{\tilde{\sigma}}}{\mathrm{d}\tau} = u^\mu \partial_\mu (\xi^\nu p_\nu)$$

$$= \frac{p^\mu \mathcal{D}_\mu (\xi_\nu p^\nu)}{m}$$

$$= \frac{p^\mu p^\nu (\mathcal{D}_\mu \xi_\nu + \mathcal{D}_\nu \xi_\mu)}{m}$$

$$= 0 \qquad (2.178)$$

and as a consequence, the existence of a Killing vector ξ^μ that satisfies the Killing equation implies that the scalar $\xi^\mu p_\mu$ is conserved along a geodetic trajectory.

At this point it is convenient to introduce the concept of the *Lie derivative* along the vector field V^μ, which acts on scalars, vectors and the metric tensor as:

$$£_V f = V^\mu \nabla_\mu f$$
$$£_V \omega_\mu = V^\nu \nabla_\nu \omega_\mu + \omega_\nu \nabla_\mu V^\nu \qquad (2.179)$$
$$£_V g_{\mu\nu} = \nabla_\mu V_\nu + \nabla_\nu V_\mu$$

where ∇_μ represents *any* symmetric torsion-free derivative operator (including, but not limited to, the covariant derivative $\mathcal{D}_\mu$ obtained from the metric tensor $g_{\mu\nu}$) [1]. Therefore, the Killing equation can be written as:

$$£_\xi g_{\mu\nu} = 0 \qquad (2.180)$$

which represents the Lie derivative of the metric tensor $g_{\mu\nu}$ along the Killing vector field ξ.

2.10 Tetrad fields

As we will discuss in chapter 5, the best way to study the dynamics of massive spin-$\frac{1}{2}$ particles in gravitational fields is through the use of local inertial frames defined at each point of spacetime. These local inertial frames are defined through a *tetrad field* $e^a{}_\mu(x)$, which is a set of four linearly independent coordinate four-vector fields [6, 23, 24]. Furthermore, the principle of general covariance requires that special relativity applies to these locally inertial frames.

In this book we will rely on the following convention: Latin indices $a, b, c \ldots$ refer to coordinates in the local inertial frame, whereas Greek indices $\mu, \nu \ldots$ correspond to the general coordinate system.

Tetrad fields are used to replace, for example, a four-vector field A^μ into four scalar fields A^a, with $a = 0, 1, 2, 3$, as follows:

$$A^a(x) = e^a{}_\mu(x)\, A^\mu(x) \qquad (2.181)$$

and vice versa:

$$A^\mu(x) = e_a{}^\mu(x)\, A^a(x) \qquad (2.182)$$

In these equations, $A^\mu(x)$ is a covariant vector field that transforms as a tensor under general coordinate transformations:

$$A^\mu(x) \rightarrow \tilde{A}^\mu(\tilde{x}) = \frac{\partial \tilde{x}^\mu}{\partial x^\nu} A^\nu(x) \qquad (2.183)$$

Physical quantities that transform with this type of tensor transformation rule are either *coordinate scalars* or *coordinate tensors*.

On the other hand, $A^a(x)$ is a set of four scalar fields ($a = 0, 1, 2, 3$) that transform as scalars under general coordinate transformations, but as a vector under local Lorentz transformations:

$$A^a(x) \rightarrow \tilde{A}^b(\tilde{x}) = \Lambda^a{}_b(x)\, A^b(x) \tag{2.184}$$

where, in general, the Lorentz transformation $\Lambda^a{}_b(x)$ can depend on the position in spacetime[8]. Physical quantities that transform with this type of transformation rule are *Lorentz scalars, Lorentz tensors,* or *Lorentz spinors.*

The Minkowski metric η^{ab} in the local inertial frame and the spacetime metric tensor $g^{\mu\nu}(x)$ are related through the tetrad field:

$$
\begin{aligned}
g^{\mu\nu}(x) &= e_a{}^{\mu}(x)\, e_b{}^{\nu}(x)\, \eta^{ab} \\
\eta^{ab} &= e^a{}_{\mu}(x)\, e^b{}_{\nu}(x)\, g^{\mu\nu}(x)
\end{aligned}
\tag{2.185}
$$

and the tetrad fields obey the following orthonormality conditions:

$$
\begin{aligned}
e^a{}_{\mu}(x)\, e_a{}^{\nu}(x) &= \delta^{\nu}_{\mu} \\
e^a{}_{\mu}(x)\, e_b{}^{\mu}(x) &= \delta^a_b
\end{aligned}
\tag{2.186}
$$

In other words, the tetrad field is not completely arbitrary, but it is required to produce the right metric tensor as described in equation (2.185). Furthermore, the lowering and raising of indices is done with the metric tensor for general coordinate indices, and with the Minkowski tensor for indices corresponding to the local inertial frame.

Therefore, tetrad fields can be used in dynamical equations to shift the dependence of spacetime curvature from the vector fields to the tetrad fields. Indeed, instead of working with A^{μ} defined in the general coordinate system, it is possible to work with $A^a e_a{}^{\mu}(x)$. As A^a is a set of four Lorentz scalar fields, then all the information about the spacetime curvature is embedded in the tetrad field $e_a{}^{\mu}(x)$. In other words, the contraction of a tensor in the general coordinates frame with a tetrad field can be understood as a *projection* to the local inertial frame. And, equivalently, the contraction of a four-vector in the local inertial frame with the tetrad field corresponds to a projection to the general coordinates system.

Alternatively, one could interpret the tetrad field as a *local orthonormal basis on the tangent space* T_p [1, 6]. This is in contrast to the *local coordinate basis of the tangent space* T_p given by the set of ordinary partial derivatives $\{\partial_{\mu}\}$. Indeed, the coordinate basis $\{\partial_{\mu}\}$ is *not* orthonormal except for the trivial case of a flat spacetime. Therefore, the tetrad field is a *non-holonomic* (i.e. non-coordinate) orthonormal basis of smooth vector fields $e^a{}_{\mu}(x)$.

---

[8] As we will see in section 5.8, as well as in the examples discussed in chapters 6 and 8, the analysis of spin-$\frac{1}{2}$ fields in curved spacetimes involves local Lorentz transformations that explicitly depend on the position in spacetime.

In this regard, it is important to remark that the tetrad field $e^a_{\ \mu}(x)$ is a *set* of four covariant vector fields:

$$e^a_{\ \mu}(x) = \left\{ e^0_{\ \mu}(x), e^1_{\ \mu}(x), e^2_{\ \mu}(x), e^3_{\ \mu}(x) \right\} \qquad (2.187)$$

and *not* a single second-rank tensor of indices a and μ. In other words, the tetrad field is a Lorentz vector and a coordinate vector. Therefore, the tetrad field transforms as:

$$e^a_{\ \mu}(x) \rightarrow \tilde{e}^a_{\ \mu}(\tilde{x}) = \frac{\partial x^\nu}{\partial \tilde{x}^\mu} \, e^a_{\ \nu}(x) \qquad (2.188)$$

under a transformation in the general coordinate system, and as:

$$e^a_{\ \mu}(x) \rightarrow \tilde{e}^a_{\ \mu}(\tilde{x}) = \Lambda^a_{\ b}(x) \, e^b_{\ \mu}(x) \qquad (2.189)$$

under a Lorentz transformation in the local inertial system.

As we will describe in more detail in chapter 5, tetrads are necessary to describe spin-$\frac{1}{2}$ particles in the presence of gravitational fields. Indeed, as we will discuss in more detail in section 5.6, it is not possible to express spinors within the context of general relativity. The origin of this impediment resides on the fact that general coordinate transformations in general relativity are described by the general linear group GLR(4) made of all regular (invertible) 4×4 matrices. And it can be shown that GLR(4) is a group without a spinor representation. However, it has been shown that every non-compact four dimensional pseudo-Riemannian manifold with a C^∞ Lorentz metric has a global spinor structure if, and only if, there is a global system of orthonormal tetrads [25]. Therefore, it is important to understand under what conditions it is possible to construct a tetrad field. Alternatively, we need to find out the conditions for the existence of non-vanishing orthonormal coordinate vectors in all points of spacetime. If such vector fields exist, then the spacetime is called *parallelizable* [19].

Needless to say, the existence of parallelizable manifolds has very strict conditions, and it is expected that not every manifold will be parallelizable. For instance, $\mathbb{S}^2$ (the two-dimensional surface of a three-dimensional ball in three-dimensional space) is not parallelizable. On the other hand, the only n-spheres that are parallelizable are $\mathbb{S}^1$, $\mathbb{S}^3$ and $\mathbb{S}^7$. In general, the existence of parallelizable spacetimes in n-spheres is similar to the problem of 'combing the hair of an n-sphere' [19].

However, in general, parallelization is a sufficient, but not necessary, condition for a manifold to accept a spinor structure [19]. The determination of the existence of a spinor structure is given by the vanishing of the *second Steiffel–Whitney class of the tangent bundles* [26–28]. Indeed, a manifold $\mathcal{M}$ accepts a global spinor structure if, and only if:

$$H^2(\mathcal{M}; \mathbb{Z}_2) = 0 \qquad (2.190)$$

expressed in terms of the cohomology group of the characteristic class[9].

---

[9] It is important to note that, even if $H^2(\mathcal{M}; \mathbb{Z}_2) \neq 0$, it is still possible to define a *generalized spin structure* [31]. The basic idea boils down to replacing the spin group by a different group that does not have strong topological restrictions [19].

It is easy to notice that this condition is weaker than parallelization. For instance, it is possible to show that:

$$H^2(\mathbb{S}^n; \mathbb{Z}_2) = 0 \quad \forall n \in \mathbb{N} \tag{2.191}$$

That is, all n-spheres accept a global spinor structure, even if some of them are not parallelizable [19].

Although these expressions that invoke sophisticated methods from algebraic topology give exact and precise conditions for the existence of spinor fields, their physical meaning is not transparent. However, it has been shown that, for every C^∞ pseudo-Riemannian four-dimensional spacetime that arises from initial conditions in the form of a Cauchy surface, it is possible to define a spinor structure [29]. Let us recall that, from the initial conditions on the Cauchy surface, it is possible to predict the entire structure of the spacetime [1, 30]. In such a case the spacetime is said to be *globally hyperbolic*. As a consequence, this result supports the idea that spacetimes without spinor structure are of no physical interest [6, 19].

Finally, it is important to mention that the metric tensor is not uniquely defined by a single choice of a tetrad field. Instead, the metric tensor is defined up to a local Lorentz transformation. Indeed, if we transform the tetrad field:

$$e_a{}^\mu(x) \to \tilde{e}_a{}^\mu(\tilde{x}) = \Lambda_a^b(x)\, e_b{}^\mu(x) \tag{2.192}$$

then the metric tensor remains invariant:

$$
\begin{aligned}
g^{\mu\nu}(x) &\to \tilde{g}^{\mu\nu}(\tilde{x}) \\
&= \tilde{e}_a{}^\mu(\tilde{x})\, \tilde{e}_b{}^\nu(\tilde{x})\, \eta^{ab} \\
&= \Lambda_a^c e_c{}^\mu(x)\, \Lambda_b^d e_d{}^\nu(x)\, \eta^{ab} \\
&= \Lambda_a^c \Lambda_b^d \eta^{ab}\, e_c{}^\mu(x)\, e_d{}^\nu(x) \\
&= e_c{}^\mu(x)\, e_d{}^\nu(x)\, \eta^{cd} \\
&= g^{\mu\nu}(x)
\end{aligned}
\tag{2.193}
$$

In other words, a tetrad representation of a specific metric is not unique, and different tetrad fields will lead to the same metric tensor if they are related by local Lorentz transformations [23].

2.11 Spin in general relativity

The spin and angular momentum of a test particle can be defined through the energy–momentum tensor $T^{\alpha\beta}$ [7]. First, we construct a tensor of rank 3 as:

$$M^{\gamma\alpha\beta} \equiv x^\alpha T^{\beta\gamma} - x^\beta T^{\alpha\gamma} \tag{2.194}$$

which satisfies:

$$\frac{\partial M^{\gamma\alpha\beta}}{\partial x^\gamma} = T^{\beta\alpha} - T^{\alpha\beta} = 0 \tag{2.195}$$

because $T^{\alpha\beta}$ is a symmetric tensor. Then, in flat space-time, the *angular momentum tensor* is given by the indefinite integral:

$$S^{\alpha\beta} = \int d^3x \, M^{0\alpha\beta} \tag{2.196}$$

and the *spin four-vector* as:

$$S_\alpha = \frac{1}{2}\epsilon_{\alpha\beta\gamma\delta}S^{\beta\gamma}u^\delta \tag{2.197}$$

where:

$$u^\alpha = \frac{p^\alpha}{\sqrt{-p_\mu p^\mu}} \tag{2.198}$$

is the four-velocity of the particle [5, 7].

Because the definition of the spin four-vector S_α includes the fully antisymmetric Levi-Civita tensor contracted with the velocity four-vector u^μ, it is clear that:

$$S_\mu u^\mu = 0 \tag{2.199}$$

Furthermore, the spin dynamics for a free particle is given by:

$$\frac{dS_\alpha}{d\tau} = 0 \tag{2.200}$$

whereas the evolution of the spin of a free-falling particle in a gravitational field is described by:

$$\frac{\mathcal{D}S_\alpha}{\mathcal{D}\tau} = \frac{dS_\alpha}{d\tau} - \Gamma^\lambda_{\alpha\beta}u^\beta S_\lambda = 0 \tag{2.201}$$

which is the equation of parallel transport for S_α. On the other hand, if in addition to gravity there are some non-gravitational external forces f^μ acting on the test particle, then, the spin evolves as:

$$\frac{\mathcal{D}S_\alpha}{\mathcal{D}\tau} = \left(S_\nu \frac{f^\nu}{m}\right)u_\alpha = S_\nu a^\nu u_\alpha \tag{2.202}$$

where u^μ and a^ν are the four-velocity and four-acceleration of the test particle, respectively. This is called the *Fermi transport equation*, which clearly reduces to the parallel transport case in the absence of forces other than gravity: $f^\mu = 0$ [1, 5]. The motion of the spin under these circumstances is known as the *Thomas precession* [1, 5].

It is worth mentioning that the Fermi transport equation is a special case of:

$$\frac{\mathcal{D}A_\mu}{\mathcal{D}\tau} = A_\nu a^\nu u_\mu - A_\nu u^\nu a_\mu \tag{2.203}$$

which is called the *Fermi–Walker transport equation* [1, 5]. Notice that this equation reduces to the Fermi transport equation when the four-vector A^μ is orthogonal to the four-speed u^μ:

$$A_\nu u^\nu = 0 \tag{2.204}$$

which is the case when A_μ is the four-spin of the test particle S_μ.

2.12 The spin–curvature coupling

So far we have assumed that the spin and the four-momentum are not directly coupled through the spacetime curvature. That is, the dynamical equations for these two physical variables are completely decoupled from each other:

$$\frac{\mathcal{D}u^\mu}{\mathcal{D}\tau} = 0$$

$$\frac{\mathcal{D}S^{\mu\nu}}{\mathcal{D}\tau} = 0 \qquad (2.205)$$

However, in general, the spin of the particle and the spacetime curvature are coupled in a non-trivial manner. Indeed, (classical) spinning particles in general relativity are required to have a finite spatial extension. As a consequence, these particles feel *tidal gravitational forces* due to their finite extension in a non-uniform gravitational field [22, 32]. Therefore, the motion of spinning particles, either classical or quantum-mechanical, does not follow geodesics [33–37].

Let us first consider the classical case (the quantum case will be discussed in detail in chapter 5). In general relativity, classical spinning test particles with finite extension obey the *Mathisson–Papapetrou equations*:

$$\frac{\mathcal{D}P^\mu}{\mathcal{D}\tau} = -\frac{1}{2}R^\mu{}_{\nu\rho\sigma}U^\nu S^{\rho\sigma}$$

$$\frac{\mathcal{D}S^{\mu\nu}}{\mathcal{D}\tau} = P^\mu U^\nu - P^\nu U^\mu \qquad (2.206)$$

where U^μ is the four-velocity of the centre of mass of the test particle and:

$$P^\mu \equiv mU^\mu - U_\sigma \frac{\mathcal{D}S^{\mu\sigma}}{\mathcal{D}\tau} \qquad (2.207)$$

is the *generalized four-momentum* of the spinning particle and:

$$s^2 = S^\mu S_\mu = \frac{1}{2}S^{\mu\nu}S_{\mu\nu}$$

$$m^2 = -P^\mu P_\nu \qquad (2.208)$$

are constants of the motion that define the spin and mass of the test particle [15, 22, 33–35, 38–40]. Notice that we keep the basic notion of a test particle: a body with a self-field that can be neglected and with dimensions much smaller than the spacetime curvature scale.

These equations do not form a closed system, so we need to add one of the following additional conditions:

$$S^{\mu 0} = 0 \qquad (2.209)$$

$$S^{\mu\nu}U_\nu = 0 \tag{2.210}$$

$$S^{\mu\nu}P_\nu = 0 \tag{2.211}$$

which are known as the *Papepetrou, Pirani and Tulczyjew conditions*, respectively.

The equations for the spin–curvature coupling are obtained through a multipole expansion, where the monopolar interaction corresponds to the standard Newtonian force [15, 33]. Thus, the equations shown above assume that we can neglect the quadrupole and higher moments of the particle–gravitational field interaction. That is, we work under the hypothesis that the dipole interaction is much smaller that the monopole interaction:

$$\frac{GMm}{r^2} \gg \frac{GM}{r^3}s \tag{2.212}$$

where M is the mass of the source of the gravitational field. Therefore, the *Möller radius* defined as the ratio between the spin and the mass of the test particle:

$$\rho \equiv \frac{s}{m} \tag{2.213}$$

is closely identified with the strength of the spin–curvature coupling. Indeed, in our approximation we have the *condition for a spinning test particle*:

$$\rho \ll r \tag{2.214}$$

where r is the distance between the test particle m and the source of the gravitational field M [15].

It can be observed that, in general, a free-falling spinning particle does not follow a geodesic:

$$\frac{D}{D\tau}(mU^\mu) = \frac{D}{D\tau}\left(U_\sigma \frac{DS^{\mu\sigma}}{D\tau}\right) - \frac{1}{2}R^\mu_{\nu\rho\sigma}U^\nu S^{\rho\sigma} \tag{2.215}$$

It is only in the spinless case that the particle follows a geodesic:

$$S^{\mu\nu} = 0 \quad \Rightarrow \quad \frac{D}{D\tau}(mU^\mu) = 0 \tag{2.216}$$

However, it is important to mention that, except for ultra-relativistic particles, the deviation from geodetic motion is relatively small. Indeed, the deviation from geodetic motion is of the order $\mathcal{O}(s/mr)$ and it is only important when the test particle travels at relativistic speed and/or in the presence of strong gravitational fields [38–42].

Similarly, in the absence of a gravitational field, the spin–curvature coupling reduces to:

$$\frac{dU^\mu}{d\tau} = 0$$
$$\frac{dS^{\mu\nu}}{d\tau} = 0 \tag{2.217}$$

which are equivalent to the free-particle equations.

2.13 Summary

This chapter presented a brief review of the special and general theories of relativity. Although most of the material is available in introductory textbooks, the use of tetrad fields and the spin–curvature coupling are often relegated as advanced topics. However, as we will explain in a later chapter, the tetrad field formalism is essential for properly defining the dynamics of spin-$\frac{1}{2}$ particles in curved spacetime. The topic of relativistic quantum fields in flat spacetime (i.e. the Minkowski metric) will be discussed in the next chapter.

Bibliography

[1] Carroll S M 2004 *Spacetime and Geometry: An Introduction to General Relativity* (Reading, MA: Addison-Wesley)

[2] Feynman R P 1995 *Feynman Lectures on Gravitation* (Reading, MA: Addison-Wesley)

[3] Hobson M P, Efstathiou G and Lasenby A N 2006 *General Relativity: An Introduction for Physicists* (Cambridge: Cambridge University Press)

[4] Landau L D and Lifshitz E M 1975 *The Classical Theory of Fields, Course of Theoretical Physics* vol 2 (Amsterdam: Elsevier)

[5] Misner C W, Thorne K S and Wheeler J A 1973 *Gravitation* (San Francisco, CA: WH Freeman)

[6] Wald R 1984 *General Relativity* (Chicago, IL: University of Chicago Press)

[7] Weinberg S 1972 *Gravitation and Cosmology: Principles and Applications of The General Theory of Relativity* (New York: Wiley)

[8] Carmeli M 1977 *Group Theory General Relativity: Representations of the Lorentz Group and Their Applications to the Gravitational Field* (London: Imperial College Press)

[9] Sexl R U and Urbantke H K 1992 *Relativity, Groups, Particles: Special Relativity and Relativistic Symmetry in Field and Particle Physics* (Berlin: Springer)

[10] Tung W K 1985 *Group Theory in Physics* (Singapore: World Scientific)

[11] Wigner E P 1939 On the unitary representations of the inhomogeneous Lorentz group *Ann. Math.* **40** 149–204

[12] Ryder L H 1996 *Quantum Field Theory* 2nd edn (Cambridge: Cambridge University Press)

[13] Goldstein H, Poole C and Safko J 2002 *Classical Mechanics* 3rd edn (Reading, MA: Addison-Wesley)

[14] Landau L D and Lifshitz E M 1975 *Mechanics, Course of Theoretical Physics* vol 1 (Amsterdam: Elsevier)

[15] Wald R 1972 Gravitational spin interaction *Phys. Rev.* D **6** 406

[16] Hehl F W, von der Heyde P and Kerlick G D 1976 General relativity with spin and torsion: foundations and prospects *Rev. Mod. Phys.* **48** 393–416

[17] Trautman A 2006 Einstein–Cartan theory *Encyclopedia of Mathematical Physics* ed J-P Francoise, G L Naber and S T Tsou vol 2 (Oxford: Elsevier) pp 189–95

[18] Audretsch J 1981 Trajectories and spin motion of massive spin-$\frac{1}{2}$ particles in gravitational fields *J. Phys. A: Math. Gen.* **14** 411–22

[19] Parker L and Toms D 2009 *Quantum Field Theory in Curved Spacetime* (Cambridge: Cambridge University Press)

[20] Griffiths J B and Podolsky J 2009 *Exact Space-Times in Einstein's General Relativity* (Cambridge: Cambridge University Press)

[21] Stephani H *et al* 2003 *Exact Solutions of Einstein's Field Equations* (Cambridge: Cambridge University Press)

[22] Semerak O 2003 Spinning particles in the Kerr field *Nonlinear Gravitodynamics*, ed R J Ruffini and C Sigismondi (Singapore: World Scientific)

[23] de Felice F and Clarke C J S 1992 *Relativity on Curved Manifolds* (Cambridge: Cambridge University Press)

[24] Zakharov A F, Zinchuk V A and Pervushin V N 2006 Tetrad formalism and reference frames in General Relativity *Phys. Part. Nuclei* **37** 104–34

[25] Geroch R 1968 Spinor structure of spacetimes in general relativity I *J. Math. Phys.* **9** 1739

[26] Clarke C J S 1971 Magnetic charge, holonomy and characteristic classes: illustrations of the methods of topology in relativity *Gen. Rel. Grav.* **2** 43–51

[27] Isham C J 1978 Spinor fields in four dimensional space-time *Proc. R. Soc. London, Ser. A, Math. Phys. Sci.* **364** 591–99

[28] Milnor J W 1963 Spin structures on manifolds *L'enseignement Math.* **9** 198–203

[29] Geroch R 1970 Spinor structure of spacetimes in general relativity II *J. Math. Phys.* **11** 343

[30] Hawking S W and Ellis G F R 1973 *The Large Scale Structure of Space-Time* (Cambridge: Cambridge University Press)

[31] Avis S J and Isham C J 1980 Generalized spin structures on four dimensional space-times *Commun. Math. Phys.* **72** 103–18

[32] Rudiger R 1981 The Dirac equation and spinning particles in general relativity *Proc. R. Soc. London, Ser. A* **377** 417–24

[33] Corinaldesi E and Papapetrou A 1951 Spinning test-particles in general relativity. II *Proc. R. Soc. London, Ser. A, Math. Phys. Sci.* **209** 259–68

[34] Papapetrou A 1951 Spinning test-particles in general relativity. I *Proc. R. Soc. London, Ser. A, Math. Phys. Sci.* **209** 248–58

[35] Plyatsko R M, Stefanyshyn O B and Fenyk M T 2011 Mathisson Papapetrou & Dixon equations in the Schwarzschild and Kerr backgrounds *Class. Quantum Grav.* **28** 195025

[36] Plyatsko R M, Stefanyshyn O B and Fenyk M T 2010 Highly relativistic spinning particle starting near $r_{ph}^{(-)}$ in a Kerr field *Phys. Rev.* D **82** 044015

[37] Ubukhov Y N, Silenko A J and Teryaev O V 2010 Spin in stationary gravitational frames and rotating frames *The Sun, The Stars, The Universe and General Relativity* ed R Ruffini and G Vereshchagin (New York: AIP) pp 112–119

[38] Plyatsko R 2005 Ultrarelativistic circular orbits of spinning particles in a Schwarzschild field *Class. Quantum Grav.* **22** 1545–51

[39] Plyatsko R 1998 Gravitational ultrarelativistic spin-orbit interaction and the weak equivalence principle *Phys. Rev.* D **58** 084031

[40] Silenko A J 2008 Classical and quantum spins in curved spacetimes *Acta Phys. Pol. B Proc. Suppl.* **1** 87–107

[41] Mashhoon B and Singh D 2006 Dynamics of extended spinning masses in a gravitational field *Phys. Rev.* D **74** 124006

[42] Singh D 2008 Perturbation method for classical spinning particle motion: I. Kerr space-time *Phys. Rev.* D **78** 104028

Chapter 3

Relativistic quantum fields

A detailed discussion of relativistic quantum mechanics and quantum field theory is outside the scope of this book, and the reader is encouraged to peruse the rich literature available on both topics [1–7].

In this chapter we will briefly review the basic concepts, equations and notation necessary to understand the basic relativistic dynamics of massive spin-$\frac{1}{2}$ particles in inertial frames[1]. To this end, we are going to take a group-theoretical approach. That is, we will derive the equations that describe the relativistic quantum dynamics of a massive spin-$\frac{1}{2}$ particle using group-theoretical arguments.

In particular, our analysis will emphasize the behaviour of quantum states under Lorentz or Poincare transformations. We will also discuss the difference between wave functions, quantum states and quantum fields according to their Lorentz or Poincare group representation. Finally, we will discuss the structure of the relativistic quantum vacuum.

3.1 The Schrödinger equation

In non-relativistic classical mechanics, the energy of a free particle is given by:

$$E = \frac{p^2}{2m} \tag{3.1}$$

where E is the energy, $p^2 = \mathbf{p} \cdot \mathbf{p}$ is the momentum squared, and m is the mass. The transition to quantum mechanics is given by making the replacement from dynamical variables to operators acting on a wave function:

$$E \to \hat{E} \equiv i\frac{\partial}{\partial t} \qquad \mathbf{p} \to \hat{\mathbf{p}} \equiv -i\nabla \tag{3.2}$$

---

[1] The detailed analysis of massless vector fields (e.g. photons) will be considered in a companion volume to be published in the near future.

which leads to the *Schrödinger equation*:

$$i\frac{\partial}{\partial t}\Psi = -\frac{\nabla^2}{2m}\Psi \tag{3.3}$$

where the wave function:

$$\Psi = \Psi(x, y, z, t) \tag{3.4}$$

is a function of the coordinates (in natural units $\hbar = c = 1$) [8–11]. Furthermore, because the Schrödinger equation is first order in time, and second order in space, the partial differential equations suggest that the Schrödinger equation is not a covariant expression.

The quantum operators and dynamical variables are characterized by the following commutation relations:

$$\left[x_i, \hat{p}_j\right] = i\delta_{ij}$$
$$\left[x_i, x_j\right] = \left[\hat{p}_i, \hat{p}_j\right] = \left[\hat{E}, \hat{p}_j\right] = \left[\hat{E}, x_j\right] = 0 \tag{3.5}$$

where the non-covariant indices i and j run from 1 to 3 (over the three spatial coordinates) [9–11].

Finally, it is important to mention that the solutions to the Schrödinger equation obey the following *continuity equation* that expresses the conservation of localization probability of the particle:

$$\frac{\partial \rho}{\partial t} + \nabla \cdot \mathbf{j} = 0 \tag{3.6}$$

where:

$$\rho \equiv |\Psi|^2 \tag{3.7}$$

is the localization probability density and:

$$\mathbf{j} \equiv \frac{i}{2m}\left(\Psi^*\nabla\Psi - \Psi\nabla\Psi^*\right) \tag{3.8}$$

is the localization probability density flux [8].

Therefore, the most general free particle solution to the Schrödinger equation is given by a plane wave such that:

$$\Psi = Ne^{i\mathbf{p}\cdot\mathbf{x}-iEt}$$
$$\rho = |N|^2 \tag{3.9}$$
$$\mathbf{j} = \frac{\mathbf{p}}{m}|N|^2$$

where N is a normalization constant [8].

3.2 The Klein–Gordon equation

We can follow a similar strategy to build relativistic quantum equations [2, 5, 8]. The relativistic energy for a free particle is given by:

$$E^2 = \mathbf{p} \cdot \mathbf{p} + m^2 \tag{3.10}$$

or equivalently, using an explicitly covariant notation:

$$p^\mu p_\mu = -m^2 \tag{3.11}$$

Using the standard replacement from classical dynamical variables to quantum operators in the previous equation leads to:

$$\left(-\frac{\partial^2}{\partial t^2} + \nabla^2 - m^2 \right) \phi(x) = 0 \tag{3.12}$$

where we have used the shorthand notation $x = x^\mu$. Alternatively, using the explicitly covariant notation:

$$\partial_\mu \equiv \frac{\partial}{\partial x^\mu} = \left(\frac{\partial}{\partial t}, \nabla \right) \qquad \partial^\mu \equiv \frac{\partial}{\partial x_\mu} = \left(-\frac{\partial}{\partial t}, \nabla \right) \tag{3.13}$$

and:

$$p_\mu = -i\partial_\mu \qquad p^\mu = -i\partial^\mu \tag{3.14}$$

we obtain:

$$\left(\partial_\mu \partial^\mu - m^2 \right) \phi(x) = 0 \tag{3.15}$$

which is the *Klein–Gordon equation* used to describe relativistic scalar (i.e. spinless or spin=0) particles such as the Higgs boson [5, 8].

Clearly, the *classical Klein–Gordon field* $\phi(x)$ that corresponds to a free Klein–Gordon particle has two energy eigenvalues, one positive and the other negative:

$$E = \pm\sqrt{\mathbf{p} \cdot \mathbf{p} + m^2} \tag{3.16}$$

Needless to say, this is a troublesome feature, as quantum free particles are required to have positive energies[2]. However, this problem is solved when we consider *quantum fields* (i.e. the quantized Klein–Gordon field), as we will see in the next section, and in section 3.7 within the context of the Dirac quantum field.

---

[2] It is important to note that relativistic classical mechanics also encounters the issue of positive and negative energy solutions separated by a finite gap $\Delta E = 2m$. However, it can be shown that, in the classical case, there is no continuous process that can take a free particle from a positive to a negative energy state [7]. Therefore, the negative energy solutions are simply discarded as *non-physical*. This is not the case in quantum mechanics: the interaction of electrons with photons could produce quantum transitions in which a positive-energy particle falls into a negative-energy state, while the energy difference is transformed into a pair of photons in the reaction $e^- e^+ \rightarrow 2\gamma$ [7].

The action of the classical Klein–Gordon field is given by:

$$S = \int \mathcal{L}_{kg} d^4x \tag{3.17}$$

and the Klein–Gordon Lagrangian density is:

$$\mathcal{L}_{kg} = -\frac{1}{2}\partial^\mu \phi \partial_\mu \phi - \frac{1}{2}m^2\phi^2 \tag{3.18}$$

The Euler–Lagrange equations for an arbitrary scalar field density Lagrangian:

$$\mathcal{L} = \mathcal{L}(\phi, \partial_\mu \phi) \tag{3.19}$$

are given by:

$$\frac{\partial}{\partial x^\mu}\left[\frac{\partial \mathcal{L}}{\partial(\partial_\mu \phi)}\right] - \frac{\partial \mathcal{L}}{\partial \phi} = 0 \tag{3.20}$$

Needless to say, the associated Euler–Lagrange equation for $\mathcal{L}_{kg}$ leads to exactly the same Klein–Gordon equation described above [5].

Furthermore, the *canonical energy–momentum tensor* is given by:

$$\pi = \frac{\partial \mathcal{L}}{\partial \dot{\phi}} = \dot{\phi} \tag{3.21}$$

where:

$$\dot{\phi} \equiv \frac{\partial \phi}{\partial t} \tag{3.22}$$

and the *canonical energy–momentum tensor* is given by:

$$\theta^{\mu\nu} = \frac{\partial \mathcal{L}}{\partial(\partial_\mu \phi)}\partial^\nu \phi - g^{\mu\nu}\mathcal{L} \tag{3.23}$$

which obeys:

$$\partial_\mu \theta^{\mu\nu} = 0 \tag{3.24}$$

The above expressions are a direct consequence of Noether's theorem as applied to the Poincare group of symmetries [5, 12]. For the specific case of the Klein–Gordon field, the canonical energy–momentum tensor is:

$$\theta^{\mu\nu} = (\partial^\mu \phi)(\partial^\nu \phi) - g^{\mu\nu}\mathcal{L}_{kg} \tag{3.25}$$

which is symmetric in the contravariant indices [5].

At this point, it is important to remark that the canonical energy–momentum tensor $\theta^{\mu\nu}$ is different from the symmetric energy–momentum tensor $T^{\mu\nu}$ from equation (2.139). Indeed, in general, the expression for $\theta^{\mu\nu}$ that corresponds to an arbitrary scalar field, equation (3.23), is not symmetric. That is, except for specific forms of the Lagrangian $\mathcal{L}$, the energy momentum tensor $\theta^{\mu\nu}$ will not be a symmetric tensor. However, $T^{\mu\nu}$ can be understood as the generally covariant generalization of

$\theta^{\mu\nu}$ in curved spacetime [13]. As such, $T^{\mu\nu}$ describes the matter and radiation fields, and their coupling to gravitation through Einstein's field equations.

Furthermore, it is worth mentioning that, in general:

$$\partial_\mu \theta^{\mu\nu} \neq 0 \tag{3.26}$$

so $\theta^{\mu\nu}$ does not necessarily produce a conserved current in curved spacetimes. In addition, in the presence of spin the tensor $\theta^{\mu\nu}$ is not symmetric, and therefore cannot be used in the right-hand side of Einstein's field equations.

In general, it is not easy to generalize $\theta^{\mu\nu}$ to curved spacetimes [14]. For instance, the *Belinfante tensor* $\Theta^{\mu\nu}$ is a rather sophisticated generalization of $\theta^{\mu\nu}$ [15]. This new energy–momentum tensor is symmetric, produces a conserved current and acts as the source of the gravitational field [7].

Also, let us remark that non-symmetric energy–momentum tensors usually appear within the context of quantum fields with spin[3]. And, as we discussed before, these fields produce spacetime torsion. Quite remarkably, the Belinfante–Rosenfeld symmetrization of the canonical energy–momentum tensor emerges as a natural feature of Einstein–Cartan theories that incorporate spacetime torsion in the description of classical gravitational fields [16].

All that said, in the context of standard general relativity, it is better simply to rely on $T^{\mu\nu}$ as the *definition* of the energy–momentum tensor [12, 17].

The explicitly covariant probability four-current for the Klein–Gordon field is defined by:

$$j^\mu \equiv i\left(\phi^* \partial^\mu \phi - \phi \partial^\mu \phi^*\right) = (\rho, \mathbf{j}) \tag{3.27}$$

and satisfies the explicitly covariant continuity equation:

$$\partial_\mu j^\mu = 0 \tag{3.28}$$

---

[3] Noether's theorem implies conservation laws for both energy-momentum and angular momentum [5]. That is, we can construct the energy momentum tensor $\theta^{\mu\nu}$ and spin tensor $s^{\alpha\beta\gamma}$ that satisfy the covariant conservation equations:

$$\partial_\nu \theta^{\mu\nu} = 0 \qquad \partial_\alpha(x^\mu \theta^{\nu\alpha} - x^\nu \theta^{\mu\alpha} + s^{\mu\nu\alpha}) = 0$$

that correspond to the conservation of energy-momentum and angular momentum in flat spacetime, respectively [16]. Clearly, if the field is not a scalar (non-zero spin), then the energy-momentum tensor does not need to be symmetric. Instead:

$$\theta^{\mu\nu} - \theta^{\nu\mu} = \partial_\alpha s^{\mu\nu\alpha}$$

In this case, the Belinfante–Rosenfeld symmetrization procedure leads to:

$$\Theta^{\mu\nu} = \theta^{\mu\nu} + \frac{1}{2}\partial_\alpha(s^{\nu\mu\alpha} + s^{\nu\alpha\mu} + s^{\mu\alpha\nu})$$

which satisfies:

$$\partial_\nu \Theta^{\mu\nu} = 0 \qquad \Theta^{\mu\nu} = \Theta^{\nu\mu}$$

That is, $\Theta^{\mu\nu}$ is conserved and symmetric. However, in such a case the tensors $\theta^{\mu\nu}$ and $s^{\mu\nu\alpha}$ that were obtained from Noether's theorem become meaningless, which is an undesirable feature of the theory.

Therefore, we obtain the following explicitly covariant expressions for the general solution (plane wave) to the free-particle Klein–Gordon equation in flat spacetime:

$$\phi = Ne^{-ipx}$$
$$j^{\mu} = 2p^{\mu}|N|^2 \tag{3.29}$$

where N is a normalization constant [8][4].

3.3 Scalar quantum fields

The quantization of the real scalar field:

$$\phi(x) \rightarrow \hat{\phi}(x) \tag{3.30}$$

is expressed through a Fourier decomposition of the field ϕ over a set of vibrational modes $\mathbf{k}$:

$$\hat{\phi}(x) = \int \frac{d^3k}{(2\pi)^{3/2}} \frac{1}{\sqrt{2\omega_{\mathbf{k}}}} \left(e^{-i\omega_{\mathbf{k}}t + i\mathbf{k}\cdot\mathbf{x}} \hat{a}_{\mathbf{k}} + e^{i\omega_{\mathbf{k}}t - i\mathbf{k}\cdot\mathbf{x}} \hat{a}_{\mathbf{k}}^{\dagger} \right) \tag{3.31}$$

where $\hat{a}$ and $\hat{a}^{\dagger}$ are the *annihilation and creation operators*, and:

$$\omega_{\mathbf{k}}^2 \equiv \mathbf{k}\cdot\mathbf{k} + m^2 \tag{3.32}$$

is the frequency or energy of the vibrational mode $\mathbf{k}$. Notice that this field expression represents a *real* scalar quantum field in the sense that:

$$\hat{\phi}(x) = \hat{\phi}^{\dagger}(x) \tag{3.33}$$

In addition, the quantum canonical field momentum operator is given by the quantized version of equation (3.21):

$$\hat{\pi}(x) = \dot{\hat{\phi}} = -i \int \frac{d^3\mathbf{k}}{(2\pi)^{3/2}} \sqrt{\frac{\omega_{\mathbf{k}}}{2}} \left(e^{-i\omega_{\mathbf{k}}t + i\mathbf{k}\cdot\mathbf{x}} \hat{a}_{\mathbf{k}} - e^{i\omega_{\mathbf{k}}t - i\mathbf{k}\cdot\mathbf{x}} \hat{a}_{\mathbf{k}}^{\dagger} \right) \tag{3.34}$$

We assume equal time commutation relations for the quantized field variables that are similar to their classical counterpart [5]:

$$\left[\hat{\phi}(\mathbf{x}, t), \hat{\pi}(\tilde{\mathbf{x}}, t) \right] = i\delta(\mathbf{x} - \tilde{\mathbf{x}})$$
$$\left[\hat{\phi}(\mathbf{x}, t), \hat{\phi}(\tilde{\mathbf{x}}, t) \right] = 0 \tag{3.35}$$
$$\left[\hat{\pi}(\mathbf{x}, t), \hat{\pi}(\tilde{\mathbf{x}}, t) \right] = 0$$

---

[4] Here and throughout the text, we use the following shorthand notation for the scalar exponent that results from the contraction of a covariant vector with a contravatiant vector:

$$pa \equiv p_{\mu}a^{\mu}$$

With the exception of those instances where the notation is misleading or confusing, the shorthand expression will be used by omitting the contracting indices.

and, as a consequence, the annihilation and creation operators, $\hat{a}$ and $\hat{a}^\dagger$, obey the commutation relations:

$$\left[\hat{a}_{\mathbf{k}}, \hat{a}_{\mathbf{q}}^\dagger\right] = \delta(\mathbf{k} - \mathbf{q}) \qquad \left[\hat{a}_{\mathbf{k}}, \hat{a}_{\mathbf{q}}\right] = \left[\hat{a}_{\mathbf{k}}^\dagger, \hat{a}_{\mathbf{q}}^\dagger\right] = 0 \tag{3.36}$$

The energy of the classical Klein–Gordon field is given by:

$$\begin{aligned} H &= \int \theta^{00} \mathrm{d}^3 x \\ &= \frac{1}{2} \int \left((\partial_0 \phi)^2 + \nabla\phi \cdot \nabla\phi + m^2 \phi^2\right) \mathrm{d}^3 x \end{aligned} \tag{3.37}$$

where θ^{00} is the energy–momentum tensor of the classical field. Therefore, the energy of the quantum Klein–Gordon field has a similar expression given by:

$$\hat{H} = \frac{1}{2} \int \left((\partial_0 \hat{\phi})^2 + \nabla\hat{\phi} \cdot \nabla\hat{\phi} + m^2 \hat{\phi}^2\right) \mathrm{d}^3 x \tag{3.38}$$

Upon substitution of the field expression for $\hat{\phi}$ and the Klein–Gordon equation for the quantum fields:

$$\partial_0^2 \hat{\phi} - \partial_i \partial^i \hat{\phi} + m^2 \hat{\phi} = 0 \tag{3.39}$$

we can write $\hat{H}$ in terms of the annihilation and creation operators as follows:

$$\begin{aligned} \hat{H} &= \frac{1}{2} \int \mathrm{d}^3 k \omega_{\mathbf{k}} \left(\hat{a}_{\mathbf{k}} \hat{a}_{\mathbf{k}}^\dagger + \hat{a}_{\mathbf{k}}^\dagger \hat{a}_{\mathbf{k}}\right) \\ &= \int \mathrm{d}^3 k \omega_{\mathbf{k}} \left[\hat{a}_{\mathbf{k}}^\dagger \hat{a}_{\mathbf{k}} + \frac{1}{2}\delta(0)\right] \end{aligned} \tag{3.40}$$

The scalar quantum field is best defined on an energy eigenvalue occupation *Fock space* [5, 18]. That is, the eigenvectors are of the form:

$$|n_{\mathbf{k}}\rangle \tag{3.41}$$

which represents a state of the quantum field with n elementary quantum field excitations of momentum $\mathbf{k}$.

The creation and annihilation operators are used to increase or decrease the number of elementary excitations on a quantum field state:

$$\begin{aligned} \hat{a}_{\mathbf{k}}^\dagger |n_{\mathbf{k}}\rangle &= \sqrt{n_{\mathbf{k}} + 1}\, |n_{\mathbf{k}} + 1\rangle \\ \hat{a}_{\mathbf{k}} |n_{\mathbf{k}}\rangle &= \sqrt{n_{\mathbf{k}}}\, |n_{\mathbf{k}} - 1\rangle \end{aligned} \tag{3.42}$$

and the *number operator*, which eigenvalue gives the number of excitations of momentum $\mathbf{k}$ in the quantum field state, is given by:

$$\hat{N}_{\mathbf{k}} \equiv \hat{a}_{\mathbf{k}}^\dagger \hat{a}_{\mathbf{k}} \quad \Rightarrow \quad \hat{N}_{\mathbf{k}} |n_{\mathbf{k}}\rangle = n_{\mathbf{k}} |n_{\mathbf{k}}\rangle \tag{3.43}$$

It can be shown that the number of elementary excitations n_k is positive or zero, but never negative:

$$n_{\mathbf{k}} \geqslant 0 \quad \forall \mathbf{k} \tag{3.44}$$

The *quantum vacuum state* $|0\rangle$ is defined as the state with no quantum field excitations:

$$\hat{N}_{\mathbf{k}}|0\rangle = 0 \quad \forall \mathbf{k} \tag{3.45}$$

Then, the creation and annihilation operators can be used to construct arbitrary states of the quantum field from the quantum vacuum state $|0\rangle$:

$$|n_{\mathbf{k}}\rangle = \frac{\left(\hat{a}_{\mathbf{k}}^{\dagger}\right)^{n_{\mathbf{k}}}}{\sqrt{n_{\mathbf{k}}!}} |0\rangle \tag{3.46}$$

Finally, using the number operator we can rewrite the Hamiltonian of the quantum field as:

$$\hat{H} = \int \mathrm{d}^3 k \omega_{\mathbf{k}} \left[\hat{N}_{\mathbf{k}} + \frac{1}{2}\delta(0)\right] \tag{3.47}$$

Because the number of excitations is never negative, the quantum field energy is never negative. In other words, even though the Klein–Gordon equation has negative energy eigenvalues, the energy of the scalar quantum field is always non-negative.

However, equation (3.47) has an infinite contribution from the ground states owing to the presence of the Dirac delta function term. Nevertheless, considering that the zero-energy level is completely arbitrary, we can safely redefine the energy of the scalar quantum field as:

$$\hat{H} = \int \mathrm{d}^3 k \omega_{\mathbf{k}} \hat{N}_{\mathbf{k}} \tag{3.48}$$

with the property:

$$\langle 0|\hat{H}|0\rangle = 0 \tag{3.49}$$

That is, once we have renormalized the energy spectrum, the quantum vacuum is the state with zero energy [19, 20].

3.4 The quantum Poincare transformations

We now need to analyse the structure and effect of a Poincare transformation on an arbitrary quantum state Ψ (which in general could be used to represent the quantum state of a scalar, vector or spinor field) [7, 21, 22]. To this end, let us consider the general Poincare transformation:

$$x^{\mu} \rightarrow \tilde{x}^{\alpha} = \Lambda^{\alpha}_{\ \beta} x^{\beta} + a^{\alpha} \tag{3.50}$$

where Λ is a Lorentz transformation and a represents a translation. If we characterize this Poincare transformation by $T(\Lambda, a)$, then this operation is equivalent to a unitary transformation on the quantum state given by:

$$\Psi \to \tilde{\Psi} = \hat{U}(\Lambda, a)\Psi \qquad (3.51)$$

which satisfies the composition rule:

$$\hat{U}(\Lambda_1, a_1)\hat{U}(\Lambda_2, a_2) = \hat{U}(\Lambda_1\Lambda_2, \Lambda_1 a_2 + a_1) \qquad (3.52)$$

Let us now consider an infinitesimal Poincare transformation:

$$\Lambda^{\alpha}_{\ \beta} = \delta^{\alpha}_{\ \beta} + \xi^{\alpha}_{\ \beta} \qquad a^{\alpha} = \epsilon^{\alpha} \qquad |\xi|, |\epsilon| \ll 1 \qquad (3.53)$$

then, its corresponding unitary matrix can be expressed as:

$$\hat{U}(1 + \xi, \epsilon) = 1 + \frac{1}{2}i\xi_{\alpha\beta}\hat{J}^{\alpha\beta} - i\epsilon_{\alpha}\hat{P}^{\alpha} + \mathcal{O}(\xi^2) + \mathcal{O}(\epsilon^2) \qquad (3.54)$$

where $\hat{J}^{\alpha\beta}$ and $\hat{P}^{\alpha}$ are the generators of the unitary transformation associated with the infinitesimal Poincare transformations described by the ξ and ϵ parameters, respectively.

From this equation we can see that if $\hat{U}$ is unitary, as required by quantum mechanics, then:

$$\hat{U}^{-1} = \hat{U}^{\dagger} \quad \Rightarrow \quad (\hat{J}^{\alpha\beta})^{\dagger} = \hat{J}^{\alpha\beta} \quad \& \quad (\hat{P}^{\alpha})^{\dagger} = \hat{P}^{\alpha} \qquad (3.55)$$

That is, $\hat{J}^{\alpha\beta}$ and $\hat{P}^{\alpha}$ are Hermitian operators. Furthermore, as $\xi_{\alpha\beta}$ is antisymmetric on the covariant indices, so is $\hat{J}^{\alpha\beta}$.

After some algebra, it can be shown that the Poincare group generators $\hat{J}^{\alpha\beta}$ and $\hat{P}^{\alpha}$ transform as:

$$\hat{U}(\Lambda, a)\, \hat{J}^{\alpha\beta}\, \hat{U}^{-1}(\Lambda, a) = \Lambda_{\mu}^{\ \alpha}\Lambda_{\nu}^{\ \beta}\left(\hat{J}^{\mu\nu} - a^{\mu}\hat{P}^{\nu} + a^{\nu}\hat{P}^{\mu}\right)$$
$$\hat{U}(\Lambda, a)\, \hat{P}^{\alpha}\, \hat{U}^{-1}(\Lambda, a) = \Lambda_{\mu}^{\ \alpha}\hat{P}^{\mu} \qquad (3.56)$$

and satisfy the commutation relations:

$$i\left[\hat{J}^{\mu\nu}, \hat{J}^{\alpha\beta}\right] = \eta^{\nu\alpha}\hat{J}^{\mu\beta} - \eta^{\mu\alpha}\hat{J}^{\nu\beta} - \eta^{\beta\mu}\hat{J}^{\alpha\nu} + \eta^{\beta\nu}\hat{J}^{\alpha\mu}$$
$$i\left[\hat{P}^{\mu}, \hat{J}^{\alpha\beta}\right] = \eta^{\mu\alpha}\hat{P}^{\beta} - \eta^{\mu\beta}\hat{P}^{\alpha} \qquad (3.57)$$
$$i\left[\hat{P}^{\mu}, \hat{P}^{\alpha}\right] = 0$$

These three expressions completely define the *Lie algebra of the Poincare group* [7, 23–25]. More formally, the linear span of the set of generators $\{\hat{P}^{\mu}, \hat{J}^{\mu\nu}\}$ forms the Lie algebra of the Poincare group.

It is convenient to rename the Poincare group generators in the following way:

$$\hat{H} = \hat{P}^0$$

$$\hat{\mathbf{P}} = (\hat{P}^1, \hat{P}^2, \hat{P}^3)$$

$$\hat{\mathbf{J}} = (\hat{J}^{23}, \hat{J}^{31}, \hat{J}^{12}) \tag{3.58}$$

$$\hat{\mathbf{K}} = (\hat{J}^{01}, \hat{J}^{02}, \hat{J}^{03})$$

as it can be shown that $\hat{\mathbf{J}}$ and $\hat{\mathbf{K}}$ are the generators for Lorentz rotations and boosts, respectively (and as a consequence $\hat{\mathbf{J}}$ corresponds to the angular momentum operator) [7]. In addition, $\hat{H}$ and $\hat{\mathbf{P}}$ correspond to translations in time and space, respectively (and consequently, $\hat{H}$ and $\hat{\mathbf{P}}$ correspond to the energy and momentum operators, respectively) [7].

The Lie algebra of the Poincare group can be used to show that the energy, momentum and angular momentum operators commute with the Hamiltonian:

$$[\hat{J}_i, \hat{H}] = [\hat{P}_i, \hat{H}] = [\hat{H}, \hat{H}] = 0 \tag{3.59}$$

where $\hat{J}_i$ and $\hat{P}_i$ represent the individual vector components of $\hat{\mathbf{J}}$ and $\hat{\mathbf{P}}$, respectively. This expression makes clear that $\hat{J}_i$, $\hat{P}_i$ and $\hat{H}$ form a set of 'good quantum numbers' to label physical quantum states (in the sense that their eigenvalues will remain constant in time). In addition, the other non-trivial commutation relations are given by:

$$[\hat{J}_i, \hat{J}_j] = i\epsilon_{ijk}\hat{J}_k$$

$$[\hat{J}_i, \hat{K}_j] = i\epsilon_{ijk}\hat{K}_k$$

$$[\hat{K}_i, \hat{K}_j] = -i\epsilon_{ijk}\hat{J}_k$$

$$[\hat{J}_i, \hat{P}_j] = i\epsilon_{ijk}\hat{P}_k \tag{3.60}$$

$$[\hat{K}_i, \hat{P}_j] = -i\hat{H}\delta_{ij}$$

$$[\hat{K}_i, \hat{H}] = -i\hat{P}_i$$

3.5 Wigner rotations

Once we have determined the algebraic structure of the quantum Poincare transformation, the next step is to analyse its specific effect on an arbitrary quantum state. To this end, let us consider the quantum state of a free particle with definite four-momentum represented by:

$$\Psi_{p,\alpha} \equiv \Psi_{p,\alpha}(x) \tag{3.61}$$

where $p \equiv p^\mu$ is the four-momentum and α is a label that represents any other possible degrees of freedom that transform in a non-trivial way under a Poincare transformation. For empirical reasons, we define *one-particle states* as those where α is a discrete number (but in the most general case, α may have a continuous spectrum) [7]. As we will show in what follows, these extra degrees of freedom will correspond to the components of the spin of the quantum particle in consideration [7, 21, 22].

Because the quantum state is considered a momentum eigenstate, we have:

$$\hat{P}^\mu \Psi_{p,\alpha} = p^\mu \Psi_{p,\alpha} \tag{3.62}$$

Let us now consider the effect of the Poincare transformation $T(\Lambda, a)$:

$$x^\mu \to \tilde{x}^\mu = \Lambda^\mu{}_\nu x^\nu + a^\mu \tag{3.63}$$

on the quantum state of the particle:

$$\Psi_{p,\alpha} \to \tilde{\Psi}_{p,\alpha} = \hat{U}(\Lambda, a)\Psi_{p,\alpha} \tag{3.64}$$

where $\hat{U}(\Lambda, a)$ is the unitary transformation that corresponds to the Poincare transformation.

For the case of a pure translation:

$$\Lambda = 0 \quad \Rightarrow \quad T(\Lambda, a) = T(a) \tag{3.65}$$

we obtain the well-known result:

$$\hat{U}(a)\Psi_{p,\alpha} = e^{-ipa}\Psi_{p,\alpha} \tag{3.66}$$

On the other hand, for the case of a pure Lorentz transformation in four-space:

$$a = 0 \quad \Rightarrow \quad T(\Lambda, a) = T(\Lambda) \tag{3.67}$$

we get:

$$\begin{aligned}
\hat{P}^\mu \hat{U}(\Lambda)\Psi_{p,\alpha} &= \left(\hat{U}(\Lambda)\hat{U}(\Lambda)^{-1}\right)\hat{P}^\mu \hat{U}(\Lambda)\Psi_{p,\alpha} \\
&= \hat{U}(\Lambda)\left(\hat{U}(\Lambda)^{-1}\hat{P}^\mu \hat{U}(\Lambda)\right)\Psi_{p,\alpha} \\
&= \hat{U}(\Lambda)\left[\left(\Lambda^\mu_\rho\right)^{-1}\hat{P}^\rho\right]\Psi_{p,\alpha} \\
&= \hat{U}(\Lambda)\left[\left(\Lambda^\mu_\rho\right)^{-1}p^\rho\right]\Psi_{p,\alpha} \\
&= \hat{U}(\Lambda)\left(\Lambda^\mu_\rho p^\rho\right)\Psi_{p,\alpha} \\
&= \left(\Lambda^\mu_\rho p^\rho\right)\hat{U}(\Lambda)\Psi_{p,\alpha} \tag{3.68}
\end{aligned}$$

Therefore, $\hat{U}(\Lambda)\Psi_{p,\alpha}$ is an eigenvector of $\hat{P}^\mu$ with eigenvalue $\Lambda^\mu_\rho p^\rho$. Notice that, to derive the previous equation, we have used the following relationship:

$$\hat{U}(\Lambda)^{-1}\hat{P}^\mu \hat{U}(\Lambda) = \left(\Lambda^\mu_\rho\right)^{-1} P^\rho \tag{3.69}$$

which is easy to derive [7].

So far we know that, under a Lorentz transformation Λ, the transformed quantum state will have momentum Λp. However, the job is not over yet, as we still need to find out the effect of the Lorentz transformation on the degrees of freedom labelled by α. We can make the reasonable assumption that the α degrees of freedom form a complete basis, in the sense that any transformation of the state $\Psi_{p,\alpha}$ with a specific value of α can be represented as a linear superposition of all the possible α states. That is, the effect of the unitary Lorentz transformation on the quantum state can be expressed as a linear superposition over the α degrees of freedom as follows:

$$\hat{U}(\Lambda)\Psi_{p,\alpha} = \sum_{\beta} C_{\alpha\beta}(\Lambda,p)\Psi_{\Lambda p,\beta} \qquad (3.70)$$

where the $C_{\alpha\beta}$ are complex numbers that still need to be determined and we have changed p to Λp as a consequence of equation (3.68) [7].

We now need to establish the structure of the $C_{\alpha\beta}$ parameters to determine fully the effect of the Lorentz transformation Λ on a quantum state $\Psi_{p,\alpha}$. To this end, we recall that, under Poincare transformations, the following two quantities remain invariant:

$$p^2 \equiv p^\mu p_\mu = \mathbf{p} \cdot \mathbf{p} - E^2 = -m^2$$
$$sign(p^0) \Leftrightarrow p^2 \leqslant 0 \qquad (3.71)$$

We can use these two invariants to classify states into specific classes. That is, for each value of p^2 and for each $sign(p^0)$ for $p^2 \leqslant 0$, we can choose a 'standard' four-momentum k^μ that identifies a specific class of quantum states [7, 21, 22]. Thus, any momentum p^μ in the class of k^μ can be expressed as:

$$p^\mu = L^\mu_{\ \nu}(p)k^\nu \qquad (3.72)$$

where L is a Lorentz transformation. Consequently, any quantum state of momentum p^μ with characteristic momentum k^μ can be defined as:

$$\Psi_{p,\alpha} \equiv N(p)\hat{U}(L(p))\Psi_{k,\alpha} \qquad (3.73)$$

where $N(p)$ is a normalization constant.

Then, if we apply the unitary quantum Poincare transformation to this state, we obtain:

$$\hat{U}(\Lambda)\Psi_{p,\alpha} = N(p)\hat{U}(\Lambda)\hat{U}(L(p))\Psi_{k,\alpha}$$
$$= N(p)\hat{U}(\Lambda L(p))\Psi_{k,\alpha}$$
$$= N(p)\hat{U}(L(\Lambda p))\hat{U}^{-1}(L(\Lambda p))\hat{U}(\Lambda L(p))\Psi_{k,\alpha}$$
$$= N(p)\hat{U}(L(\Lambda p))\hat{U}(L^{-1}(\Lambda p))\hat{U}(\Lambda L(p))\Psi_{k,\alpha}$$
$$= N(p)\hat{U}(L(\Lambda p))\hat{U}(L^{-1}(\Lambda p)\Lambda L(p))\Psi_{k,\alpha}$$
$$= N(p)\hat{U}(L(\Lambda p))\hat{U}(W)\Psi_{k,\alpha} \qquad (3.74)$$

where:

$$W \equiv W(\Lambda, p) = L^{-1}(\Lambda p)\Lambda L(p) \tag{3.75}$$

transforms the standard four-momentum as:

$$k^\mu \xrightarrow{L} p^\mu \xrightarrow{\Lambda} \Lambda^\mu_{\ \nu} p^\nu \xrightarrow{L^{-1}} k^\mu \quad \Rightarrow \quad W^\mu_{\ \nu} k^\nu = k^\mu \tag{3.76}$$

and therefore leaves the standard four-momentum invariant. This transformation is often called the *Wigner rotation*. The set of Wigner rotations forms a group known as the *little group*, which is a subgroup of the Poincare group. Its effect on the quantum state can be written as a rotation in Hilbert space over the α degrees of freedom:

$$\hat{U}(W)\Psi_{k,\alpha} = \sum_\beta D_{\alpha\beta}\Psi_{k,\beta} \tag{3.77}$$

where the $D_{\alpha\beta}$ coefficients furnish a representation of the little group [7].

Then, from equation (3.74) we have:

$$\begin{aligned}
\hat{U}(\Lambda)\Psi_{p,\alpha} &= N(p)\hat{U}(L(\Lambda p))\hat{U}(W)\Psi_{k,\alpha} \\
&= N(p)\hat{U}(L(\Lambda p)) \sum_\beta D_{\alpha\beta}\Psi_{k,\beta} \\
&= N(p) \sum_\beta D_{\alpha\beta}\hat{U}(L(\Lambda p))\Psi_{k,\beta} \\
&= \frac{N(p)}{N(\Lambda p)} \sum_\beta D_{\alpha\beta}\Psi_{\Lambda p,\beta}
\end{aligned} \tag{3.78}$$

where we use the fact that:

$$\Psi_{\Lambda p,\beta} = N(\Lambda p)\hat{U}(L(\Lambda p))\Psi_{k,\beta} \tag{3.79}$$

by virtue of equation (3.73). Notice that in general:

$$D_{\alpha\beta} = D_{\alpha\beta}(\Lambda) \tag{3.80}$$

the matrix elements of the Wigner rotation $D_{\alpha\beta}$ are a function of the Lorentz transformation Λ.

Therefore, to understand the effect of a Poincare transformation on a quantum state, it is necessary to determine the $D_{\alpha\beta}$ coefficients that form a representation of the little group. Notice that, although the $C_{\alpha\beta}$ coefficients were specific for a particular quantum state, the $D_{\alpha\beta}$ coefficients are general for *all* the quantum states within the same class characterized by the standard four-momentum k^μ. This is known as the *method of induced representations* [7, 21, 22].

Table 3.1 presents a classification of all possible physical and non-physical quantum states in terms of the invariant conditions and the standard four-momentum of a free particle [7, 21, 22]. The meaning of this table is clear: the quantum state of elementary particles is intrinsically connected to the irreducible

Table 3.1. Classification of quantum states in terms of invariant conditions and the standard four-momentum [7].

Case	Condition	k^μ	Little Group	Physical State
(a)	$p^2 = -m^2 < 0; p^0 > 0$	$(m, 0, 0, 0)$	SO(3)	Massive particles
(b)	$p^2 = -m^2 < 0; p^0 < 0$	$(-m, 0, 0, 0)$	SO(3)	Non-physical
(c)	$p^2 = 0; p^0 > 0$	$(k, k, 0, 0)$	ISO(2)	Massless particles
(d)	$p^2 = 0; p^0 < 0$	$(-k, k, 0, 0)$	ISO(2)	Non-physical
(e)	$p^2 > 0$	$(0, k, 0, 0)$	SO(2, 1)	Virtual Particles
(f)	$p^\mu = 0$	$(0, 0, 0, 0)$	SO(3, 1)	Vacuum

representations of the Poincare group. In particular, each type of elementary particle can be assigned to a unique representation of the Poincare group.

Cases (b) and (d) represent *non-physical states*. Case (b) represents particles with a negative mass, and consequently, free particles with negative energy. In a similar manner, case (d) represents massless free particles with negative energy. These states have not been observed in nature[5]. Therefore, these non-physical states need to be considered as mathematical artefacts and should not be accounted for in any discussion of actual quantum fields (as far as current experimental results show).

Case (e) represents *virtual particles* that are off the mass shell and often have space-like momentum. These particles only exist as intermediary carriers of some physical interaction and are never found free [8]. As such, for this classification of free particles, this case can also be considered as non-physical.

Case (f) represents the *quantum vacuum*. We can observe that the quantum vacuum is the *only* quantum state that is left invariant under the full 3 + 1 Lorentz group SO(3, 1).

Cases (a) and (c) represent physical particles; they will be discussed in further detail in the following subsections.

3.5.1 Massive particles

Case (a) represents *massive particles* (e.g. electrons). It can be observed from table 3.1 that the extra degrees of freedom α make the state $\Psi_{p,\alpha}$ transform as a member of SO(3), the special orthogonal group. Notice that it is well known that SO(3) is related to SU(2), the spin group (also known as the angular momentum group) [25, 26]. Indeed, there is a homomorphism h such that:

$$h : \mathrm{SU}(2) \to \mathrm{SO}(3) \tag{3.81}$$

and as a consequence, the Lie algebras of SO(3) and SU(2) are isomorphic. This will be discussed in further detail in the next section.

Therefore, the extra degrees of freedom labelled by α have to correspond to spin degrees of freedom. In other words, it is possible to define spin as those extra degrees

---

[5] As we discussed in section 3.3, and as we will show in section 3.7, even if the classical Dirac and Klein–Gordon fields have positive and negative energy eigenvalues, the states in a quantum field always have positive energies.

of freedom of a free massive particle that are affected by a Lorentz transformation in a non-trivial way. That is, the spin degrees of freedom dictate how a physical quantum state transforms under the Lorentz or Poincare groups.

The irreducible unitary representations of SU(2) span a Hilbert space of $2j + 1$ dimensions, where $j = 0, \frac{1}{2}, 1, \ldots$ The value of j is what is usually referred as the *spin* of the massive particle. Thus, for each value of the spin of a massive particle given by j, there is a degeneracy of $2j + 1$ states labelled by σ. The σ label is usually referred as the *spin projection*. Thus, the representations that form the canonical basis for states of definite momentum $\mathbf{p}$ and non-zero mass are denoted by:

$$|\mathbf{p}, j, \sigma\rangle \tag{3.82}$$

which satisfy:

$$\hat{\mathbf{J}}^2|\mathbf{p}, j, \sigma\rangle = j(j + 1)|\mathbf{p}, j, \sigma\rangle$$

$$\hat{J}_3|\mathbf{p}, j, \sigma\rangle = \sigma|\mathbf{p}, j, \sigma\rangle$$

$$\hat{P}_k|\mathbf{p}, j, \sigma\rangle = p_k|\mathbf{p}, j, \sigma\rangle \tag{3.83}$$

$$(\hat{J}_1 \pm i\hat{J}_2)|\mathbf{p}, j, \sigma\rangle = \sqrt{j(j + 1) \mp \sigma - \sigma^2}\,|\mathbf{p}, j, \sigma \pm 1\rangle$$

and $\sigma = -j, -j + 1, \ldots, j - 1, j$. That is, the quantum states of massive particles are completely described by the momentum $\mathbf{p}$, spin j and spin projection σ eigenvectors. Notice that we have arbitrarily chosen the three-axis as the quantization axis for the spin projection operator $\hat{J}_3$.

Furthermore, under Poincare operations $T(\Lambda, a)$ the quantum states transform as:

$$\hat{U}(a)|\mathbf{p}, j, \sigma\rangle = e^{-ipa}|\mathbf{p}, j, \sigma\rangle$$

$$\hat{U}(\Lambda)|\mathbf{p}, j, \sigma\rangle = \sum_{\tilde{\sigma}=-j}^{j} D_{\tilde{\sigma}\sigma}^{(j)}|\Lambda\mathbf{p}, j, \tilde{\sigma}\rangle \tag{3.84}$$

Notice that, in general, a Lorentz transformation Λ leads to a superposition of states over all possible values of the spin projection σ.

It is also convenient to introduce the *Pauli–Lubanski pseudo-vector*:

$$\hat{W}_\mu \equiv -\frac{1}{2}\epsilon_{\mu\nu\alpha\beta}\hat{J}^{\nu\alpha}\hat{P}^\beta \tag{3.85}$$

which eigenstates satisfy:

$$\hat{W}_\mu\hat{W}^\mu|\mathbf{p}, j, \sigma\rangle = -j(j + 1)m^2|\mathbf{p}, j, \sigma\rangle \tag{3.86}$$

where m is the mass of the particle. This operator is important because it can be shown that the irreducible unitary representations of the Poincare group can be characterized by the eigenvalues of:

$$C_1 \equiv -\hat{P}^2 \qquad C_2 \equiv -\hat{W}^2/\mathbf{p}^2 \tag{3.87}$$

which are the only two independent *Casimir invariants* of the Poincare group (i.e. C_1 and C_2 are invariant under Poincare transformations) [24, 25]. That is, irreducible representations of the Poincare group carry *invariant labels* (C_1, C_2) associated with their mass and spin:

$$(C_1, C_2) \leftrightarrow (m, j) \tag{3.88}$$

3.5.2 Massless particles

Case (c) represents *massless particles* (e.g. photons). It is interesting to observe that massive and massless particles have different little groups. Indeed, massless particles transform as representations of ISO(2), the Euclidean geometry group of rotations and translations in two dimensions [24, 25]. That is, the spin of the electron and the spin of the photon may be 'similar' concepts, but they are formally described by a *completely different group structure*.

Indeed, the irreducible representations of ISO(2) are denoted by:

$$|\mathbf{p}, \lambda\rangle \tag{3.89}$$

and satisfy:

$$\hat{W}_\mu \hat{W}^\mu |\mathbf{p}, \lambda\rangle = 0$$
$$\hat{W}_k |\mathbf{p}, \lambda\rangle = \lambda p_k |\mathbf{p}, \lambda\rangle \quad k = 0, 3 \tag{3.90}$$
$$\hat{W}_k |\mathbf{p}, \lambda\rangle = 0 \qquad k = 1, 2$$

for $\lambda = 0, \pm\frac{1}{2}, \pm 1, \ldots$.

It can be shown that all the other meaningful and physically independent operators have zero eigenvalues [7]. This is a requirement needed to avoid a continuous degree of freedom in the spin structure. Indeed, experiments suggest that the spin variable has a discrete spectrum.

The λ eigenvalue labels the *helicity* of the massless particle. That is, λ corresponds to the component of the spin in the direction of motion:

$$\hat{\mathbf{J}} \cdot \frac{\mathbf{p}}{|\mathbf{p}|} |\mathbf{p}, \lambda\rangle = \lambda |\mathbf{p}, \lambda\rangle \tag{3.91}$$

Besides the momentum $\mathbf{p}$, the helicity is the only other 'good quantum' number that can be used to label the quantum state of massless particles.

Under Poincare operations $T(\Lambda, a)$, these states transform as:

$$\hat{U}(a)|\mathbf{p}, \lambda\rangle = e^{-ipa}|\mathbf{p}, \lambda\rangle$$
$$\hat{U}(\Lambda)|\mathbf{p}, \lambda\rangle = e^{i\lambda a(\Lambda, \mathbf{p})}|\Lambda\mathbf{p}, \lambda\rangle \tag{3.92}$$

where $\alpha = \alpha(\Lambda, \mathbf{p})$ is the angle of rotation around the three-axis as specified by the little group.

It can be observed from these transformation rules that the helicity is a Lorentz invariant. That is, all inertial observers will see the massless particle with exactly the same helicity. In contrast, notice how the Lorentz transformation of the quantum state of a massive particle in equation (3.84) is a superposition of the $2j + 1$ possible values of the spin projection σ.

Furthermore, one could argue that each value of λ corresponds to different types of massless particle. However, particles of opposite helicity may be related by the symmetry of space inversion (parity). In this context, parity transformations are of the type:

$$\mathbf{x} \rightarrow -\mathbf{x} \tag{3.93}$$

where $\mathbf{x}$ is the spatial coordinates.

For example, electromagnetic and gravitational forces obey space inversion symmetry. Therefore, it can be shown that inversion symmetry implies that both helicity states with $\lambda = \pm 1$ correspond to *photons* (the spin-1 carriers of the electromagnetic interaction) of positive and negative helicity, whereas both helicity states with $\lambda = \pm 2$ correspond to *gravitons* (the spin-2 carriers of the gravitational interaction) of positive and negative helicity. That is, the quantum states with opposite helicities correspond to exactly the same particle because of the underlying parity symmetry [7]. In addition, notice that there are no photons with helicity $\lambda = 0$, and neither gravitons with helicities $\lambda = 0, \pm 1$ (but of course, we can have massive spin-1 vector particles with spin projection $\sigma = 0, \pm 1$ or massive spin-2 vector particles with spin projection $\sigma = 0, \pm 1, \pm 2$).

On the other hand, massless neutrinos interact through the weak force, which does not respect the space inversion symmetry [8]. Therefore, helicity states with $\lambda = -\frac{1}{2}$ corresponds to neutrinos, and helicity states with $\lambda = +\frac{1}{2}$ corresponds to antineutrinos. Formally, massless neutrinos and anti-neutrinos are different types of elementary particle.

Then, the value of the spin for massless particles is simply defined as:

$$s \equiv |\lambda| \tag{3.94}$$

and we cannot use the second Casimir invariant C_2, nor the Pauli–Lubanski pseudo-vector, to characterize the states.

That being said, it is also important to mention that even though helicity is a Lorentz invariant, a general quantum state may be affected under a Lorentz transformation that does not change the momentum. Indeed, notice how the Lorentz transformation of a quantum state depends on the helicity λ in the exponent of equation (3.92). Although this leads to a global phase that has no effect for helicity eigenstates, a superposition of helicity states may change in a non-trivial manner.

Clearly, the different group-theoretical properties associated with massive and massless particles indicate that the spin of a massive particle is different from the spin of a massless particle. Indeed:

- The helicity λ of a massless particle is a Lorentz invariant, while the spin projection σ of a massive particle is transformed among all possible $2j + 1$ possible spin projection states. Then, it makes sense to talk about the helicity of massless particles, because all inertial observers will agree on its value. On the other hand, even though we can define a helicity for massive states, this helicity will not be a Lorentz invariant. Indeed, it only makes sense to talk about the *helicity of a massive particle state* rather than the helicity of a massive particle.

- Also, a massive particle with spin j has $2j + 1$ spin projection states labelled by σ that correspond to exactly the same type of particle. On the other hand, a massless particle of helicity λ corresponds to the same type of particle with opposite helicity $-\lambda$ only if they are related through a space inversion symmetry.

Needless to say, the reason for the discrepancy between the spin of a massive and massless particles comes down to the fact that massless particles travel at the speed of light. Thus, it is permissible to describe spin as the angular momentum of a massive particle observed in the rest frame. However, such an inertial frame cannot be found for massless particles.

As such, from a formal point of view, it is necessary to avoid the use of the word 'spin' to describe the state of massless particles, as the correct quantum number should be the 'helicity'. However, if we limit our discussion to massive spin-$\frac{1}{2}$ particles (e.g. electrons) and massless spin-1 particles (e.g. photons), as most textbooks on quantum information do, then the number of degrees of freedom will coincide for both classes. In such a case, there may be specific situations in which the different group-theoretical structure of the spin will not play a major role on the dynamics of the system, and spin and helicity become terms that can be interchanged without leading to confusion.

3.6 The Dirac equation

The *Dirac equation* describes relativistic spin-$\frac{1}{2}$ particles. Most textbooks derive this equation as a linearization of the Klein–Gordon equation [2, 8]. That is, by requiring a Hamiltonian of the form:

$$\hat{H}\psi = \left(\boldsymbol{\alpha} \cdot \hat{\mathbf{P}} + \beta m\right)\psi \tag{3.95}$$

such that:

$$\hat{H}^2\psi = \left(\hat{\mathbf{P}}^2 + m^2\right)\psi \tag{3.96}$$

and:

$$\gamma^\mu \equiv (\beta, \beta\boldsymbol{\alpha}) \tag{3.97}$$

are a representation of the *Dirac matrices*. The Hamiltonian in equation (3.95) is known as the *Dirac Hamiltonian*.

In this section, however, we will derive the Dirac equation by following simple, but powerful, group-theoretical arguments [5]. The argument consists of first exploring the connection between the special orthogonal group in three dimensions

SO(3) and the special unitary group in two dimensions SU(2). Whereas SO(3) represents rotations in three-dimensional space, SU(2) is used to represent unitary quantum operations on two-level states (i.e. qubits). As we will show, these two groups are related and therefore rotations can be represented as unitary transformations.

Subsequently, we will study the relation between SL(2,$\mathbb{C}$), the special linear group of complex 2×2 matrices with determinant equal to 1, and SO$^+$(3, 1), the proper, orthochronous Lorentz group. As we will find out, while rotations can be expressed as unitary transformations, this is not the case when we add boosts. Therefore, applying a boost to a quantum field results in non-unitary transformations. As we will explain, that these transformations are non-unitary is not a reason for concern, as we just have to be careful to understand properly what entities are being transformed and under what representation.

The relationship between SL(2,$\mathbb{C}$) and SO$^+$(3,1) will allow us to understand how spin-$\frac{1}{2}$ quantum fields transform under a Lorentz transformation. As we will find out, it is precisely the Dirac equation that describes the effect of a Lorentz transformation on a quantum field[6].

3.6.1 SO(3) and SU(2)

As a first step, we will discuss the relation between the groups SO(3) and SU(2) in the context of relativistic quantum mechanics.

SO(3)

Let us first consider the rotation group in three dimensions SO(3) [25]. Recall that SO(3) consists of direct isometries (i.e. isometries that preserve orientation) and leaves the origin fixed. Then, a three-dimensional vector:

$$\mathbf{r} = (x, y, z) \qquad (3.98)$$

is transformed as:

$$\mathbf{r} \rightarrow \tilde{\mathbf{r}} = \hat{R}\,\mathbf{r} \qquad (3.99)$$

with the property:

$$\tilde{\mathbf{r}}^T \cdot \tilde{\mathbf{r}} = \mathbf{r}^T \cdot \mathbf{r} \qquad (3.100)$$

or equivalently:

$$x^2 + y^2 + z^2 = \tilde{x}^2 + \tilde{y}^2 + \tilde{z}^2 \qquad (3.101)$$

---

[6] It is important to note that, even though we will restrict our discussion to spin-$\frac{1}{2}$ particles, the method that we will present in this section is rather general, and it could be easily extended to obtain the dynamic equations that describe particles of arbitrary spin [27–29].

which makes explicit that the magnitude of the vector $\mathbf{r}$ is unchanged under SO(3) transformations. For a rotation given by an angle θ over the z axis, we can write $\hat{R}_z(\theta)$ as:

$$\hat{R}_z(\theta) = \begin{pmatrix} \cos\theta & \sin\theta & 0 \\ -\sin\theta & \cos\theta & 0 \\ 0 & 0 & 1 \end{pmatrix} \tag{3.102}$$

while an infinitesimal rotation by an angle $\delta\theta$ is given by:

$$R_z(\delta\theta) = 1 + i J_z \delta\theta \tag{3.103}$$

where J_z is the infinitesimal generator of rotations over z:

$$J_z = \frac{1}{i} \frac{dR_z(\theta)}{d\theta} \bigg|_{\theta=0} = \begin{pmatrix} 0 & -i & 0 \\ i & 0 & 0 \\ 0 & 0 & 0 \end{pmatrix} \tag{3.104}$$

We can find similar expressions for J_x and J_y, which obey the commutation relation that defines the generators of SO(3):

$$[J_x, J_y] = i J_z \tag{3.105}$$

and their fully anti-symmetric cyclic permutations [25].

SU(2)

Let us now consider the case of SU(2), the group of unitary matrices with unit determinant that are used to represent operations on quantum states:

$$\hat{U} \in \mathrm{SU}(2) \quad \Rightarrow \quad \hat{U}\hat{U}^{\dagger} = 1 \quad \text{and} \quad \det \hat{U} = 1 \tag{3.106}$$

In general, we can write an arbitrary element of SU(2) as a 2×2 matrix of the form:

$$\hat{U} = \begin{pmatrix} a & b \\ -b^* & a^* \end{pmatrix} \tag{3.107}$$

where $a, b \in \mathbb{C}$ and satisfy the normalization condition:

$$|a|^2 + |b|^2 = 1 \tag{3.108}$$

The three Pauli matrices $\{\sigma_x, \sigma_y, \sigma_z\}$ are the generators of SU(2) and satisfy the commutation relation:

$$\left[\frac{\sigma_x}{2}, \frac{\sigma_y}{2}\right] = i\frac{\sigma_z}{2} \tag{3.109}$$

and their fully anti-symmetric cyclic permutations [25].

In their most basic representation, SU(2) matrices operate on two-dimensional objects such as:

$$\eta = \begin{pmatrix} \eta_1 \\ \eta_2 \end{pmatrix} \tag{3.110}$$

in the following way:

$$\eta \to \tilde{\eta} = \hat{U}\eta \quad \eta^\dagger \to \tilde{\eta}^\dagger = \eta^\dagger \hat{U}^\dagger \tag{3.111}$$

Then:

$$\eta^\dagger \eta \to \eta^\dagger \hat{U}^\dagger \hat{U}\eta = \eta^\dagger \eta \tag{3.112}$$

and as a consequence:

$$\eta^\dagger \eta = |\eta_1|^2 + |\eta_2|^2 \tag{3.113}$$

is an invariant under SU(2) transformations. The η entities that transform under SU(2) are called *spinors* [25].

In a similar manner, the outer product of two-spinors is a Hermitian matrix which we denote by $\mathcal{Z}$:

$$\mathcal{Z} \equiv \eta\eta^\dagger = \begin{pmatrix} |\eta_1|^2 & \eta_1\eta_2^* \\ \eta_2\eta_1^* & |\eta_2|^2 \end{pmatrix} \tag{3.114}$$

that transforms as:

$$\eta\eta^\dagger \to \hat{U}\eta\eta^\dagger \hat{U}^\dagger \tag{3.115}$$

As we have seen, η and $\eta^\dagger$ transform in a different way under SU(2). However:

$$\begin{pmatrix} \eta_1 \\ \eta_2 \end{pmatrix} \quad \text{and} \quad \begin{pmatrix} -\eta_2^* \\ \eta_1^* \end{pmatrix} \tag{3.116}$$

transform in exactly the same way. Indeed:

$$\begin{pmatrix} \eta_1 \\ \eta_2 \end{pmatrix} \to \begin{pmatrix} \tilde{\eta}_1 \\ \tilde{\eta}_2 \end{pmatrix} = \begin{pmatrix} a & b \\ -b^* & a^* \end{pmatrix}\begin{pmatrix} \eta_1 \\ \eta_2 \end{pmatrix}$$

$$= \begin{pmatrix} a\eta_1 + b\eta_2 \\ -b^*\eta_1 + a^*\eta_2 \end{pmatrix} \tag{3.117}$$

and:

$$\begin{pmatrix} -\eta_2^* \\ \eta_1^* \end{pmatrix} \to \begin{pmatrix} -\tilde{\eta}_2^* \\ \tilde{\eta}_1^* \end{pmatrix} = \begin{pmatrix} a(-\eta_2^*) + b(\eta_1^*) \\ -b^*(-\eta_2^*) + a^*(\eta_1^*) \end{pmatrix}$$

$$= \begin{pmatrix} a & b \\ -b^* & a^* \end{pmatrix}\begin{pmatrix} -\eta_2^* \\ \eta_1^* \end{pmatrix} \tag{3.118}$$

which means that $-\eta_2^*$ undergoes the same transformation rule as η_1, and η_1^* undergoes the same transformation rule as η_2. Therefore, we write this similarity as:

$$\begin{pmatrix} \eta_1 \\ \eta_2 \end{pmatrix} \sim \begin{pmatrix} -\eta_2^* \\ \eta_1^* \end{pmatrix} \tag{3.119}$$

where the '$\sim$' symbol can be read as 'transforms as'.

Let us now define the matrix:

$$\Upsilon = \begin{pmatrix} 0 & -1 \\ 1 & 0 \end{pmatrix} \tag{3.120}$$

which can be used to conveniently write:

$$\begin{pmatrix} -\eta_2^* \\ \eta_1^* \end{pmatrix} = \begin{pmatrix} 0 & -1 \\ 1 & 0 \end{pmatrix} \begin{pmatrix} \eta_1^* \\ \eta_2^* \end{pmatrix} = \Upsilon \begin{pmatrix} \eta_1^* \\ \eta_2^* \end{pmatrix} \tag{3.121}$$

Therefore, we have that:

$$\eta \sim \Upsilon \eta^* \tag{3.122}$$

We also observe that:

$$\eta^\dagger \sim (\Upsilon \eta^*)^\dagger = (\Upsilon \eta)^T = (-\eta_2, \eta_1) \tag{3.123}$$

and therefore:

$$\eta \eta^\dagger \sim \begin{pmatrix} \eta_1 \\ \eta_2 \end{pmatrix} (-\eta_2 \quad \eta_1) = \begin{pmatrix} -\eta_1 \eta_2 & \eta_1^2 \\ -\eta_2^2 & \eta_1 \eta_2 \end{pmatrix} \tag{3.124}$$

If now we define:

$$\mathcal{W} \equiv \begin{pmatrix} \eta_1 \eta_2 & -\eta_1^2 \\ \eta_2^2 & -\eta_1 \eta_2 \end{pmatrix} \tag{3.125}$$

we have that:

$$\eta \eta^\dagger \sim -\mathcal{W} \quad \text{and} \quad Tr(\mathcal{W}) = 0 \tag{3.126}$$

Consequently $\mathcal{W}$ transforms as:

$$\mathcal{W} \to \hat{U} \mathcal{W} \hat{U}^\dagger \tag{3.127}$$

under an SU(2) transformation.

Relation between SO(3) and SU(2)

As we will see next, the transformation groups SO(3) and SU(2) are related. To this end, let us first define $\mathcal{Y}$ as:

$$\mathcal{Y} \equiv \boldsymbol{\sigma} \cdot \mathbf{r} = x\sigma_x + y\sigma_y + z\sigma_z = \begin{pmatrix} z & x - iy \\ x + iy & -z \end{pmatrix} \qquad (3.128)$$

which is a traceless 2×2 matrix where $\mathbf{r}$ is a vector in three-dimensional Cartesian coordinates and $\boldsymbol{\sigma}$ is the vector of Pauli matrices. If $\mathcal{Y}$ transforms as $\mathcal{W}$ under SU(2):

$$\mathcal{Y} \sim \mathcal{W} \quad \Rightarrow \quad \mathcal{Y} \rightarrow \tilde{\mathcal{Y}} = \hat{U}\mathcal{Y}\hat{U}^\dagger \qquad (3.129)$$

then we notice that:

$$\det(\tilde{\mathcal{Y}}) = \det(\hat{U}\mathcal{Y}\hat{U}^\dagger) = \det(\hat{U})\det(\mathcal{Y})\det(\hat{U}^\dagger) = \det(\mathcal{Y}) \qquad (3.130)$$

and therefore:

$$x^2 + y^2 + z^2 = \tilde{x}^2 + \tilde{y}^2 + \tilde{z}^2 \qquad (3.131)$$

That is, the SU(2) unitary operation $\hat{U}$ acting on $\mathcal{Y}$ is equivalent to an orthogonal transformation SO(3) of the Cartesian vector $\mathbf{r} = (x, y, z)$.

Consequently, an SU(2) transformation on the spinor η is equivalent to an SO(3) transformation of the Cartesian vector $\mathbf{r}$ if we make the following identification of variables:

$$\begin{aligned} z &= \eta_1\eta_2 \\ x - iy &= -\eta_1^2 \\ x + iy &= \eta_2^2 \end{aligned} \qquad (3.132)$$

or equivalently:

$$\begin{aligned} x &= \frac{1}{2}\left(\eta_2^2 - \eta_1^2\right) \\ y &= \frac{1}{2i}\left(\eta_1^2 + \eta_2^2\right) \\ z &= \eta_1\eta_2 \end{aligned} \qquad (3.133)$$

However, this representation is not unique[7].

---

[7] Indeed, the traditional mapping to the *Bloch sphere* considers the matrix $\mathcal{X}$ that transforms as $\mathcal{Z}$:

$$\mathcal{X} = \frac{1 + \boldsymbol{\sigma} \cdot \mathbf{r}}{2} = \frac{1}{2}\begin{pmatrix} 1 + z & x - iy \\ x + iy & 1 - z \end{pmatrix} \sim \mathcal{Z}$$

and therefore we have to make an identification of variables based on the expressions:

$$\frac{1 + z}{2} = |\eta_1|^2 \quad \frac{1 - z}{2} = |\eta_2|^2 \quad \frac{x + iy}{2} = \eta_2\eta_1^* \quad \frac{x - iy}{2} = \eta_1\eta_2^*$$

Notice that in this case, $\mathcal{Z}$ has trace 1, whereas $\mathcal{W}$ is traceless.

As an example, let us consider the SU(2) transformation characterized by $(a = e^{i\alpha/2}, b = 0)$. It is easy to show the following equivalence between the associated SU(2) and SO(3) transformations:

$$\hat{U} = \begin{pmatrix} e^{i\alpha/2} & 0 \\ 0 & e^{-i\alpha/2} \end{pmatrix} \leftrightarrow \hat{R} = \begin{pmatrix} \cos\alpha & \sin\alpha & 0 \\ -\sin\alpha & \cos\alpha & 0 \\ 0 & 0 & 1 \end{pmatrix} \qquad (3.134)$$

where the $\leftrightarrow$ symbol can be read as 'equivalent to'. Then:

$$\hat{U} = e^{i\sigma_z\alpha/2} \leftrightarrow \hat{R} = e^{i\hat{J}_z\alpha} \qquad (3.135)$$

and in general:

$$\hat{U} = e^{i\sigma\cdot\theta/2} \leftrightarrow \hat{R} = e^{i\hat{\mathbf{J}}\cdot\theta} \qquad (3.136)$$

where $\mathbf{J} = (J_x, J_y, J_z)$ are the generators of SO(3). Therefore, the effect of spatial rotations on a quantum field can be described by SU(2) transformations, and vice versa.

This relationship between SO(3) and SU(2) means that both groups have a similar algebraic structure [24, 25]. Specifically, their generators obey exactly the same commutation relations:

$$\text{SU(2)} : \left[\frac{\sigma_i}{2}, \frac{\sigma_j}{2}\right] = i\epsilon_{ijk}\frac{\sigma_k}{2} \qquad \text{SO(3)} : \left[J_i, J_j\right] = i\epsilon_{ijk}J_k \qquad (3.137)$$

There is, however, a topological distinction. For the case of SO(3), increasing the rotation by an angle of 2π leads to:

$$\theta \to \tilde{\theta} = \theta + 2\pi \;\Rightarrow\; \hat{R}(\theta) \to \hat{R}(\tilde{\theta}) = \hat{R}(\theta) \qquad (3.138)$$

On the other hand, in the case of SU(2) we obtain:

$$\theta \to \tilde{\theta} = \theta + 2\pi \;\Rightarrow\; \hat{U}(\theta) \to \hat{U}(\tilde{\theta}) = -\hat{U}(\theta) \qquad (3.139)$$

Therefore, we need to increase the rotation by a 4π angle so the SU(2) transformation returns to itself. This means that the connection between SO(3) and SU(2) is a *homomorphism* from SU(2) to SO(3) [25].

3.6.2 SL(2,ℂ) and SO⁺(3,1)

We have just seen how SO(3) spatial rotations can be implemented as SU(2) unitary transformations. Here we will show how the incorporation of boosts imply non-unitary transformations. Indeed, similar to SO(3) and SU(2), SO⁺(3,1) and SL(2,ℂ) are related by a homomorphism. However, in general, SL(2,ℂ), which operates on quantum fields, does not imply unitary transformations.

First of all, let us recall that the Lorentz group has six generators: three for spatial rotations $\hat{\mathbf{J}}$ and three for boosts $\hat{\mathbf{K}}$. In particular, the commutation relations that define these generators were given in equation (3.60). Clearly, these commutation relations are satisfied if we make the following choices:

$$\hat{\mathbf{J}} = \frac{\sigma}{2} \qquad \hat{\mathbf{K}} = \pm i \frac{\sigma}{2} \qquad (3.140)$$

Then, we can consider that there are two types of spinor, according to the sign chosen for $\hat{\mathbf{K}}$.

More explicitly, let us define the generators $\hat{\mathbf{A}}$ and $\hat{\mathbf{B}}$ as follows:

$$\hat{\mathbf{A}} = \frac{1}{2}\left(\hat{\mathbf{J}} + i\hat{\mathbf{K}}\right) \qquad \hat{\mathbf{B}} = \frac{1}{2}\left(\hat{\mathbf{J}} - i\hat{\mathbf{K}}\right) \qquad (3.141)$$

Then, the commutation relations for these two new generators are given by:

$$\left[\hat{A}_i, \hat{A}_j\right] = i\epsilon_{ijk}\hat{A}_k$$

$$\left[\hat{B}_i, \hat{B}_j\right] = i\epsilon_{ijk}\hat{B}_k \qquad (3.142)$$

$$\left[\hat{A}_i, \hat{B}_j\right] = 0$$

That is, $\hat{\mathbf{A}}$ and $\hat{\mathbf{B}}$ each generates an SU(2) group and they are completely decoupled from each other:

$$SO^+(3,1) \simeq SU(2)_A \times SU(2)_B \qquad (3.143)$$

where the $\simeq$ symbol means that the Lie algebras of the groups are isomorphic. That is, the Lie algebra of the Lorentz group can be reduced to the direct product of two subalgebras upon performing the transformation in equation (3.141).

Therefore, we can label states that transform under the Lorentz group $SO^+(3,1)$ by using two indices, each corresponding to $SU(2)_A$ or $SU(2)_B$:

$$(j_A, j_B) \qquad (3.144)$$

For instance:

$$
\begin{aligned}
(j,0) &\quad\Rightarrow\quad \hat{\mathbf{J}} = +i\hat{\mathbf{K}} \quad\Rightarrow\quad \hat{\mathbf{B}} = 0 \\
(0,j) &\quad\Rightarrow\quad \hat{\mathbf{J}} = -i\hat{\mathbf{K}} \quad\Rightarrow\quad \hat{\mathbf{A}} = 0
\end{aligned}
\qquad (3.145)
$$

For the spin-$\frac{1}{2}$ case, we can use these labels to classify the spinors into two distinct types:

$$
\begin{aligned}
\text{Type I}: &\quad \left(\frac{1}{2}, 0\right) \leftrightarrow \xi \\
\text{Type II}: &\quad \left(0, \frac{1}{2}\right) \leftrightarrow \eta
\end{aligned}
\qquad (3.146)
$$

which we represent by ξ and η.

The type I spinor transforms under a $\boldsymbol{\theta}$ rotation and a $\boldsymbol{\varphi}$ boost as:

$$\begin{aligned}
\xi \rightarrow \tilde{\xi} &= e^{i\mathbf{J}\cdot\boldsymbol{\theta}+i\mathbf{K}\cdot\boldsymbol{\varphi}}\,\xi \\
&= e^{i\boldsymbol{\sigma}\cdot\boldsymbol{\theta}/2+\boldsymbol{\sigma}\cdot\boldsymbol{\varphi}/2}\,\xi \\
&= e^{i\boldsymbol{\sigma}\cdot(\boldsymbol{\theta}-i\boldsymbol{\varphi})/2}\,\xi \\
&= \mathcal{M}\xi
\end{aligned} \tag{3.147}$$

On the other hand, the type II spinor transforms under a $\boldsymbol{\theta}$ rotation and a $\boldsymbol{\varphi}$ boost as:

$$\begin{aligned}
\eta \rightarrow \tilde{\eta} &= e^{i\mathbf{J}\cdot\boldsymbol{\theta}+i\mathbf{K}\cdot\boldsymbol{\varphi}}\,\eta \\
&= e^{i\boldsymbol{\sigma}\cdot\boldsymbol{\theta}/2-\boldsymbol{\sigma}\cdot\boldsymbol{\varphi}/2}\,\eta \\
&= e^{i\boldsymbol{\sigma}\cdot(\boldsymbol{\theta}+i\boldsymbol{\varphi})/2}\,\eta \\
&= \mathcal{N}\eta
\end{aligned} \tag{3.148}$$

Notice that $\mathcal{M}$ and $\mathcal{N}$ are *not* unitary transformations. Indeed, the pure boost component of $\mathcal{N}$ and $\mathcal{M}$ is given by a term of the form:

$$e^{\pm\boldsymbol{\sigma}\cdot\boldsymbol{\varphi}/2} \tag{3.149}$$

which is not unitary. In other words, because of the imaginary factor i in equation (3.141), the set of all four generators $\{\mathbf{J}, \mathbf{K}\}$ and $\{\mathbf{A}, \mathbf{B}\}$ cannot be *simultaneously Hermitian* [25]. Indeed, it can be observed that while $\{\mathbf{A}, \mathbf{B}\}$ are Hermitian, $\{\mathbf{J}, \mathbf{K}\}$ are not.

Furthermore, it can be shown that the Lorentz group is not compact [25]. For example, let us first consider the case of rotations. To complete a full rotation, the angle parameter θ has to vary from 0 to 2π. On the other hand, the boost can take the speed v from the minimal value of 0 to the maximal value of 1 (but never actually reaching the speed of light, $c = 1$ in natural units). However, the angular variable φ is related to the speed v through a hyperbolic tangent:

$$\tanh\varphi = v \tag{3.150}$$

and therefore the range of values of φ are in the open interval $[0, \infty)$, which makes the transformation group not to be compact[8].

In addition, it can be proved that there are only two types of representation for a non-compact group: infinite-dimensional unitary representations and finite-dimensional non-unitary representations [24, 25]. In the present case, we are talking about finite-dimensional representations (the number of spinor components is taken to be 2, which is a finite number), and as a consequence they are not unitary.

These observations do not contradict the fact that the Lorentz group $SO^+(3,1)$ is *equivalent* to the product of two SU(2) groups. This is a true statement even though

---

[8] Notice that the group SO(3) is compact because the range of values of the parameter that defines the rotation transformation is on a compact set (the circle). Indeed, the angle takes on values that lie on the interval $[0, 2\pi]$, which is a compact space. On the other hand, the range of values of the hyperbolic angle φ in the Lorentz boost lies on the open interval $[0, \infty)$, which is not a compact space [25].

$SU(2)_A$ and $SU(2)_B$ are each represented with unitary matrices. The discrepancy emerges because the transformations shown above are performed using the $(\hat{\mathbf{J}}, \hat{\mathbf{K}})$ generators (in terms of rotations and boosts), which are functions of $(\hat{\mathbf{A}}, \hat{\mathbf{B}})$. In other words, in the (j_A, j_B) representation, the $\hat{\mathbf{K}}$ generator is not Hermitian. Indeed, even though both groups have isomorphic Lie algebras and share the same irreducible representations, the unitarity of the transformation is not preserved under the transformation (that identifies operators between both groups) given by equation (3.141). The issue of non-unitarity will be discussed in further detail later in this chapter.

It is also important to notice that, because $\hat{\mathbf{A}}$ and $\hat{\mathbf{B}}$ generate two separate SU(2) groups, $\mathcal{M}$ and $\mathcal{N}$ are inequivalent representations of the Lorentz group. That is:

$$\nexists \mathcal{S} \quad \text{such that}: \quad \mathcal{N} = \mathcal{S}\mathcal{M}\mathcal{S}^{-1} \tag{3.151}$$

However, they are both related by:

$$\mathcal{N} = \Upsilon \mathcal{M}^{*} \Upsilon^{-1} \tag{3.152}$$

where Υ can be written as:

$$\Upsilon = -i\sigma_2 \tag{3.153}$$

It can be shown that $\mathcal{M}$ and $\mathcal{N}$ can be represented using 2×2 complex matrices with unit determinant (using Euler's formula). Therefore, $\{\mathcal{M}, \mathcal{N}\}$ forms a representation of SL(2,$\mathbb{C}$), the special linear group of complex 2×2 matrices with determinant equal to one [25]. Notice that the difference between SL(2,$\mathbb{C}$) and SU(2) is that SL(2,$\mathbb{C}$) does not necessarily imply unitary transformations.

Consequently, there is a *homomorphism* from SL(2,$\mathbb{C}$) to the Lorentz group SO^{+}(3,1). That is, there is more than one element in SL(2,$\mathbb{C}$) that corresponds to each element of SO^{+}(3,1). If $A \in$ SL(2,$\mathbb{C}$), then the Lorentz transformation associated with A and $-A$ is the same: $L(A) = L(-A)$. So, more precisely:

$$\text{SO}^{+}(3,1) \simeq \text{SL}(2,\mathbb{C})/Z_2 \tag{3.154}$$

where Z_2 is the cyclic group of order 2.

To summarize, we can represent the SO^{+}(3,1) Lorentz group using the representations of the following two groups:

$$\text{SO}^{+}(3,1) \simeq \begin{cases} \text{SU}(2) \times \text{SU}(2) \\ \text{SL}(2,\mathbb{C})/Z_2 \end{cases} \tag{3.155}$$

where, as before, the symbol $\simeq$ means that the groups have isomorphic Lie algebras.

3.6.3 Four-spinors

In the previous section we defined two distinct types of spinor that obey different transformation rules under a Lorentz transformation:

$$\begin{aligned} \text{Type I}: & \quad \xi \to \mathcal{M}\xi \\ \text{Type II}: & \quad \eta \to \mathcal{N}\eta \end{aligned} \tag{3.156}$$

However, these two-spinors, ξ and η, can be combined into a single spinor of four components usually referred as a *four-spinor* [5].

To construct these four-spinors formally, we need to introduce the space inversion or *parity operation* defined as:

$$\mathbf{r} \to \hat{\mathcal{P}}\mathbf{r} = -\mathbf{r} \tag{3.157}$$

which affects the velocity in a similar manner:

$$\mathbf{v} \to \hat{\mathcal{P}}\mathbf{v} = -\mathbf{v} \tag{3.158}$$

Consequently, $\hat{\mathbf{K}}$ and $\hat{\mathbf{J}}$ transform in different ways under a parity operation:

$$\hat{\mathbf{K}} \to \hat{\mathcal{P}}\hat{\mathbf{K}} = -\hat{\mathbf{K}} \qquad \hat{\mathbf{J}} \to \hat{\mathcal{P}}\hat{\mathbf{J}} = +\hat{\mathbf{J}} \tag{3.159}$$

which is a consequence of $\hat{\mathbf{J}}$ being an *axial vector*. Then, these relations imply that:

$$(j,0) \overset{\hat{\mathcal{P}}}{\leftrightarrow} (0,j)$$
$$\xi \overset{\hat{\mathcal{P}}}{\leftrightarrow} \eta \tag{3.160}$$

That is, parity transforms ξ spinors into η spinors, and vice versa. Therefore, if we take into account the parity operation, then η and ξ can be incorporated into a single mathematical entity, the four-spinor ψ:

$$\psi \equiv \begin{pmatrix} \xi \\ \eta \end{pmatrix} \tag{3.161}$$

which transforms as:

$$\psi = \begin{pmatrix} \xi \\ \eta \end{pmatrix} \to \begin{pmatrix} \tilde{\xi} \\ \tilde{\eta} \end{pmatrix} = \begin{pmatrix} \mathcal{M} & 0 \\ 0 & \mathcal{N} \end{pmatrix} \begin{pmatrix} \xi \\ \eta \end{pmatrix}$$
$$= \begin{pmatrix} D(\Lambda) & 0 \\ 0 & \overline{D}(\Lambda) \end{pmatrix} \begin{pmatrix} \xi \\ \eta \end{pmatrix} \tag{3.162}$$

where:

$$\overline{D}(\Lambda) = \Upsilon D^*(\Lambda) \Upsilon^{-1} \tag{3.163}$$

and $D(\Lambda)$ denotes a representation of the Lorentz transformation $\Lambda_\mu^{\ \nu}$. On the other hand, under a parity transformation:

$$\begin{pmatrix} \xi \\ \eta \end{pmatrix} \overset{\hat{\mathcal{P}}}{\to} \begin{pmatrix} 0 & 1 \\ 1 & 0 \end{pmatrix} \begin{pmatrix} \xi \\ \eta \end{pmatrix} = \begin{pmatrix} \eta \\ \xi \end{pmatrix} \tag{3.164}$$

Therefore, ψ is a non-unitary irreducible representation of the Lorentz group extended by parity and denoted by:

$$SO^+(3,1) \times \mathcal{R} = O^+(3,1) \tag{3.165}$$

where $\mathcal{R}$ denotes the *reflections group*. Furthermore, as we will see next, the dynamical relation between the two spinor types η and ξ is given by the Dirac equation. At such a point it will become clear why we are required to introduce the parity operation to relate both types of spinors.

3.6.4 Particle dynamics

It is convenient to rename the spinors ξ and η as:

$$\xi \to \psi_R \qquad \eta \to \psi_L \tag{3.166}$$

Under a pure Lorentz boost, $\theta = 0$ and $\varphi \neq 0$, and the two spinors are transformed as:

$$
\begin{aligned}
\tilde{\psi}_R &= e^{\boldsymbol{\sigma} \cdot \boldsymbol{\varphi}/2} \psi_R \\
&= (\cosh(\varphi/2) + \boldsymbol{\sigma} \cdot \mathbf{n} \sinh(\varphi/2)) \psi_R \\
\tilde{\psi}_L &= e^{-\boldsymbol{\sigma} \cdot \boldsymbol{\varphi}/2} \psi_L \\
&= (\cosh(\varphi/2) - \boldsymbol{\sigma} \cdot \mathbf{n} \sinh(\varphi/2)) \psi_L
\end{aligned}
\tag{3.167}
$$

If we assume that the boost is the Lorentz transformation necessary to move from a frame where the particle is at rest, $\mathbf{v} = 0$, to another in which the particle is in motion, $\mathbf{v} \neq 0$, we can rewrite the transformation as:

$$
\begin{aligned}
\psi_R(\mathbf{p}) &= (\cosh(\varphi/2) + \boldsymbol{\sigma} \cdot \hat{\mathbf{p}} \sinh(\varphi/2)) \psi_R(0) \\
\psi_L(\mathbf{p}) &= (\cosh(\varphi/2) - \boldsymbol{\sigma} \cdot \hat{\mathbf{p}} \sinh(\varphi/2)) \psi_L(0)
\end{aligned}
\tag{3.168}
$$

where:

$$\mathbf{p} = \gamma m \mathbf{v} \neq 0 \tag{3.169}$$

is the spatial component of the four-momentum of the particle as observed from the frame of reference in which the particle is not at rest.

We can rewrite the Lorentz transformation with the standard kinematic parameters:

$$\gamma = \cosh \varphi \qquad \gamma\beta = \sinh \varphi \tag{3.170}$$

where:

$$\beta = v \qquad \gamma = \frac{1}{\sqrt{1 - v^2}} \tag{3.171}$$

3-29

And then, the spinor transformation equations look like:

$$\psi_R(\mathbf{p}) = \left(\sqrt{\frac{\gamma+1}{2}} + \boldsymbol{\sigma} \cdot \hat{\mathbf{p}}\sqrt{\frac{\gamma-1}{2}}\right)\psi_R(0)$$

$$\psi_L(\mathbf{p}) = \left(\sqrt{\frac{\gamma+1}{2}} - \boldsymbol{\sigma} \cdot \hat{\mathbf{p}}\sqrt{\frac{\gamma-1}{2}}\right)\psi_L(0)$$

(3.172)

Furthermore, if we assume that the particle has energy E and mass m:

$$E = m\gamma$$

(3.173)

then we have the following expressions:

$$\psi_R(\mathbf{p}) = \frac{E + m + \boldsymbol{\sigma} \cdot \mathbf{p}}{\sqrt{2m(E+m)}}\psi_R(0)$$

$$\psi_L(\mathbf{p}) = \frac{E + m - \boldsymbol{\sigma} \cdot \mathbf{p}}{\sqrt{2m(E+m)}}\psi_L(0)$$

(3.174)

It is important to notice that the only distinction between the two spinor components ξ and η is dynamical. That is, we cannot tell them apart on a frame in which the particle is at rest. Indeed, the mathematical expressions that define these components are able to distinguish them only because each one of these spinors transforms in a different manner under a Lorentz boost. As we will see later, the two-spinors ψ_R and ψ_L correspond to the right-handed and left-handed helicity components of the four-spinor ψ. Therefore, it is impossible to tell them apart in the rest frame. This means that:

$$\psi_R(0) = \psi_L(0)$$

(3.175)

and therefore:

$$\psi_R(\mathbf{p}) = \frac{E + \boldsymbol{\sigma} \cdot \mathbf{p}}{m}\psi_L(\mathbf{p})$$

$$\psi_L(\mathbf{p}) = \frac{E - \boldsymbol{\sigma} \cdot \mathbf{p}}{m}\psi_R(\mathbf{p})$$

(3.176)

which can be rewritten in matrix expression as:

$$\begin{pmatrix} -m & E + \boldsymbol{\sigma} \cdot \mathbf{p} \\ E - \boldsymbol{\sigma} \cdot \mathbf{p} & -m \end{pmatrix}\begin{pmatrix} \psi_R(\mathbf{p}) \\ \psi_L(\mathbf{p}) \end{pmatrix} = 0$$

(3.177)

At this point it is convenient to define the 4×4 *Weyl representation of the Dirac matrices* as:

$$\gamma^0 = \begin{pmatrix} 0 & \mathbb{I} \\ \mathbb{I} & 0 \end{pmatrix} \quad \boldsymbol{\gamma} = \begin{pmatrix} 0 & -\boldsymbol{\sigma} \\ \boldsymbol{\sigma} & 0 \end{pmatrix}$$

(3.178)

where $\mathbb{I}$ is the 2×2 identity matrix. Therefore, equation (3.177) can be rewritten as:

$$\left(-m\mathbb{I} + E\gamma^0 - \mathbf{p} \cdot \boldsymbol{\gamma}\right)\psi(\mathbf{p}) = 0 \qquad (3.179)$$

Introducing the explicitly covariant notation:

$$p_\mu = (-E, \mathbf{p}) \qquad \gamma^\mu = (\gamma^0, \boldsymbol{\gamma}) \qquad (3.180)$$

we finally obtain:

$$\left(\gamma^\mu p_\mu + m\right)\psi(\mathbf{p}) = 0 \qquad (3.181)$$

Alternatively, using wave functions in coordinate space we obtain:

$$\left(i\gamma^\mu \partial_\mu - m\right)\psi(\mathbf{r}) = 0 \qquad (3.182)$$

which is known as the *Dirac equation* and describes the dynamics of free spin-$\frac{1}{2}$ relativistic particles (e.g. electrons and quarks) [5, 8]. Furthermore, it can be shown that each spinor component ψ_i satisfies the Klein–Gordon equation:

$$\left(\partial^\mu \partial_\mu - m^2\right)\psi_i = 0 \qquad (3.183)$$

as a direct consequence of the algebraic structure of the Dirac matrices.

At this point it is important to recall that the representation of the Dirac matrices is not unique. However, they need to satisfy the anti-commutation relation:

$$\gamma^\mu \gamma^\nu + \gamma^\nu \gamma^\mu = -2\eta^{\mu\nu} \qquad (3.184)$$

In addition, if we recall the generic form of the gamma matrices that emerges from the Dirac Hamiltonian:

$$\gamma^\mu = (\beta, \beta\boldsymbol{\alpha}) \qquad (3.185)$$

then the α_i and β are complex 4×4 matrices that anti-commute with each other and:

$$\alpha_1^2 = \alpha_2^2 = \alpha_3^2 = \beta^2 = \mathbb{I} \qquad (3.186)$$

where $\mathbb{I}$ is the 4×4 identity matrix. Consequently, the only requirement is for α_i and β to be Hermitian, traceless matrices of even dimensionality, with eigenvalues ± 1.

Furthermore, if γ^μ is a four-vector of 4×4 matrices that satisfies equation (3.184), and if S is a 4×4 unitary matrix, then:

$$\tilde{\gamma}^\mu = S\gamma^\mu S^{-1} \qquad (3.187)$$

is also a valid four-vector of 4×4 Dirac matrices (albeit in a different representation). In this case, the Dirac equation is satisfied by the spinor:

$$\tilde{\psi} = S\psi \qquad (3.188)$$

Finally, the Hermitian conjugation of the Dirac matrices is given by:

$$\gamma^{\mu\dagger} = \gamma^0 \gamma^\mu \gamma^0 \qquad (3.189)$$

And the *adjoint spinor* to ψ is defined as:

$$\overline{\psi} = \psi^\dagger \gamma^0 \tag{3.190}$$

and obeys the Dirac equation

$$i\partial_\mu \overline{\psi} \gamma^\mu + m\overline{\psi} = 0 \tag{3.191}$$

and can be used to define the four-current:

$$j^\mu = \overline{\psi} \gamma^\mu \psi \tag{3.192}$$

that satisfies the covariant continuity equation:

$$\partial_\mu j^\mu = 0 \tag{3.193}$$

It is relatively simple to corroborate that the Dirac Lagrangian is given by:

$$\mathcal{L} = i\overline{\psi}\gamma^\mu \partial_\mu \psi - m\overline{\psi}\psi \tag{3.194}$$

and the canonical momentum is:

$$\pi = \frac{\partial \mathcal{L}}{\partial \dot{\psi}} = i\psi^\dagger \tag{3.195}$$

Let us now discuss why it is important to require that the relativistic equation for a spin-$\frac{1}{2}$ particle contains two independent spinors related by a parity transformation. First, the relativistic wave equation has to be a linear differential relation between spinor components. To obtain a fully covariant free particle equation, we are limited to the four-momentum as the only operator that can appear in the equation [1]. Then, the most general form that the differential operator can take in the relativistic equation is:

$$A_\mu \hat{p}^\mu \tag{3.196}$$

where A_μ is some four-vector. Therefore, for two-spinors, the most general relativistic equations are of the form:

$$a_\mu^1 \hat{p}^\mu \xi + a_\mu^2 \hat{p}^\mu \eta = c_1 \eta + c_2 \xi$$
$$a_\mu^3 \hat{p}^\mu \xi + a_\mu^4 \hat{p}^\mu \eta = c_3 \eta + c_4 \xi \tag{3.197}$$

where a_μ^i are four-vectors and c_i are constants. We can always transform the spinors ξ and η in such a way that this system of differential equations can be rewritten as:

$$a_\mu^5 \hat{p}^\mu \xi = c_5 \eta$$
$$a_\mu^6 \hat{p}^\mu \eta = c_5 \xi \tag{3.198}$$

Therefore, we can combine these two equations into:

$$a_\mu^5 \hat{p}^\mu \frac{a_\nu^6 \hat{p}^\nu}{c_5} \eta = c_5 \eta$$

$$a_\mu^6 \hat{p}^\mu \frac{a_\nu^5 \hat{p}^\nu}{c_5} \xi = c_5 \xi \tag{3.199}$$

which can be rewritten as:

$$\left(a_\nu^5 \hat{p}^\nu a_\mu^6 \hat{p}^\mu - c_5^2\right)\xi = 0$$
$$\left(a_\nu^5 \hat{p}^\nu a_\mu^6 \hat{p}^\mu - c_5^2\right)\eta = 0$$

(3.200)

If we make the reasonable assumption that each of the spinor components satisfies the Klein–Gordon equation, then we have to make the following identification:

$$c_5 = m$$
$$a_\nu^5 \hat{p}^\nu a_\mu^6 \hat{p}^\mu = -\hat{p}^2$$

(3.201)

so we can write:

$$\left(\hat{p}^2 + m^2\right)\xi = 0$$
$$\left(\hat{p}^2 + m^2\right)\eta = 0$$

(3.202)

Therefore, if the mass of the spin-$\frac{1}{2}$ particle is different from zero, then we are required to have two distinct two-spinors, ξ and η, in the relativistic wave equation [1]. Indeed, equation (3.199) assumed that $m = c_5 \neq 0$. To summarize, the Dirac equation for a massive particle is compatible with the Klein–Gordon equation if, and only if, the quantum field is made of two two-spinors. Or, equivalently, to guarantee relativistic covariance, a non-zero mass parameter in the Dirac equation requires the simultaneous consideration of two two-spinors [1].

Let us now consider the reason why both spinors need to be related by the parity transformation [18]. To this end, let us rewrite equation (3.176) as follows:

$$\xi = \frac{E + \boldsymbol{\sigma} \cdot \mathbf{p}}{m}\eta$$

$$\eta = \frac{E - \boldsymbol{\sigma} \cdot \mathbf{p}}{m}\xi$$

(3.203)

If we require that the laws of physics are invariant under a parity transformation, then the relativistic wave equations need to be invariant under a parity transformation[9]. Let us assume for a moment that ξ and η are not related by parity. Then, under a parity transformation, we have:

$$\xi \to \tilde{\xi} \qquad \eta \to \tilde{\eta}$$

(3.204)

---

[9] Invariance under parity transformations is equivalent to requesting an invariance between left and right. This is a reasonable assumption for *free particles*. However, it is known that electroweak interactions violate parity as only left-handed particles are involved in the weak interactions [7, 8, 18].

where $\tilde{\xi}$ and $\tilde{\eta}$ are some spinors. Therefore, under a parity transformation:

$$\tilde{\xi} = \frac{E - \boldsymbol{\sigma} \cdot \mathbf{p}}{m} \tilde{\eta}$$

$$\tilde{\eta} = \frac{E + \boldsymbol{\sigma} \cdot \mathbf{p}}{m} \tilde{\xi}$$

(3.205)

Clearly, these relativistic equations are invariant under a parity transformation if:

$$\tilde{\xi} = \eta \quad \Rightarrow \quad \xi \to \eta$$
$$\tilde{\eta} = \xi \quad \Rightarrow \quad \eta \to \xi$$

(3.206)

which is exactly the same transformation relation that we had obtained before. In other words, the Dirac equation is invariant under parity transformations if, and only if, both two-spinor components η and ξ are transformed into each other by a parity transformation [18].

3.6.5 Free particle spinors

We now need to find free particle solutions to the Dirac equation [6, 8, 30]. In the ultra-relativistic limit:

$$v \approx 1 \quad \Rightarrow \quad E^2 = \mathbf{p} \cdot \mathbf{p} + m^2 \approx \mathbf{p} \cdot \mathbf{p} \quad \Rightarrow \quad E \approx |\mathbf{p}| \gg m \qquad (3.207)$$

the Dirac equation as expressed in equation (3.176) reduces to:

$$(E + \boldsymbol{\sigma} \cdot \mathbf{p})\psi_L \approx 0$$
$$(E - \boldsymbol{\sigma} \cdot \mathbf{p})\psi_R \approx 0$$

(3.208)

which explicitly decouples the two-spinorial components of the four-spinor. Therefore, we can write the two components of the Dirac equation as:

$$\boldsymbol{\sigma} \cdot \hat{\mathbf{p}} \psi_L \approx -\psi_L$$
$$\boldsymbol{\sigma} \cdot \hat{\mathbf{p}} \psi_R \approx \psi_R$$

(3.209)

where we have defined:

$$\hat{\mathbf{p}} \equiv \frac{\mathbf{p}}{|\mathbf{p}|} \approx \frac{\mathbf{p}}{E} \qquad (3.210)$$

which is a unit vector in the direction of motion. Equations (3.209) are known as the *Weyl equations* and ψ_R and ψ_L are called the right-handed and left-handed *Weyl spinors*, respectively [8].

As we will discuss in more detail later, the operator $\boldsymbol{\sigma} \cdot \hat{\mathbf{p}}$ corresponds to the helicity of the particle: the projection of the spin in the direction of motion[10]. Therefore, the

---

[10] Notice that, as we already explained in the derivation of the Dirac equation, in this representation the spinors at rest provide no information about the dynamics of the particle, and as a consequence $\psi_R = \psi_L$ when $\mathbf{p} = 0$.

Weyl representation of the Dirac matrices diagonalizes the helicity in the ultra-relativistic limit [8]. Although this is a convenient representation of the Dirac matrices for studying high-energy physics, most quantum information applications take place in the small v limit. Consequently, we need to determine what is the best representation of the Dirac matrices that can be used to represent spinors at rest.

To this end, let us observe that the Dirac equation for spinors at rest looks like:

$$(\gamma^0 p_0 + m)\psi = 0 \quad \Rightarrow \quad p_0\psi = -m\gamma^0\psi \tag{3.211}$$

Thus, it would be convenient to use a representation of the Dirac matrices where γ^0 is in diagonal form [5]. This is called the *standard representation of the Dirac matrices* (also known as the *Dirac–Pauli representation*), given by:

$$\gamma^0 = \begin{pmatrix} \mathbb{I} & 0 \\ 0 & -\mathbb{I} \end{pmatrix} \qquad \gamma = \begin{pmatrix} 0 & \sigma \\ -\sigma & 0 \end{pmatrix} \tag{3.212}$$

Furthermore, let us first note that each component of the four-spinor ψ satisfies the Klein–Gordon equation:

$$(\partial_\mu \partial^\mu - m^2)\psi_i = 0 \qquad i = 1, 2, 3, 4 \tag{3.213}$$

Therefore, we should look for four-momentum plane wave eigensolutions of the form:

$$\psi = u(\mathbf{p})e^{-ipx} \tag{3.214}$$

where $u(\mathbf{p})$ is a four-spinor that does not depend on the coordinates [5, 8]. Notice that we have introduced the shorthand expression:

$$px \equiv p_\mu x^\mu = -\omega_p t + \mathbf{p} \cdot \mathbf{x} \tag{3.215}$$

for notational convenience. As usual, the frequency ω_p and the momentum $\mathbf{p}$ are related by the expression:

$$\omega_p^2 = \mathbf{p} \cdot \mathbf{p} + m^2 \tag{3.216}$$

The spinor components of the proposed solution can be found by looking at the energy eigenstates of the Dirac Hamiltonian:

$$\hat{H}u(\mathbf{p}) = (\boldsymbol{\alpha} \cdot \mathbf{p} + \beta m)u(\mathbf{p}) = Eu(\mathbf{p}) \tag{3.217}$$

It can be shown that there are four independent solutions, two with positive energy ($E > 0$) and two with negative energy ($E < 0$). The positive energy solutions correspond to *particles*, whereas the negative energy solutions represent *antiparticles* [8].

The four-spinors that correspond to the two positive energy solutions ($s = 1, 2$) are given by:

$$u^{(s)}(\mathbf{p}) = N \begin{pmatrix} \chi^{(s)} \\ \dfrac{\sigma \cdot \mathbf{p}}{E + m}\chi^{(s)} \end{pmatrix} \qquad E > 0 \tag{3.218}$$

where N is a normalization constant and $\chi^{(s)}$ are constant two-spinors defined as:

$$\chi^{(1)} = \begin{pmatrix} 1 \\ 0 \end{pmatrix} \quad \chi^{(2)} = \begin{pmatrix} 0 \\ 1 \end{pmatrix} \tag{3.219}$$

Similarly, the negative energy solutions are found to be:

$$u^{(s+2)}(\mathbf{p}) = N \begin{pmatrix} \dfrac{-\boldsymbol{\sigma} \cdot \mathbf{p}}{|E| + m} \chi^{(s)} \\[2mm] \chi^{(s)} \end{pmatrix} \quad E < 0 \tag{3.220}$$

The integration of the density over the unit volume is given by:

$$\int \rho \, dV = \int \bar{\psi} \gamma^0 \psi \, dV = \int \psi^\dagger \psi \, dV = u^\dagger u = |N|^2 \frac{2E}{E+m} \tag{3.221}$$

and if we choose the normalization:

$$\int \rho \, dV = 2E \quad \Rightarrow \quad N = \sqrt{E+m} \tag{3.222}$$

Alternatively, we could have chosen the following normalization conditions for the spinors:

$$\int \rho \, dV = \frac{E}{m} \quad \Rightarrow \quad N = \sqrt{\frac{E+m}{2m}}$$

$$\int \rho \, dV = 1 \quad \Rightarrow \quad N = \sqrt{\frac{E+m}{2E}} \tag{3.223}$$

For the former normalization condition, we can explicitly write the positive energy four-spinor solutions as:

$$u^{(1)}(\mathbf{p}) = \sqrt{E+m} \begin{pmatrix} 1 \\[2mm] 0 \\[2mm] \dfrac{p_3}{E+m} \\[3mm] \dfrac{p_1 + \mathrm{i} p_2}{E+m} \end{pmatrix} \tag{3.224}$$

and:

$$u^{(2)}(\mathbf{p}) = \sqrt{E+m} \begin{pmatrix} 0 \\ 1 \\ \dfrac{p_1 - ip_2}{E+m} \\ \dfrac{-p_3}{E+m} \end{pmatrix} \tag{3.225}$$

On the other hand, the negative energy four-spinor solutions are given by:

$$u^{(3)}(\mathbf{p}) = \sqrt{E+m} \begin{pmatrix} \dfrac{-p_3}{|E|+m} \\ -\dfrac{p_1 + ip_2}{|E|+m} \\ 1 \\ 0 \end{pmatrix} \tag{3.226}$$

and:

$$u^{(4)}(\mathbf{p}) = \sqrt{E+m} \begin{pmatrix} -\dfrac{p_1 - ip_2}{|E|+m} \\ \dfrac{p_3}{|E|+m} \\ 0 \\ 1 \end{pmatrix} \tag{3.227}$$

Therefore, a spin-$\frac{1}{2}$ particle such as the electron is described by the positive energy spinors:

$$u^{(1,2)}(\mathbf{p})e^{-ipx} = u^{(1,2)}(\mathbf{p})e^{iEt-i\mathbf{p}\cdot\mathbf{x}} \tag{3.228}$$

In addition, according to the *Feynman–Stückelberg interpretation of the negative energy solutions*, an antiparticle of energy $E > 0$ and momentum $\mathbf{p}$ can be represented by a Dirac spinor of energy $-E < 0$ and momentum $-\mathbf{p}$:

$$u^{(3,4)}(-\mathbf{p})e^{-iEt+i\mathbf{p}\cdot\mathbf{x}} \equiv v^{(2,1)}(\mathbf{p})e^{-iEt+i\mathbf{p}\cdot\mathbf{x}} = v^{(2,1)}(\mathbf{p})e^{ipx} \qquad (3.229)$$

defined in terms of a positive energy $p^0 = E > 0$ momentum four-vector and we introduced the definition of the *antiparticle spinor* $v^{(s)}$. However, as we will see in a later section, the quantization of the Dirac field leads to strictly non-negative energies of the quantum field. In addition, in the context of Dirac fields, antiparticles and particles carry a charge of equal magnitude but different sign.

The most general solution to the free-particle Dirac equation, which represents a superposition of particle and antiparticle states, can be written as:

$$\psi = \sum_{s=1}^{2} \alpha_s u^{(s)}(\mathbf{p})e^{-ipx} + \sum_{s=1}^{2} \beta_s v^{(s)}(\mathbf{p})e^{ipx} \qquad (3.230)$$

where α_i and β_i are complex parameters normalized to unity:

$$\alpha_1\alpha_1^* + \alpha_2\alpha_2^* + \beta_1\beta_1^* + \beta_2\beta_2^* = 1 \qquad (3.231)$$

Solely relying on the postulates of quantum mechanics, there is nothing that forbids a general type of coherent superposition such as the one given above. However, within the context of particle physics, it is known that some coherent superpositions of pure states are not found in nature [3, 26, 31, 32]. For example, a superposition of states with different charge has never been observed, nor with a mixture of integer and fractional spin. That is, the charge and spin quantum numbers have completely determinate values. Furthermore, it is known that the interference due to the mixing of positive and energy solutions in bound-state Dirac spinors leads to *Zitterbewegung*, a rapid motion of the electrons that violates Newton's second law [2, 6].

Any statement that forbids unphysical superposition of states (which are not observed in experiments) is known as a *superselection rule* [3, 26, 31, 32]. Therefore, we will invoke the charge superselection rule and work only with positive energy solutions (charge q particle states), or only with negative energy solutions (charge $-q$ antiparticle states).

In this book we will concentrate on the positive energy solutions to the Dirac equation, but similar results can be derived for the negative energy solutions. Thus, the general positive energy solution to the Dirac equation in flat spacetime is simply given by:

$$\psi_+ = \sum_{s=1}^{2} \alpha_s u^{(s)}(\mathbf{p})e^{-ipx} \qquad (3.232)$$

where α_1 and α_2 are complex parameters normalized as:

$$\alpha_1\alpha_1^* + \alpha_2\alpha_2^* = 1 \qquad (3.233)$$

3.6.6 Spin and helicity

As shown in section 3.5, the spin can be defined as those extra degrees of freedom that transform in a non-trivial way under a Lorentz transformation. In the case of the Dirac equation, the four-spinor ψ has four degrees of freedom, one for each component. However, the energy eigenstates have only two degrees of freedom, two for positive energy solutions and two for negative energy solutions. Therefore, the spin of the particle has to be related to the two degrees of freedom that correspond to a single energy eigenvalue. That is, spin breaks the degeneracy of the energy in the solutions of the Dirac equation.

Indeed, let us define the operator:

$$\hat{\mathbf{\Sigma}} \equiv \begin{pmatrix} \sigma & 0 \\ 0 & \sigma \end{pmatrix} \tag{3.234}$$

and a unit vector in the direction of the momentum:

$$\hat{\mathbf{p}} \equiv \frac{\mathbf{p}}{|\mathbf{p}|} \tag{3.235}$$

Then, it can be shown that the following commutation relations hold for a free particle:

$$[\hat{H}, \hat{\mathbf{\Sigma}} \cdot \hat{\mathbf{p}}] = 0 \qquad [\hat{\mathbf{P}}, \hat{\mathbf{\Sigma}} \cdot \hat{\mathbf{p}}] = 0 \tag{3.236}$$

where $\hat{H}$ is the Dirac Hamiltonian. That is, the operator $\hat{\mathbf{\Sigma}} \cdot \hat{\mathbf{p}}$ commutes with the Hamiltonian and momentum operators $\hat{H}$ and $\hat{\mathbf{P}}$, respectively. Consequently, the eigenvalues of $\hat{\mathbf{\Sigma}} \cdot \hat{\mathbf{p}}$ are good quantum numbers that can be used to label the proposed free particle solutions and split the energy degeneracy of the eigenstates.

Furthermore, it can be shown that, if we define the orbital angular momentum operator in the traditional manner as:

$$\hat{\mathbf{L}} = \mathbf{r} \times \hat{\mathbf{P}} \tag{3.237}$$

then:

$$[\hat{H}, \hat{\mathbf{L}}] = -i(\boldsymbol{\alpha} \times \hat{\mathbf{P}}) \neq 0 \tag{3.238}$$

where $\boldsymbol{\alpha}$ is the vector of matrices that appears in the Dirac Hamiltonian. Consequently:

$$\frac{d\hat{\mathbf{L}}}{dt} = i[\hat{H}, \hat{\mathbf{L}}] \neq 0 \tag{3.239}$$

which implies that the orbital angular momentum is *not* conserved [8]. However, the *total angular momentum* defined as:

$$\hat{\mathbf{J}} \equiv \hat{\mathbf{L}} + \frac{1}{2}\hat{\mathbf{\Sigma}} \tag{3.240}$$

is a conserved quantity:

$$\frac{d\hat{\mathbf{J}}}{dt} = i[\hat{H}, \hat{\mathbf{J}}] = 0 \tag{3.241}$$

In other words, $\hat{\Sigma}$ is an operator related to the *intrinsic* angular momentum of the free particle (i.e. its spin). Therefore, the helicity (the spin component in the direction of motion):

$$\hat{\lambda} \equiv \frac{1}{2}\sigma \cdot \hat{\mathbf{P}} \tag{3.242}$$

is the right quantum number to break the degeneracy of the free particle spinors [8].

For example, for a (positive energy) free Dirac particle moving in the three-direction:

$$\mathbf{p} = (0,0,p) \tag{3.243}$$

we have:

$$\hat{\lambda}\,\chi^{(s)} = \frac{1}{2}\sigma_3\,\chi^{(s)} = \lambda\chi^{(s)} = \begin{cases} +\dfrac{1}{2}\chi^{(1)} \\[2mm] -\dfrac{1}{2}\chi^{(2)} \end{cases} \tag{3.244}$$

and therefore $\chi^{(1)}$ is a spinor associated with positive helicity states (the spin points in the direction of motion), whereas $\chi^{(2)}$ is a spinor associated with negative helicity states (the spin is anti-parallel to the direction of motion).

On the other hand, if the free particle is moving along the one-direction:

$$\mathbf{p} = (p,0,0) \tag{3.245}$$

we have:

$$\hat{\lambda}\chi^{(1)} = \frac{1}{2}\sigma_1\chi^{(1)} = \frac{1}{2}\chi^{(2)}$$

$$\hat{\lambda}\chi^{(2)} = \frac{1}{2}\sigma_1\chi^{(2)} = \frac{1}{2}\chi^{(1)} \tag{3.246}$$

which shows that $\chi^{(1)}$ and $\chi^{(2)}$ are not eigenstates of the helicity for a particle moving along the one-direction. However the linear combination of spinors:

$$\chi^{(+)} = \frac{\chi^{(1)} + \chi^{(2)}}{\sqrt{2}}$$

$$\chi^{(-)} = \frac{\chi^{(1)} - \chi^{(2)}}{\sqrt{2}} \tag{3.247}$$

are helicity eigenstates:

$$\hat{\lambda}\chi^{(\pm)} = \frac{1}{2}\sigma_1\chi^{(\pm)} = \lambda\chi^{(\pm)} = \begin{cases} +\dfrac{1}{2}\chi^{(+)} \\[2mm] -\dfrac{1}{2}\chi^{(-)} \end{cases} \tag{3.248}$$

for a free particle moving along the one-direction.

In the case where the particle is at rest, then the helicity operator $\hat{\lambda}$ is replaced by a spin projection operator σ_i [33]. Indeed, let us consider a particle with helicity λ and momentum aligned to the three-direction $\mathbf{p} = (0, 0, p)$ such that:

$$\hat{\lambda}\Psi_{\mathbf{p},\lambda} = \lambda\Psi_{\mathbf{p},\lambda} \tag{3.249}$$

If we apply a Lorentz boost along the three-direction, in such a way that the particle is seen in its rest frame, then we have:

$$\hat{B}_3(-\mathbf{p})\Psi_{\mathbf{p},\lambda} = \Psi_{\mathbf{0},s_3} \tag{3.250}$$

where $\Psi_{\mathbf{0},s_3}$ is the state of the particle in the rest frame with a spin quantization along the three-axis given by:

$$\hat{J}_3\Psi_{\mathbf{0},s_3} = s_3\Psi_{\mathbf{0},s_3} \tag{3.251}$$

and therefore both states $\Psi_{\mathbf{p},\lambda}$ and $\Psi_{\mathbf{0},s_3}$ are simply connected through a Lorentz transformation. Indeed, as helicity is the projection of the spin on the direction of motion, there is ambiguity if the particle is at rest.

3.7 Dirac quantum fields

In a similar fashion as with the scalar quantum field, the Dirac quantum field is constructed through a Fourier decomposition of modes:

$$\hat{\psi}(x) = \int \frac{\mathrm{d}^3k}{(2\pi)^{3/2}} \sum_{\alpha=1,2} \left[\hat{b}_\alpha(k)u^{(\alpha)}(k)e^{ikx} + \hat{d}_\alpha^\dagger(x)v^{(\alpha)}(k)e^{-ikx} \right]$$

$$\hat{\bar{\psi}}(x) = \int \frac{\mathrm{d}^3k}{(2\pi)^{3/2}} \sum_{\alpha=1,2} \left[\hat{b}_\alpha^\dagger(k)\bar{u}^{(\alpha)}(k)e^{-ikx} + \hat{d}_\alpha(x)\bar{v}^{(\alpha)}(k)e^{ikx} \right] \tag{3.252}$$

In this case, however, there are two distinct types of annihilation operator denoted by $\hat{b}_\alpha(k)$ and $\hat{d}_\alpha(k)$.

Furthermore, the Hamiltonian for the quantum Dirac field can be expressed as:

$$\hat{H} = \int \mathrm{d}^3x : \hat{\psi}^\dagger(x)i\frac{\partial}{\partial t}\hat{\psi}(x) :$$

$$= \int \mathrm{d}^3k\, \omega_{\mathbf{k}} \sum_\alpha \left[\hat{b}_\alpha^\dagger(k)\hat{b}_\alpha(k) + \hat{d}_\alpha^\dagger(k)\hat{d}_\alpha(k) \right] \tag{3.253}$$

where we have used the following anti-commutation relationships:

$$\left\{ \hat{b}_\alpha(k), \hat{b}_{\alpha'}^\dagger(k') \right\} = \left\{ \hat{d}_\alpha(k), \hat{d}_{\alpha'}^\dagger(k') \right\} = \delta^3(\mathbf{k} - \mathbf{k}')\delta_{\alpha,\alpha'}$$

$$\left\{ \hat{b}_\alpha(k), \hat{b}_{\alpha'}(k') \right\} = \left\{ \hat{b}_\alpha^\dagger(k), \hat{b}_{\alpha'}^\dagger(k') \right\} = 0 \tag{3.254}$$

$$\left\{ \hat{d}_\alpha(k), \hat{d}_{\alpha'}(k') \right\} = \left\{ \hat{d}_\alpha^\dagger(k), \hat{d}_{\alpha'}^\dagger(k') \right\} = 0$$

to obtain positive definite states of energy [5, 18]. In addition, in equation (3.253) we have inserted the *normal ordering operator* (denoted by the symbols ': :'), which places all annihilation operators to the right of all the creation operators. Furthermore, these anti-commutation relations imply that the Dirac quantum fields satisfy the following equal time anti-commutation relationships:

$$\left\{ \hat{\psi}_i(\mathbf{x}, t), \hat{\psi}_j^\dagger(\tilde{\mathbf{x}}, t) \right\} = \delta(\mathbf{x} - \tilde{\mathbf{x}})\, \delta_{ij}$$

$$\left\{ \hat{\psi}_i(\mathbf{x}, t), \hat{\psi}_j(\tilde{\mathbf{x}}, t) \right\} = \left\{ \hat{\psi}_i^\dagger(\mathbf{x}, t), \hat{\psi}_j^\dagger(\tilde{\mathbf{x}}, t) \right\} = 0$$

(3.255)

And we can recall that, for the case of the Dirac field, the canonical momentum operator is given by:

$$\hat{\pi} = i\hat{\psi}^\dagger$$

(3.256)

As we have seen, the Dirac quantum field involves two pairs of particle creation and annihilation operators denoted by $(\hat{b}, \hat{b}^\dagger)$ and $(\hat{d}, \hat{d}^\dagger)$, respectively. Let us define the number operators for particles of type b and type d as:

$$\hat{N}_b = \hat{b}^\dagger \hat{b}$$

$$\hat{N}_d = \hat{d}^\dagger \hat{d}$$

(3.257)

where $\hat{N}_b$ is known as the *particle number operator* and $\hat{N}_d$ is the *antiparticle number operator*. Then, the Hamiltonian of the quantum Dirac field can be written as:

$$\hat{H} = \int \mathrm{d}^3k\, \omega_\mathbf{k} \left[\hat{N}_b(k) + \hat{N}_d(k) \right]$$

(3.258)

where for simplicity we have omitted the summation over spin polarization states. Then, the total energy of the quantum field is the sum of all the positive definite energies corresponding to all the particles and anti-particles in the system. That is, the energy of the Dirac quantum field is non-negative.

In the Fourier decomposition of the quantum field $\hat{\psi}$, the terms with e^{ikx} are known as the *positive energy solutions* or *positive frequency solutions*, whereas the terms with e^{-ikx} are known as the *negative energy solutions* or *negative frequency solutions* [18]. This distinction is based on the sign of the eigenvalue of the *energy operator*:

$$\hat{E}\mathrm{e}^{ikx} = i\frac{\partial}{\partial t}\mathrm{e}^{ikx} = \omega_\mathbf{k}\mathrm{e}^{ikx}$$

$$\hat{E}\mathrm{e}^{-ikx} = i\frac{\partial}{\partial t}\mathrm{e}^{-ikx} = -\omega_\mathbf{k}\mathrm{e}^{-ikx}$$

(3.259)

with:

$$\omega_\mathbf{k}^2 = \mathbf{k} \cdot \mathbf{k} + m^2 > 0$$

(3.260)

Notice that the energy operator $\hat{E}$ is different from the quantum field Hamiltonian operator $\hat{H}$. Furthermore, even though the field is made of positive and negative energy frequencies (as eigenvalues of the energy operator $\hat{E}$), the total energy of the quantum field is positive definite (with reference to the eigenvalues of the quantum field Hamiltonian operator $\hat{H}$ in the occupation Fock space) [5, 18]. Consequently, the quantization of the Dirac field removes the problem of negative energy solutions that are obtained when only the Dirac equation eigenvalues are considered.

In addition, it can be shown that the Dirac particles have an associated electric charge, while the antiparticles have a corresponding negative charge, so the total electric charge of the Dirac field is given by:

$$\hat{Q} \propto \int \mathrm{d}^3 k \left[\hat{N}_b(k) - \hat{N}_d(k) \right] \tag{3.261}$$

That is, the states:

$$\hat{b}_\alpha^\dagger(k)|0\rangle \qquad \hat{d}_\alpha^\dagger(k)|0\rangle \tag{3.262}$$

represent particles of momentum k^μ, spin polarization α, mass $-k^\mu k_\mu$ and opposite charge, respectively [5, 18].

Therefore, we have the following four cases of interest that determine the action of the four quantum field operators:

- For $\hat{\psi}$, the operator $\hat{d}_\alpha^\dagger(k)$ that corresponds to the negative energy solutions creates anti-particles of momentum k^μ, charge $-q$ and spin polarization α;
- For $\hat{\psi}$, the operator $\hat{b}_\alpha(k)$ that corresponds to the positive energy solutions destroys particles of momentum k^μ, charge q and spin polarization α;
- For $\hat{\bar{\psi}}$, the operator $\hat{b}_\alpha^\dagger(k)$ that corresponds to the negative energy solutions creates particles of momentum k^μ, charge q and spin polarization α;
- For $\hat{\bar{\psi}}$, the operator $\hat{d}_\alpha(k)$ that corresponds to the positive energy solutions destroys anti-particles of momentum k^μ, charge $-q$ and spin polarization α.

Once more, notice that despite the negative frequencies present in the Fourier decomposition associated with the negative energy solutions of the Dirac equation, all the quantum excitations generated by the Dirac field have non-negative energy.

Furthermore, there is a fundamental reason why, for example, the particle annihilation operator is associated with the positive energy solutions [1]. Indeed, let us consider the Dirac field in the Heisenberg picture:

$$\hat{\psi} \propto f(r)\hat{b}(t) \tag{3.263}$$

where the operator $\hat{b}(t)$ explicitly depends on time. The operator $\hat{b}(t)$, in terms of the operator $\hat{b}$ in the Schrödinger picture, is given by:

$$\hat{b}(t) = \mathrm{e}^{i\hat{H}t}\,\hat{b}\,\mathrm{e}^{-i\hat{H}t} \tag{3.264}$$

The action of this annihilation operator on an arbitrary state is to reduce the energy from some initial energy E_i to some final energy E_f:

$$E_i \rightarrow E_f \qquad \omega \equiv E_i - E_f > 0 \qquad (3.265)$$

in the following way:

$$
\begin{aligned}
\mathrm{e}^{i\hat{H}t}\, \hat{b}\, \mathrm{e}^{-i\hat{H}t}|E_i\rangle &= \mathrm{e}^{i\hat{H}t}\, \hat{b}\, \mathrm{e}^{-iE_i t}|E_i\rangle \\
&= \mathrm{e}^{i\hat{H}t}\, \mathrm{e}^{-iE_i t}|E_f\rangle \\
&= \mathrm{e}^{iE_f t}\, \mathrm{e}^{-iE_i t}|E_f\rangle \\
&= \mathrm{e}^{i(E_f - E_i)t}|E_f\rangle \\
&= \mathrm{e}^{-i\omega t}|E_f\rangle \qquad\qquad (3.266)
\end{aligned}
$$

which according to the energy operator corresponds to a positive energy solution. A similar rationale can be argued for the other operators.

Therefore, the energy operator gives a natural symmetry that can be used to characterize the difference between particles (related to the $E > 0$ eigenvalues of the Dirac equation) and antiparticles (related to the $E < 0$ eigenvalues of the Dirac equation) [34]. As we will discuss in chapter 5, this is not the case if the Dirac field is interacting with a gravitational field.

3.8 Group representations in quantum field theory

In section 3.6 we showed that, if we start with an arbitrary four-spinor ψ, then we can deduce its relativistic wave equation just by looking at how its four components transform under a Lorentz transformation. However, we found that some of the transformations involved are not unitary. This is an expected result, considering that the Lorentz group is non-compact, and therefore its unitary representations are infinite-dimensional, and its finite representations are not unitary.

This appears to be a problematical result, as we know that quantum mechanics requires unitary operators. However, we need to be very careful to understand what the physical quantities are that are being transformed under the Lorentz or Poincare transformations, and under what representation.

First, we have to make clear that not all physical quantities in quantum mechanics need to be transformed with unitary operations [24, 25]. For instance, *physical variables* such as the four-position x^μ and the four-momentum p^μ transform as non-unitary, finite dimensional representations of the Lorentz group. In addition, *classical and quantum fields* such as spinors ψ_a, the classical electromagnetic potential A^μ and their quantum counterparts, $\hat{\psi}_a$ and $\hat{A}^\mu$ also transform as non-unitary finite dimensional representations of the Lorentz group. In these examples, the spinor component index a and the spacetime index μ label the finite dimensional components of the physical entity that are being transformed in a non-trivial way under a Lorentz transformation. Consequently, these labels are called *Lorentz indices*.

On the other hand, *quantum states* require unitary operations. That is, unitarity is required for physical entities such as:

$$|p^{\mu}, j, \sigma\rangle \tag{3.267}$$

where the three labels represent the eigenvectors of the momentum, spin and spin projection operators, respectively. These labels are called *Poincare indices*.

This is a subtle, but important difference. Quantum states are the physical entities with observable quantities that are measured in an experiment and require unitary operations. On the other hand, wave functions and fields are not physical observables (as they cannot be observed directly by experimentation), so unitarity is not required.

As stated before, we need to be very careful to determine not only how a given physical entity transforms under a Lorentz or Poincare transformation. It is also important to determine what representation of the Lorentz or Poincare group is being used to represent them. In what follows we will discuss these issues in further detail.

3.8.1 Non-unitary representations of the Lorentz group and quantum fields

Let us recall that the Lorentz group can be decomposed into two independent SU(2) subgroups:

$$\mathrm{SO}^{+}(3,1) \simeq \mathrm{SU}(2)_A \times \mathrm{SU}(2)_B \tag{3.268}$$

Then, we can use the labels:

$$(u, k; v, l) \tag{3.269}$$

to denote the physical entities that undergo a non-trivial Lorentz transformation. More specifically, (u, k) and (v, l) are labels that correspond to the A and B subgroups, respectively. These entities behave as if they were spin-like states, in the sense that:

$$
\begin{aligned}
\hat{A}^2 |u, k\rangle &= u(u+1)|u, k\rangle \\[4pt]
\hat{A}_3 |u, k\rangle &= k|u, k\rangle \\[4pt]
\hat{B}^2 |v, l\rangle &= v(v+1)|v, l\rangle \\[4pt]
\hat{B}_3 |v, l\rangle &= l|v, l\rangle
\end{aligned}
\tag{3.270}
$$

which is equivalent to having a quantum system with two spin variables (A and B). Then, a representation of the full Lorentz group can be written as:

$$|u, k\rangle \otimes |v, l\rangle \tag{3.271}$$

where u and v are the spin numbers, and k and l correspond to the spin projection numbers.

For example, let us recall that in our derivation of the Dirac equation we used:

$$u = v = \frac{1}{2}$$

$$k, l = -\frac{1}{2}, \frac{1}{2} \tag{3.272}$$

to represent two types of spinors:

$$\xi \leftrightarrow \left| \frac{1}{2}, \pm \frac{1}{2} \right\rangle \otimes |0, 0\rangle$$

$$\eta \leftrightarrow |0, 0\rangle \otimes \left| \frac{1}{2}, \pm \frac{1}{2} \right\rangle \tag{3.273}$$

that are transformed with the non-unitary operators $\mathcal{M}$ and $\mathcal{N}$ as:

$$\xi \rightarrow \mathcal{M}\xi$$

$$\eta \rightarrow \mathcal{N}\eta \tag{3.274}$$

Therefore, these spinors correspond to finite, non-unitary representations of the Lorentz group. Furthermore, the Dirac equation relates the components of these spinors as they undergo Lorentz transformations:

$$(i\gamma^\mu \partial_\mu - m)\psi_\alpha(r) = 0 \tag{3.275}$$

where $\psi_\alpha(x)$ is the four-spinor that results from combining ξ and η, and α is the Lorentz index that labels the four-spinor components.

In addition, by definition, the Dirac quantum field $\hat{\psi}_\alpha(x)$ satisfies the Dirac equation and carries a Lorentz index α. Consequently, the quantum field transforms as a finite-dimensional, non-unitary representation of the Lorentz group.

3.8.2 Unitary representations of the Poincare group and quantum states

Let us recall that the Poincare group is the Lorentz group extended by translations in four-dimensional spacetime:

$$\mathrm{ISO}^+(3, 1) = \mathbb{R}^{3,1} \rtimes \mathrm{SO}^+(3, 1) \tag{3.276}$$

and its infinite-dimensional unitary representation is generated by:

$$\hat{P}^\mu \qquad \hat{S}^2 \qquad \hat{S}_3 \tag{3.277}$$

which correspond to the momentum, spin squared and spin projection operators, respectively. As any pair of these operators commute with each other, there exists a unique set of simultaneous eigenstates, and the physical state of the quantum system is completely specified by the associated eigenvalues [11]. Therefore, the states that transform under the Poincare group can be labelled using the eigenvectors of these operators as follows:

$$|p^\mu, j, \sigma\rangle \tag{3.278}$$

which correspond to a particle of momentum p^μ, spin j and spin projection σ. These states are known as *Poincare states* and they satisfy:

$$\hat{P}^\mu \,|p^\mu, j, \sigma\rangle = p^\mu \,|p^\mu, j, \sigma\rangle$$

$$\hat{S}^2 \,|p^\mu, j, \sigma\rangle = j(j+1)\,|p^\mu, j, \sigma\rangle \qquad (3.279)$$

$$\hat{S}_3 \,|p^\mu, j, \sigma\rangle = \sigma\,|p^\mu, j, \sigma\rangle$$

Clearly, the label p^μ can take an infinitely large number of values, which makes explicit that this representation is infinite-dimensional. Also, these states are transformed using unitary operators.

Furthermore, p^μ is used to determine the momentum square $p^\mu p_\mu$ and the time-like component p^0, which in turn defines the little group. And j and σ label the rotation parameters of the little group. In addition, these three operators commute with the Hamiltonian of the system, which means that they are 'good' quantum numbers to label the quantum states (i.e. the associated eigenvectors remain the same as time evolves).

The representation of the quantum state is related to the representation of the creation and annihilation operators. Indeed, let us rewrite the Dirac quantum field as:

$$\hat{\Psi}_\alpha(x) = \int \frac{\mathrm{d}^3 k}{(2\pi)^{3/2}} \sum_{\sigma=1,2} (\hat{b}(k,\sigma) u_\alpha(k,\sigma) e^{ikx} + \hat{d}^\dagger(k,\sigma) v_\alpha(k,\sigma) e^{-ikx}) \qquad (3.280)$$

where α is the Lorentz index that labels the components of the four-spinor, and σ is the spin projection eigenvalue (and we have assumed that the spin has a fixed value $j = \frac{1}{2}$ for the Dirac spinors, therefore, we can omit this label in all the equations). As discussed before, the quantum field $\hat{\Psi}_\alpha(x)$ only carries a Lorentz index (α), and as a consequence, it transforms as a finite-dimensional non-unitary representation of the Lorentz group.

On the other hand, the creation and annihilation operators $\hat{b}(k,\sigma)$ and $\hat{d}(k,\sigma)$ carry Poincare indices k and σ, do not carry Lorentz indices and therefore transform as the infinite-dimensional unitary representations of the Poincare group. In other words, the creation operators create Poincare states:

$$\hat{b}^\dagger(k,\sigma)|0\rangle \propto |k,\sigma\rangle \qquad (3.281)$$

while the annihilation operators destroy Poincare states [25].

3.8.3 Unitary/non-unitary representations and wave functions

In quantum field theory, the *wave function* is defined as the transition amplitude of the quantum field between a given quantum state and the quantum vacuum:

$$\phi = \langle 0|\hat{\Psi}_\alpha(x)|\phi\rangle$$

$$= \langle 0|\hat{\Psi}_\alpha(x)|\mathbf{k},\sigma\rangle$$

$$= u_\alpha(\mathbf{k},\sigma) e^{ikx} \qquad (3.282)$$

where ϕ is the wave function that corresponds to the quantum state $|\phi\rangle$ of momentum $\mathbf{k}$ and spin polarization σ [5]. Therefore, wave functions carry both Lorentz and Poincare indices [25].

To summarize:

- **Quantum fields** obey the Dirac equation and conform to finite-dimensional non-unitary representations of the Lorentz group;
- **Quantum states** are physical states that transform as the infinite-dimensional unitary representations of the Poincare group;
- **Wave functions** are the complex numbers that connect the finite-dimensional non-unitary representations of the Lorentz group with the infinite-dimensional unitary representations of the Poincare group.

This makes evident that quantum fields, quantum states and wave functions are different types of physical entity, described by different mathematical objects, with different transformation rules under the Lorentz and Poincare groups [25].

3.9 Representations of quantum fields with arbitrary spin

Let us consider the properties of quantum fields of arbitrary spin under Lorentz transformations. If we have the infinitesimal Lorentz transformation:

$$
\begin{aligned}
x^\alpha \to \tilde{x}^\alpha &= \Lambda^\alpha_{\ \beta} x^\beta \\
&= (\delta^\alpha_\beta + \xi^\alpha_{\ \beta}) x^\beta
\end{aligned}
\tag{3.283}
$$

where:

$$
\begin{aligned}
\xi_{\alpha\beta} &= -\xi_{\beta\alpha} \\
|\xi^\alpha_{\ \beta}| &\ll 1
\end{aligned}
\tag{3.284}
$$

then, a general multicomponent field $T^{ab\cdots}$ will transform as:

$$
T^{ab\cdots} \to T^{a'b'\cdots} = [D(\Lambda)]^{a'b'\cdots}_{a\ b\cdots}\ T^{ab\cdots}
\tag{3.285}
$$

where:

$$
D(\Lambda) \equiv 1 + \frac{1}{2}\xi^{\alpha\beta}\Sigma_{\alpha\beta}
\tag{3.286}
$$

is an adequate matrix representation of the Lorentz transformation Λ, and $\Sigma_{\alpha\beta}$ are antisymmetric tensors:

$$
\Sigma_{\alpha\beta} = -\Sigma_{\beta\alpha}
\tag{3.287}
$$

It can be shown that, if these transformations form a group, then the $\Sigma_{\alpha\beta}$ tensors are required to satisfy the following commutation relation:

$$
[\Sigma_{\alpha\beta}, \Sigma_{\gamma\delta}] = \eta_{\gamma\beta}\Sigma_{\alpha\delta} - \eta_{\alpha\gamma}\Sigma_{\beta\delta} + \eta_{\delta\beta}\Sigma_{\gamma\alpha} - \eta_{\delta\alpha}\Sigma_{\gamma\beta}
\tag{3.288}
$$

and are called the *generators of the Lorentz group* [13]. Indeed, the above condition is necessary to preserve the group multiplication rule expressed by:

$$D(\Lambda_1)D(\Lambda_2) = D(\Lambda_1\Lambda_2) \qquad (3.289)$$

where Λ_1 and Λ_2 are two different Lorentz transformations.

It is now convenient to define the three-vectors **a** and **b** as:

$$a_i \equiv \frac{1}{2}(-i\epsilon_i{}^{kl}\Sigma_{kl} + \Sigma_{i0}) \qquad (3.290)$$

and:

$$b_i \equiv \frac{1}{2}(-i\epsilon_i{}^{kl}\Sigma_{kl} - \Sigma_{i0}) \qquad (3.291)$$

which satisfy the following relations:

$$\mathbf{a} \times \mathbf{a} = i\mathbf{a}$$
$$\mathbf{b} \times \mathbf{b} = i\mathbf{b} \qquad (3.292)$$
$$[a_i, b_j] = 0$$

which are equivalent to:

$$[a_i, a_j] = i\epsilon_{ijk}\, a_k$$
$$[b_i, b_j] = i\epsilon_{ijk}\, b_k \qquad (3.293)$$
$$[a_i, b_j] = 0$$

Therefore, **a** and **b** can be interpreted as the generators for two independent SU(2) groups (notice the similarity between this result and the discussion in section 3.8). Indeed, as we discussed before, the Lorentz group can be decomposed into:

$$SO^+(3,1) \simeq SU(2)_A \times SU(2)_B \qquad (3.294)$$

Consequently, each representation of the Lorentz group can be labelled by A or B, where these labels are defined by the eigenvalues of the squared operators on a specific field:

$$\mathbf{a}^2\, T_A = A(A+1)\, T_A$$
$$\mathbf{b}^2\, T_B = B(B+1)\, T_B \qquad (3.295)$$

where $\mathbf{a}^2$ and $\mathbf{b}^2$ act on vector spaces of dimension $2A+1$ and $2B+1$, respectively. Thus, representations of the Lorentz group can be written as the direct product of the two separate irreducible representations as:

$$(A, B) \qquad (3.296)$$

which act in a vector space of $(2A + 1)(2B + 1)$ dimensions [34].

In what follows we will further discuss the specific cases that correspond to the most widely used quantum fields: scalar fields, vector fields, tensor fields and Dirac fields.

3.9.1 Scalar fields

Under a Lorentz transformation, a scalar field has the trivial transformation rule:

$$\phi(x) \rightarrow \tilde{\phi}(\tilde{x}) = \phi(\tilde{x}) \tag{3.297}$$

and as a consequence, from equation (3.286) we conclude that:

$$\Sigma_{\alpha\beta} = 0 \tag{3.298}$$

This expression for the Σ tensor completely defines the value of **a** and **b** through equations (3.290) and (3.291), respectively. Therefore, in the case of scalar fields we have that:

$$\mathbf{a}^2 = \mathbf{b}^2 = 0 \tag{3.299}$$

This means that the scalar field transforms as the $(0, 0)$ irreducible representation of the Lorentz group. That is, a scalar field has spin zero $(0 + 0 = 0)$.

3.9.2 Vector fields

Under an infinitesimal Lorentz transformation, a vector field A^a transforms as:

$$A^a \rightarrow \tilde{A}^a = \Lambda^a_{\ b} A^b = (\delta^a_b + \xi^a_{\ b}) A^b \tag{3.300}$$

where, for the sake of clarity, we have used a Latin index to label the field components. Therefore:

$$\delta^a_b + \xi^a_{\ b} = \left[1 + \frac{1}{2} \xi^{\alpha\beta} \Sigma_{\alpha\beta} \right]^a_{\ b} \tag{3.301}$$

which implies:

$$\xi^a_{\ b} = \frac{1}{2} \xi^{\alpha\beta} \left[\Sigma_{\alpha\beta} \right]^a_{\ b} \tag{3.302}$$

and as a result:

$$\left[\Sigma_{\alpha\beta} \right]^a_{\ b} = \delta^a_\alpha \eta_{\beta b} - \delta^a_\beta \eta_{\alpha b} \tag{3.303}$$

(notice that in this case, because the field is a four-vector, the Latin indices run over exactly the same set of Greek indices). As before, this expression for the Σ tensor completely defines the value of **a** and **b**. Then, for the case of vector fields, this expression for Σ implies that:

$$\mathbf{a}^2 = \mathbf{b}^2 = \frac{1}{2} \left(\frac{1}{2} + 1 \right) \tag{3.304}$$

and consequently the vector field transforms as the $(\frac{1}{2}, \frac{1}{2})$ irreducible representation of the Lorentz group. Therefore, a vector field has spin $\frac{1}{2} + \frac{1}{2} = 1$.

3.9.3 Tensor fields

Second-rank tensors are used to represent the electromagnetic field tensor $F^{\mu\nu}$ and the energy-momentum tensor $T^{\mu\nu}$. Under a Lorentz transformation, a second-rank tensor $Q^{\alpha\beta}$ transforms as:

$$Q^{\alpha\beta} \to \tilde{Q}^{\alpha\beta} = \Lambda^{\alpha}_{\mu} \Lambda^{\beta}_{\nu} Q^{\mu\nu} \tag{3.305}$$

Therefore, the matrix representation of the Lorentz transformation for a second-rank tensor is the product of two matrix representations of the Lorentz transformation for a vector field. Then, the product representation is given by:

$$\left(\frac{1}{2}, \frac{1}{2}\right) \otimes \left(\frac{1}{2}, \frac{1}{2}\right) \tag{3.306}$$

which reduces to four distinct irreducible representations:

$$\left(\frac{1}{2}, \frac{1}{2}\right) \otimes \left(\frac{1}{2}, \frac{1}{2}\right) = \begin{cases} (1, 1) \\ (1, 0) \\ (0, 1) \\ (0, 0) \end{cases} \tag{3.307}$$

Therefore, the second-rank tensor field has one component with spin $1 + 1 = 2$, two components with spin $1 + 0 = 1$ and one component with spin $0 + 0 = 0$ [34]. In particular, the six independent anti-symmetric components $A^{\mu\nu}$:

$$A^{\mu\nu} \equiv Q^{\mu\nu} - Q^{\nu\mu} \tag{3.308}$$

are invariant under Lorentz transformations, transform under the $(1, 0) \oplus (0, 1)$ representation of the group and correspond to the spin-1 components. The ten symmetric components $S^{\mu\nu}$:

$$S^{\mu\nu} \equiv Q^{\mu\nu} + Q^{\nu\mu} \tag{3.309}$$

can be separated into nine traceless components and the trace itself S^{μ}_{μ}. All these components are irreducible representations of the Lorentz group. Furthermore, the traceless symmetric second-rank tensors transform as the $(1,1)$ representation and correspond to the spin-2 component. On the other hand, the trace S^{μ}_{μ} transforms as the $(0, 0)$ representation and corresponds to the spin-0 component.

3.9.4 Dirac fields

Under a Lorentz transformation Λ, the Dirac field ψ transforms as:

$$\psi \to \tilde{\psi} = \hat{U}(\Lambda)\psi \tag{3.310}$$

where $\hat{U}(\Lambda)$ is a spinor representation of a Lorentz group transformation (which was discussed in detail in section 3.6). Then, it can be shown that in this case we require:

$$\Sigma_{\alpha\beta} = \frac{1}{4} \left[\gamma_\alpha, \gamma_\beta \right] \tag{3.311}$$

which is a 4 × 4 matrix in four-spinor space (the dimension of the Dirac matrices), and in addition has the structure of a 4 × 4 matrix in spacetime (because of the two covariant indices) [13, 34]. Once again, this expression for the Σ matrix completely defines the value of **a** and **b**. Consequently, it can be shown that the Dirac fields are associated with the $(\frac{1}{2}, 0)$ and $(0, \frac{1}{2})$ irreducible representation of the Lorentz group [34]. Therefore, Dirac fields have spin $\frac{1}{2} + 0 = \frac{1}{2}$.

3.10 The quantum vacuum in flat spacetime

In this section we will discuss the quantum vacuum in flat spacetime. We will restrict our attention to the case of a real scalar quantum field, but these ideas can be easily generalized to more complex cases. Let us write the general expression for the real scalar quantum field as:

$$\hat{\phi}(x) = \frac{1}{\sqrt{2}} \int \frac{\mathrm{d}^3 k}{(2\pi)^{3/2}} \left(\mathrm{e}^{\mathrm{i}k\cdot x} v_{\mathbf{k}}^* \hat{a}_{\mathbf{k}} + \mathrm{e}^{-\mathrm{i}k\cdot x} v_{\mathbf{k}} \hat{a}_{\mathbf{k}}^\dagger \right) \tag{3.312}$$

where:

$$v_{\mathbf{k}} = v_{\mathbf{k}}(t) \tag{3.313}$$

is a function of time, but independent of the spatial coordinates. As we discussed before, the vacuum is defined as the state with no elementary quantum excitations. That is:

$$\hat{N}|0\rangle = 0 \tag{3.314}$$

However, we need to confirm that $|0\rangle$ is the real vacuum, namely the state of minimum energy [35].

First, let us start with the classical scalar field Hamiltonian:

$$H = \int \theta^{00} \mathrm{d}^3 x$$

$$= \frac{1}{2} \int \left(|\partial_0 \phi|^2 + \nabla\phi^* \cdot \nabla\phi + m^2 |\phi|^2 \right) \mathrm{d}^3 x \tag{3.315}$$

which has the following quantum counterpart:

$$\hat{H} = \frac{1}{2} \int \left(|\partial_0 \hat{\phi}|^2 + \nabla\hat{\phi}^* \cdot \nabla\hat{\phi} + m^2 |\hat{\phi}|^2 \right) \mathrm{d}^3 x \tag{3.316}$$

As all the terms in the integrand are positive definite, the minimum of the energy occurs when the integrand is minimal. In addition, because the quantum field $\hat{\phi}$ obeys the Klein–Gordon equation, we have:

$$\partial_0^2 \hat{\phi} - \partial_i \partial^i \hat{\phi} + m^2 \hat{\phi} = 0 \tag{3.317}$$

Using the Fourier decomposition of $\hat{\phi}$ on the Klein–Gordon equation leads to the following expression for each value of the momentum parameter $\mathbf{k}$:

$$e^{i\mathbf{k}\cdot\mathbf{x}}\hat{a}_{\mathbf{k}}\partial_0^2 v_{\mathbf{k}}^* + k^2 e^{i\mathbf{k}\cdot\mathbf{x}}\hat{a}_{\mathbf{k}} v_{\mathbf{k}}^* + m^2 e^{i\mathbf{k}\cdot\mathbf{x}}\hat{a}_{\mathbf{k}} v_{\mathbf{k}}^*$$
$$+ e^{-i\mathbf{k}\cdot\mathbf{x}}\hat{a}_{\mathbf{k}}^\dagger \partial_0^2 v_{\mathbf{k}} + k^2 e^{-i\mathbf{k}\cdot\mathbf{x}}\hat{a}_{\mathbf{k}}^\dagger v_{\mathbf{k}} + m^2 e^{-i\mathbf{k}\cdot\mathbf{x}}\hat{a}_{\mathbf{k}}^\dagger v_{\mathbf{k}} = 0 \tag{3.318}$$

which can be rewritten as:

$$\ddot{v}_{\mathbf{k}}^* + k^2 v_{\mathbf{k}}^* + m^2 v_{\mathbf{k}}^* = 0$$
$$\ddot{v}_{\mathbf{k}} + k^2 v_{\mathbf{k}} + m^2 v_{\mathbf{k}} = 0 \tag{3.319}$$

or equivalently:

$$\ddot{v}_{\mathbf{k}} + \omega_{\mathbf{k}}^2 v_{\mathbf{k}} = 0 \tag{3.320}$$

where as usual:

$$\omega_{\mathbf{k}}^2 \equiv \mathbf{k} \cdot \mathbf{k} + m^2 \tag{3.321}$$

corresponds to the energy associated with the quantum excitation mode with momentum $\mathbf{k}$. Notice that in this case, as in all isotropic models, the frequency and mode functions solely depend on the magnitude of the momentum $k = |\mathbf{k}|$ and are independent of the direction. Thus, in the isotropic case, we can write $v_{\mathbf{k}}$ as v_k, and $\omega_{\mathbf{k}}$ as ω_k without leading to confusion.

Then, a solution to these differential equations is given by:

$$v_{\mathbf{k}} = e^{i\omega_k t} \tag{3.322}$$

However, any other function of the type:

$$u_{\mathbf{k}} = \alpha_{\mathbf{k}} v_{\mathbf{k}} + \beta_{\mathbf{k}} v_{\mathbf{k}}^* \qquad \alpha_{\mathbf{k}}, \beta_{\mathbf{k}} \in \mathbb{C} \tag{3.323}$$

will also be a solution to the differential equation (3.319). Then, we need to determine which of these infinitely many solutions corresponds to the *true vacuum* of the scalar quantum field.

First, let us confirm our suspicion that u and v correspond to different vacuum states of the same scalar quantum field. Thus, let us assume that $\hat{\chi}$ is the scalar quantum field with a vacuum described by the u function generated through the linear transformation of v with a specific choice of $\alpha_{\mathbf{k}}$ and $\beta_{\mathbf{k}}$. Then:

$$\hat{\chi}(x) = \frac{1}{\sqrt{2}} \int \frac{d^3 k}{(2\pi)^{3/2}} \left(e^{i\mathbf{k}\cdot\mathbf{x}} u_{\mathbf{k}}^* \hat{b}_{\mathbf{k}} + e^{-i\mathbf{k}\cdot\mathbf{x}} u_{\mathbf{k}} \hat{b}_{\mathbf{k}}^\dagger \right) \tag{3.324}$$

where we have introduced a new set of annihilation and creation operators $\hat{b}$ and $\hat{b}^\dagger$.

Then, the vacuum states for each of the quantum fields, $\hat{\phi}$ and $\hat{\chi}$, are given by:

$$\hat{a}_{\mathbf{k}}|0\rangle_a = 0$$
$$\hat{b}_{\mathbf{k}}|0\rangle_b = 0 \tag{3.325}$$

where the subindex a refers to the annihilation operators of the quantum field $\hat{\phi}$, and b to the annihilation operators of the quantum field $\hat{\chi}$. More specifically, the operator $\hat{a}^{\dagger}$ creates 'a' particles, while the operator $\hat{b}^{\dagger}$ creates 'b' particles; and similarly for the annihilation operators.

However, $\hat{\phi}$ and $\hat{\chi}$ are simply different Fourier expansions to exactly the same quantum field. That is,

$$\hat{\phi} = \hat{\chi} \tag{3.326}$$

which implies that:

$$u_{\mathbf{k}}^* \hat{b}_{\mathbf{k}} + u_{\mathbf{k}} \hat{b}_{-\mathbf{k}}^{\dagger} = v_{\mathbf{k}}^* \hat{a}_{\mathbf{k}} + v_{\mathbf{k}} \hat{a}_{-\mathbf{k}}^{\dagger} \tag{3.327}$$

and using the linear transformation that connects the u and v functions, equation (3.323), we get:

$$(\alpha_{\mathbf{k}}^* v_{\mathbf{k}}^* + \beta_{\mathbf{k}}^* v_{\mathbf{k}}) \hat{b}_{\mathbf{k}} + (\alpha_{\mathbf{k}} v_{\mathbf{k}} + \beta_{\mathbf{k}} v_{\mathbf{k}}^*) \hat{b}_{-\mathbf{k}}^{\dagger} = v_{\mathbf{k}}^* \hat{a}_{\mathbf{k}} + v_{\mathbf{k}} \hat{a}_{-\mathbf{k}}^{\dagger} \tag{3.328}$$

This expression is equivalent to:

$$\hat{a}_{\mathbf{k}} = \alpha_{\mathbf{k}}^* \hat{b}_{\mathbf{k}} + \beta_{\mathbf{k}} \hat{b}_{-\mathbf{k}}^{\dagger}$$
$$\hat{a}_{-\mathbf{k}}^{\dagger} = \beta_{\mathbf{k}}^* \hat{b}_{\mathbf{k}} + \alpha_{\mathbf{k}} \hat{b}_{-\mathbf{k}}^{\dagger} \tag{3.329}$$

These transformations between the $\hat{a}$ and $\hat{b}$ operators are known as the *Bogolyubov transformations* [13, 34, 35]. If we demand that the new set of operators $\hat{b}$ and $\hat{b}^{\dagger}$ obey the same commutation relations as their $\hat{a}$ and $\hat{a}^{\dagger}$ counterparts, then the Bogolyubov transformations need to be normalized in such a way that:

$$|\alpha_{\mathbf{k}}|^2 - |\beta_{\mathbf{k}}|^2 = 1 \tag{3.330}$$

Let us now define number operators that correspond to the $\hat{a}$ and $\hat{b}$ operators:

$$\hat{N}_{\mathbf{k}}^{(a)} = \hat{a}_{\mathbf{k}}^{\dagger} \hat{a}_{\mathbf{k}}$$
$$\hat{N}_{\mathbf{k}}^{(b)} = \hat{b}_{\mathbf{k}}^{\dagger} \hat{b}_{\mathbf{k}} \tag{3.331}$$

That is, these operators give how many particles of type 'a' or type 'b' are present in a given quantum state. Then, the expectation value of the number of a particles in the vacuum of b is given by:

$$\begin{aligned}
{}_b\langle 0|\hat{N}^{(a)}|0\rangle_b &= {}_b\langle 0|\hat{a}_{\mathbf{k}}^{\dagger} \hat{a}_{\mathbf{k}}|0\rangle_b \\
&= {}_b\langle 0|(\beta_{\mathbf{k}}^* \hat{b}_{-\mathbf{k}} + \alpha_{\mathbf{k}} \hat{b}_{\mathbf{k}}^{\dagger})(\alpha_{\mathbf{k}}^* \hat{b}_{\mathbf{k}} + \beta_{\mathbf{k}} \hat{b}_{-\mathbf{k}}^{\dagger})|0\rangle_b \\
&= {}_b\langle 0|\hat{b}_{-\mathbf{k}} \hat{b}_{-\mathbf{k}}^{\dagger}|0\rangle_b \, \beta_{\mathbf{k}} \beta_{\mathbf{k}}^* \\
&= |\beta_k|^2
\end{aligned} \tag{3.332}$$

Therefore, the vacuum of b has a $|\beta_{\mathbf{k}}|^2$ particle density of a particles. In addition, in a similar manner, the vacuum of a has a $|\alpha_{\mathbf{k}}|^2$ particle density of b particles.

This means that the true quantum vacuum cannot be found just by looking at the minimum of the Hamiltonian $\hat{H}$. Instead, the real vacuum has to be found by minimizing the expectation value of the energy in the vacuum state $\langle 0|\hat{H}|0\rangle$. To this end, the Hamiltonian of the quantum field, in terms of 'a'-type particles, can be written as:

$$\hat{H} = \frac{1}{4}\int d^3k(\hat{a}_{\mathbf{k}}\hat{a}_{-\mathbf{k}}F_{\mathbf{k}}^* + \hat{a}_{\mathbf{k}}^\dagger\hat{a}_{-\mathbf{k}}^\dagger F_{\mathbf{k}})$$
$$+\frac{1}{4}\int d^3k(2\hat{a}_{\mathbf{k}}^\dagger\hat{a}_{\mathbf{k}} + \delta^{(0)}(0))E_{\mathbf{k}} \tag{3.333}$$

where:

$$E_{\mathbf{k}} = |\dot{v}_{\mathbf{k}}|^2 + \omega_{\mathbf{k}}^2|v_{\mathbf{k}}|^2$$
$$F_{\mathbf{k}} = \dot{v}_{\mathbf{k}}^2 + \omega_{\mathbf{k}}^2 v_{\mathbf{k}}^2 \tag{3.334}$$

Then, the vacuum energy of the quantum field $\hat{\phi}$ is:

$$_a\langle 0|\hat{H}|0\rangle_a = \frac{\delta^3(0)}{4}\int d^3k\, E_{\mathbf{k}} \tag{3.335}$$

The $\delta^3(0)$ factor is a consequence of using an infinitely large integration volume and can be safely renormalized [4, 5]. Then, it can be observed that, if we are only interested in finding the minimum of this expression, we only need to find the minimum of the function $E_{\mathbf{k}}$.

To find the minimum of the $E_{\mathbf{k}}$ function, let us first rewrite $v_{\mathbf{k}}$ in polar form:

$$v_{\mathbf{k}} = r_{\mathbf{k}}e^{i\alpha} \tag{3.336}$$

where $r_{\mathbf{k}}, \alpha \in \mathbb{R}$. Then, the derivative of $v_{\mathbf{k}}$ becomes:

$$\dot{v}_{\mathbf{k}} = \dot{r}_{\mathbf{k}}e^{i\alpha} + i\dot{\alpha}r_{\mathbf{k}}e^{i\alpha} \tag{3.337}$$

which means that the $E_{\mathbf{k}}$ function can be written as:

$$E_{\mathbf{k}} = \dot{r}_{\mathbf{k}}^2 + \frac{1}{r_{\mathbf{k}}^2} + \omega_{\mathbf{k}}^2 r_{\mathbf{k}}^2 \tag{3.338}$$

For minimization purposes, we can consider $E_{\mathbf{k}}$ as a function on two independent variables, $r_{\mathbf{k}}$ and $\dot{r}_{\mathbf{k}}$. Then, the minimum is found at:

$$\frac{\partial E_{\mathbf{k}}}{\partial \dot{r}_{\mathbf{k}}} = 0 \quad \Rightarrow \quad \dot{r}_{\mathbf{k}} = 0$$

$$\frac{\partial E_{\mathbf{k}}}{\partial r_{\mathbf{k}}} = 0 \quad \Rightarrow \quad -\frac{1}{r_{\mathbf{k}}^3} + \omega_{\mathbf{k}}^2 r_{\mathbf{k}} = 0 \tag{3.339}$$

Therefore, the minimum is found at:

$$\dot{r}_{\mathbf{k}} = 0$$
$$r_{\mathbf{k}} = \frac{1}{\sqrt{\omega_{\mathbf{k}}}} \tag{3.340}$$

That is, the $v_{\mathbf{k}}$ function associated with the true quantum vacuum state is given by:

$$v_{\mathbf{k}} = \frac{e^{i\alpha}}{\sqrt{\omega_{\mathbf{k}}}} \tag{3.341}$$

In addition, let us recall that we also have the commutation condition:

$$\left[\hat{\phi}(\mathbf{x},t), \hat{\pi}(\mathbf{y},t)\right] = \left[\hat{\phi}(\mathbf{x},t), \dot{\hat{\phi}}(\mathbf{y},t)\right] = i\delta(\mathbf{x} - \mathbf{y}) \tag{3.342}$$

which implies the normalization condition:

$$\frac{\dot{v}_{\mathbf{k}} v_{\mathbf{k}}^* - v_{\mathbf{k}} \dot{v}_{\mathbf{k}}^*}{2i} = 1 \tag{3.343}$$

and therefore:

$$r_{\mathbf{k}}^2 \dot{\alpha} = 1 \quad \Rightarrow \quad \alpha = \omega_{\mathbf{k}} t \tag{3.344}$$

So, finally we get the following expression:

$$v_{\mathbf{k}} = \frac{e^{i\omega_{\mathbf{k}} t}}{\sqrt{\omega_{\mathbf{k}}}} \tag{3.345}$$

which corresponds to the true quantum vacuum. Then, the expression for the quantum field with the true vacuum state is given by:

$$\hat{\phi}(x) = \frac{1}{\sqrt{2}} \int \frac{d^3 k}{(2\pi)^{3/2} \sqrt{\omega_{\mathbf{k}}}} (\hat{a}_{\mathbf{k}} e^{i\mathbf{k}\cdot\mathbf{x} - i\omega_{\mathbf{k}} t} + \hat{a}_{\mathbf{k}}^\dagger e^{-i\mathbf{k}\cdot\mathbf{x} + i\omega_{\mathbf{k}} t}) \tag{3.346}$$

which is the same expression as equation (3.31). This exercise may appear to be excessive. However, as we will see in chapter 5, this energy minimization process is crucial to understanding the structure of the quantum vacuum in curved spacetime.

It is important to remark that equation (3.346) not only describes the quantum field with the true vacuum state, but it also distinguishes between positive and negative frequencies. Indeed, the quantum field modes are positive or negative with respect to time as eigenfunctions of the $i\partial_t$ energy operator:

$$\hat{E} v_{\mathbf{k}} = i\partial_t v_{\mathbf{k}} = -\omega_{\mathbf{k}} v_{\mathbf{k}}$$
$$\hat{E} v_{\mathbf{k}}^* = i\partial_t v_{\mathbf{k}}^* = \omega_{\mathbf{k}} v_{\mathbf{k}}^* \tag{3.347}$$

for $\omega_{\mathbf{k}} \geqslant 0$.

Finally, it is important to recall from our previous discussions that the vacuum is invariant under Lorentz transformations:

$$\tilde{p}^{\mu} = L^{\mu}_{\ \nu}p^{\nu} = 0 \Leftrightarrow p^{\nu} = 0 \tag{3.348}$$

That is, all inertial observers see the same state with minimal energy.

3.11 Summary

In this chapter we presented a rather detailed review of the relativistic dynamics of massive spin-$\frac{1}{2}$ particles. Furthermore, we showed how a group-theoretical approach determines the dynamics of quantum fields without ambiguity. In other words, the group structure of the Lorentz and Poincare transformations are enough to describe the nature of spin and the dynamics of quantum particles of arbitrary spin. Even though our discussion was limited to scalar and spin-$\frac{1}{2}$ particles, the arguments can be extended to the case of quantum fields of arbitrary spin. Furthermore, as a consequence of these group-theoretical expressions, we described the difference between the concept of spin for massless and massive particles, as well as the difference between quantum fields, quantum states and wave functions. Finally, we discussed the structure of the quantum vacuum for the scalar field. In particular, we showed how to compute the right expression for the quantum vacuum. Needless to say, similar arguments can be followed when considering the more complex case of the quantum vacuum for the Dirac and vector fields.

Bibliography

[1] Berestetski V B, Lifshitz E M and Piaevski L P 1982 *Quantum Electrodynamics, Landau and Lifshitz Course of Theoretical Physics* vol 4 2nd edn (Oxford: Pergamon Press)

[2] Bjorken J D and Drell S D 1964 *Relativistic Quantum Mechanics* (New York: McGraw-Hill)

[3] Bogolubov N N *et al* 1990 *General Principles of Quantum Field Theory* (Dordrecht: Kluwer Academic Press)

[4] Itzykson C and Zuber J B 1980 *Quantum Field Theory* (New York: McGraw Hill)

[5] Ryder L H 1996 *Quantum Field Theory* 2nd edn (Cambridge: Cambridge University Press)

[6] Thaller B 1992 *The Dirac Equation* (Berlin: Springer)

[7] Weinberg S 1995 *The Quantum Theory of Fields* vol 1 (Cambridge: Cambridge University Press)

[8] Halzen F and Martin A D 1984 *Quarks and Leptons: An Introductory Course in Modern Particle Physics* (New York: Wiley)

[9] Landau L D and Lifshitz E M 1981 *Quantum Mechanics: Non-Relativistic Theory, Course of Theoretical Physics* vol 3 3rd edn (Amsterdam: Elsevier)

[10] Merzbacher E 1997 *Quantum Mechanics* 3rd edn (New York: Wiley)

[11] Messiah A 1999 *Quantum Mechanics* (New York: Dover)

[12] Wald R 1984 *General Relativity* (Chicago, IL: University of Chicago Press)

[13] Parker L and Toms D 2009 *Quantum Field Theory in Curved Spacetime* (Cambridge: Cambridge University Press)

[14] Kuchař K 1976 Dynamics of tensor fields in hyperspace III *J. Math. Phys.* **17** 801–20

[15] Belinfante F 1939 *Physica* **6** 887

[16] Trautman A 2006 Einstein–Cartan theory *Encyclopedia of Mathematical Physics* ed J-P Francoise, G L Naber and S T Tsou vol 2 (Oxford: Elsevier) pp 189–95

[17] Carroll S M 2004 *Spacetime and Geometry: An Introduction to General Relativity* (Reading, MA: Addison-Wesley)

[18] Aitchison I J R and Hey A J G 2003 *Gauge Theories in Particle Physics* vol 1 (London: Taylor & Francis)

[19] Aitchison I J R 1985 Nothing's plenty: the vacuum in modern quantum field theory *Contemp. Phys.* **26** 333

[20] Milonni P W 1994 *The Quantum Vacuum* (New York: Academic)

[21] Halpern F R 1968 *Special Relativity and Quantum Mechanics* (Englewood Cliffs, NJ: Prentice Hall)

[22] Ohnuki Y 1988 *Unitary Representations of the Poincare Group and Relativistic Wave Equations* (Singapore: World Scientific)

[23] Carmeli M 1977 *Group Theory and General Relativity: Representations of the Lorentz Group and Their Applications to the Gravitational Field* (London: Imperial College Press)

[24] Sexl R U and Urbantke H K 1992 *Relativity, Groups, Particles: Special Relativity and Relativistic Symmetry in Field and Particle Physics* (Berlin: Springer)

[25] Tung W K 1985 *Group Theory in Physics* (Singapore: World Scientific)

[26] Tung W K 1967 Relativistic wave equations and field theory for arbitrary spin *Phys. Rev.* **156** 1385

[27] Weinberg S 1964 Feynman rules for any spin *Phys. Rev.* **133** B1318–32

[28] Weinberg S 1964 Feynman Rules for any spin. II. Massless particles *Phys. Rev.* **134** B882–96

[29] Bagrov V G and Gitman D M 1990 *Exact Solutions of Relativistic Wave Equations* (Dordrecht: Kluwer Academic)

[30] Araki H 1999 *Mathematical Theory of Quantum Fields* (Oxford: Oxford University)

[31] Haag R 1996 *Local Quantum Physics: Fields, Particles, Algebras* 2nd edn (Berlin: Springer)

[32] Streater R F and Wightman A S 1964 *PCT, Spin and Statistics, and All That* (Princeton, NJ: Princeton University Press)

[33] Troshin S M 1994 *Spin Phenomena in Particle Interactions* (Singapore: World Scientific)

[34] Birrell N D and Davis P C W 1982 *Quantum Fields in Curved Space* (Cambridge: Cambridge University Press)

[35] Mukhanov V F and Winitzki S 2007 *Introduction to Quantum Effects in Gravity* (Cambridge: Cambridge University Press)

Chapter 4

Quantum information in inertial frames

In this chapter we will explore some issues related to relativistic quantum information in the context of inertial frames of reference. This chapter does not offer a detailed review of the topic, so the reader is encouraged to peruse the relevant literature [1–5]. We will begin by analysing how a qubit state is affected by a Lorentz transformation and we will limit most of our discussion to the case of spin-$\frac{1}{2}$ particles described by momentum eigenstates of the Dirac equation. We will use this analysis to study the performance of a simple steganographic quantum communication channel, as well as the teleportation channel. We will conclude by discussing the effects of a Lorentz transformation on a state made of the superposition of two momentum eigenstates.

4.1 Qubit transformations

Let us consider a four-spinor *field* ψ and an infinitesimal Lorentz transformation:

$$\Lambda^\alpha_{\ \beta} = \delta^\alpha_\beta + \xi^\alpha_{\ \beta} \quad |\xi| \ll 1 \tag{4.1}$$

We recall that parameters that combine spatial and temporal components such as ξ^{0i} correspond to boosts, whereas parameters than only involve spatial components such as ξ^{ij} correspond to rotations ($i, j = 1, 2, 3$). Then, the Lorentz transformation on ψ is of the form:

$$D(\Lambda) \approx 1 + i\xi^{\alpha\beta}\Sigma_{\alpha\beta} + \mathcal{O}(\xi^2) \tag{4.2}$$

where:

$$\Sigma^{\alpha\beta} = -\frac{i}{8}\left[\gamma^\alpha, \gamma^\beta\right] \tag{4.3}$$

Notice that we are using slightly different expressions than the ones used in equations (3.286) and (3.311).

doi:10.1088/978-1-627-05330-3ch4

It is easy to show that, in the Dirac–Pauli representation of the gamma matrices, the commutator is given by:

$$[\gamma^i, \gamma^j] = -2i\epsilon_{ijk}\begin{pmatrix} \sigma_k & 0 \\ 0 & \sigma_k \end{pmatrix} \tag{4.4}$$

for the spatial components ($i = 1, 2, 3$). Also:

$$[\gamma^0, \gamma^i] = 2\begin{pmatrix} 0 & \sigma_i \\ \sigma_i & 0 \end{pmatrix} \tag{4.5}$$

for the commutator that involves a time-like gamma matrix component. As we have already discussed, we notice that boosts lead to non-unitary transformations. Indeed, for boosts, the tensor $\Sigma_{\alpha\beta}$ is not Hermitian.

Now, let us recall from equations (3.218) and (3.220) that the general solution of the Dirac equation is of the form:

$$\psi = \begin{pmatrix} \alpha\chi \\ \beta\chi \end{pmatrix} \tag{4.6}$$

where χ is a two-spinor and $\alpha, \beta \in \mathbb{C}$. Then, the unitary Lorentz transformation on this state leads to:

$$\psi \rightarrow \tilde{\psi} = \hat{U}(\Lambda)\psi \tag{4.7}$$

For example, for an infinitesimal rotation around the one-axis and ignoring higher terms:

$$\begin{aligned} \tilde{\psi} &= \left(1 + 2i\xi^{23}\Sigma_{23}\right)\psi \\ &= \left(1 - i\frac{\xi^{23}}{2}\begin{pmatrix} \sigma_1 & 0 \\ 0 & \sigma_1 \end{pmatrix}\right)\psi \\ &= \begin{pmatrix} \alpha\left(1 - i\frac{\xi^{23}}{2}\sigma_1\right)\chi \\ \beta\left(1 - i\frac{\xi^{23}}{2}\sigma_1\right)\chi \end{pmatrix} \end{aligned} \tag{4.8}$$

And for a finite rotation, we have:

$$\tilde{\psi} = \begin{pmatrix} \alpha e^{-i\xi\sigma_1/2}\chi \\ \beta e^{-i\xi\sigma_1/2}\chi \end{pmatrix} \tag{4.9}$$

and can be observed that the transformation acts in an equal and individual manner on both components of the four-spinor.

Similarly, a finite boost along the one-direction transforms the four-spinor into:

$$\tilde{\psi} = \begin{pmatrix} \beta e^{-\xi\sigma_1/2}\chi \\ \alpha e^{-\xi\sigma_1/2}\chi \end{pmatrix} \tag{4.10}$$

and as expected, it is not a unitary transformation.

For some applications to quantum information science, it is convenient to use states rather than fields, as these transform with unitary operators. From equation (3.78) we know that the Lorentz transformation of the spinor state $\psi_{p,\alpha}$ (with momentum p^μ and spin projection components labelled α) is given by:

$$\hat{U}(\Lambda)\psi_{p,\alpha} \propto \sum_\beta \mathcal{D}_{\alpha\beta}\psi_{\Lambda p,\beta} \qquad (4.11)$$

where $\mathcal{D}_{\alpha\beta}$ form a representation of the Wigner's little group.

In the case of spin-$\frac{1}{2}$ particles, we recall that there are two degrees of freedom embedded in the β index (corresponding to the two spin projections). For notational convenience, we can call these states $+\frac{1}{2}$ and $-\frac{1}{2}$. As discussed in the previous chapter, the Dirac state of a massive spin-$\frac{1}{2}$ particle with momentum $\mathbf{p}$ and positive helicity is represented by $|\mathbf{p},+1/2\rangle$, whereas the same particle with negative helicity is given by $|\mathbf{p},-1/2\rangle$. We make strong emphasis that these Dirac states are momentum–helicity eigenstates.

Then, the Lorentz transformation Λ acts on the spin up $(+\frac{1}{2})$ state as:

$$\hat{U}(\Lambda)|p,1/2\rangle = \cos\left(\frac{\Omega_p}{2}\right)|\Lambda p,1/2\rangle + \sin\left(\frac{\Omega_p}{2}\right)|\Lambda p,-1/2\rangle \qquad (4.12)$$

where Ω_p is the *Wigner angle*. That is, in the new frame, the particle has momentum Λp and the spin projection is rotated by an angle Ω_p. In addition, a similar result is found if the particle has negative helicity:

$$\hat{U}(\Lambda)|p,-1/2\rangle = -\sin\left(\frac{\Omega_p}{2}\right)|\Lambda p,1/2\rangle + \cos\left(\frac{\Omega_p}{2}\right)|\Lambda p,-1/2\rangle \qquad (4.13)$$

and we will completely ignore the anti-particle states.

Let us write the general superposition of spin states as:

$$|\psi\rangle = \alpha|\Lambda p,1/2\rangle + \beta|\Lambda p,-1/2\rangle = \begin{pmatrix}\alpha\\\beta\end{pmatrix} \qquad (4.14)$$

where $\alpha,\beta \in \mathbb{C}$ and:

$$\alpha^*\alpha + \beta^*\beta = 1 \qquad (4.15)$$

Then, we can rewrite the Wigner rotations using Euler's formula as:

$$\begin{aligned}\mathcal{D} &= e^{i\Omega_p\sigma_2/2}\\ &= \cos\left(\frac{\Omega_p}{2}\right) + i\sigma_2\sin\left(\frac{\Omega_p}{2}\right)\\ &= \begin{pmatrix}\cos\left(\frac{\Omega_p}{2}\right) & \sin\left(\frac{\Omega_p}{2}\right)\\ -\sin\left(\frac{\Omega_p}{2}\right) & \cos\left(\frac{\Omega_p}{2}\right)\end{pmatrix}\end{aligned} \qquad (4.16)$$

and the state $|\psi\rangle$ is quantized over the three-axis:

$$\sigma_3 \begin{pmatrix} 1 \\ 0 \end{pmatrix} = + \begin{pmatrix} 1 \\ 0 \end{pmatrix} \qquad \sigma_3 \begin{pmatrix} 0 \\ 1 \end{pmatrix} = - \begin{pmatrix} 0 \\ 1 \end{pmatrix} \tag{4.17}$$

These equations form the ground base to analyse relativistic effects on quantum states, and will be used in the rest of the present chapter, as well as in chapter 9, where we will discuss gravitational effects on quantum states.

4.2 Relativistic dynamics

In this section we will discuss the effect of Lorentz transformations on spin-$\frac{1}{2}$ qubits. To this end, we will analyse the structure of the Wigner rotation, as well as its effects on momentum eigenstates.

Let us assume a massive spin-$\frac{1}{2}$ qubit with four-momentum p^μ with respect to some inertial frame. For simplicity, we can always choose an inertial frame in such a way that the spatial component of the four-momentum is aligned with the three-axis:

$$\mathbf{p} = p\,\hat{\mathbf{k}} \;\Rightarrow\; p^\mu = (p^0, \mathbf{p}) = (m\gamma, 0, 0, mv\gamma) \tag{4.18}$$

where the four-velocity is given by:

$$\frac{\mathrm{d}x^\mu}{\mathrm{d}\tau} = (\gamma, 0, 0, v\gamma) \tag{4.19}$$

and the standard velocity is:

$$v \equiv \frac{\mathrm{d}x^3}{\mathrm{d}x^0} = \frac{\mathrm{d}x^3}{\mathrm{d}t} \tag{4.20}$$

and it is assumed to be different from zero.

Now, let us suppose we apply a Lorentz transformation to the system, in such a way that:

$$p^\mu \to \tilde{p}^\mu = \Lambda^\mu{}_\nu p^\nu \tag{4.21}$$

where p^μ is a four-vector with spatial components on the three-direction and $\Lambda^\mu{}_\nu$ is a boost on the one-direction with velocity $\tilde{v}$.

The derivation of the Wigner angle for an arbitrary Lorentz transformation is substantially long, and the reader is referred to the literature [1]. In what follows we will limit ourselves to present the final result. If we define:

$$\tanh(-\eta) = v \qquad \tanh(-\xi) = \tilde{v} \tag{4.22}$$

then the magnitude of Ω_p is given by:

$$\tan \Omega_p = \frac{\sinh \eta \, \sinh \xi}{\cosh \eta + \cosh \xi} = \frac{\sqrt{v^2}\ \sqrt{\tilde{v}^2}}{\sqrt{1 - v^2} + \sqrt{1 - \tilde{v}^2}} \tag{4.23}$$

Notice that the parameter η is associated to a Lorentz boost that takes the particle from its rest position to a kinematic state of momentum $\mathbf{p}$, while the parameter ξ is associated with the Lorentz transformation Λ. Then, the non-trivial spin rotation is the result of two non-parallel Lorentz transformations (associated to the parameters η and ξ, respectively). If both Lorentz transformations are parallel to each other, then the rotation angle is zero regardless of the magnitude of the speeds. This behaviour is because if both boosts are parallel to each other, then this is equivalent to a single boost in that direction. In this case, we know that a single Lorentz transformation merely describes the relativistic dynamics of the particle, as we discussed in the previous chapter. In such a case, the application of a single Lorentz transformation to a spinor quantum field will basically lead to an expression for the Dirac equation.

If both relevant speeds are of equal magnitude, $\tilde{v} = v$, then the value of Ω_p is given by:

$$\tan \Omega_p = \frac{1}{2} \frac{v^2}{\sqrt{1 - v^2}} \tag{4.24}$$

which can be rewritten as:

$$\tan \Omega_p = \frac{1}{2} \beta^2 \gamma^2 \tag{4.25}$$

using the standard relativistic kinematic variables.

Let us assume Alice is in the original inertial frame, and Bob is in the boosted inertial frame. We can observe that, in the non-relativistic case:

$$v \ll 1 \Rightarrow \tan \Omega_p \approx 0 \Rightarrow \Omega_p \approx 0 \tag{4.26}$$

which means that Bob will see the particle with momentum Λp and with a negligible change in the spin projection:

$$|\phi\rangle_{\text{Bob}}^{v \ll 1} = \hat{U}(\Lambda)|p, 1/2\rangle \approx |\Lambda p, 1/2\rangle \tag{4.27}$$

On the other hand, in the ultra-relativistic limit:

$$v \approx 1 \Rightarrow \tan \Omega_p \approx \infty \Rightarrow \Omega_p \approx \pi/2 \tag{4.28}$$

and Bob will observe the particle with momentum Λp and a spin projection state with nearly equal probability of being up or down:

$$|\phi\rangle_{\text{Bob}}^{v \approx 1} = \hat{U}(\Lambda)|p, 1/2\rangle \approx \frac{1}{\sqrt{2}} (|\Lambda p, 1/2\rangle + |\Lambda p, -1/2\rangle) \tag{4.29}$$

4.3 Steganographic quantum channel

Let us suppose now that Alice and Bob use qubits to transmit classical information *steganographically*. A steganographic or hidden channel has the potential to provide secure communications without encryption [6]. In a sense, the secret message is *hidden* in a transmission, and except for the authorized parties, nobody else suspects the existence of the message. As such, this technique is often described as a provider of *security through obscurity*.

To form a steganographic quantum channel, Alice and Bob encode classical information by choosing an orthonormal basis of the form:

$$|\tilde{0}\rangle = \alpha|0\rangle + \beta|1\rangle \qquad |\tilde{1}\rangle = -\beta^*|0\rangle + \alpha^*|1\rangle \qquad (4.30)$$

where α and β are complex numbers, whereas the tilde in $|\tilde{0}\rangle$ and $|\tilde{1}\rangle$ is used to represent the *logical states* '0' and '1', respectively. These logical states are the ones used to encode the secret message. On the other hand, $|0\rangle$ and $|1\rangle$ represent the *physical states* of the particle (in this case, spin up or spin down).

Suppose Alice sends a logical '0' encoded in $|\tilde{0}\rangle$, and Bob receives the state $\hat{U}|\tilde{0}\rangle$. The unitary operator $\hat{U}$ may represent the effect of a noisy environment or any other factor that may affect the measurement result. Bob performs a measurement in the steganographic basis and will obtain either a '0' or a '1'. Then, the probability that Alice sends state '*j*' and Bob measures state '*i*' is given by:

$$P(i|j) = \left|\langle \tilde{i}|\hat{U}|\tilde{j}\rangle\right|^2 \qquad (4.31)$$

and the *capacity* of the channel is given by:

$$C = 1 + \frac{1}{2}\sum_{i,j=0}^{1} P(i|j)\log P(i|j) \qquad (4.32)$$

It can be observed that this is a *binary symmetric channel*. That is, the success probability P and the error probability ϵ are:

$$P \equiv P(0|0) = P(1|1) \qquad \epsilon \equiv P(0|1) = P(1|0) \qquad \Rightarrow P + \epsilon = 1 \qquad (4.33)$$

whereas the channel capacity is given by:

$$C = 1 + P\log P + \epsilon\log\epsilon \qquad (4.34)$$

4.3.1 Relativistic communications

Let us suppose that Alice and Bob are on the inertial frames described above, where Bob needs to apply two orthogonal boosts to obtain the rest frame of the qubit sent by Alice. For simplicity, we can also assume that there is no environmental noise affecting the state of the qubits. We can identify a computational basis using Dirac eigenstates in the following way:

$$|0\rangle \leftrightarrow |p,+\rangle \qquad |1\rangle \leftrightarrow |p,-\rangle \qquad (4.35)$$

In matrix form, the effect of the Lorentz transformation on a qubit state is given by:

$$|0\rangle \rightarrow \hat{U}|0\rangle = \begin{pmatrix} a & -b \\ b & a \end{pmatrix}\begin{pmatrix} 1 \\ 0 \end{pmatrix} = \begin{pmatrix} a \\ b \end{pmatrix} = a|0\rangle + b|1\rangle \qquad (4.36)$$

and:

$$|1\rangle \rightarrow \hat{U}|1\rangle = \begin{pmatrix} a & -b \\ b & a \end{pmatrix} \begin{pmatrix} 0 \\ 1 \end{pmatrix} = \begin{pmatrix} -b \\ a \end{pmatrix} = -b|0\rangle + a|1\rangle \qquad (4.37)$$

where we have defined:

$$a \equiv \cos\left(\Omega_p/2\right) \qquad b \equiv \sin\left(\Omega_p/2\right) \qquad \Rightarrow a^2 + b^2 = 1 \qquad (4.38)$$

and we have neglected the functional dependency on the momentum, which is a correct assumption for as long as we are only considering momentum eigenstates.

In Alice's frame of reference, she prepares either $|\tilde{0}\rangle$ or $|\tilde{1}\rangle$. However, Bob will measure states rotated by an angle Ω_p:

$$
\begin{aligned}
|\tilde{0}\rangle_b = \hat{U}|\tilde{0}\rangle &= \alpha(a|0\rangle + b|1\rangle) + \beta(-b|0\rangle + a|1\rangle) \\
&= (\alpha a - \beta b)|0\rangle + (\alpha b + a\beta)|1\rangle \\
|\tilde{1}\rangle_b = \hat{U}|\tilde{1}\rangle &= -\beta^*(a|0\rangle + b|1\rangle) + \alpha^*(-b|0\rangle + a|1\rangle) \\
&= -(\beta^* a + \alpha^* b)|0\rangle - (\beta^* b - \alpha^* a)|1\rangle
\end{aligned}
\qquad (4.39)
$$

Furthermore, if we assume that both boosts have the same magnitude in the speed parameter (but different direction), $v = \tilde{v}$, then the rotation angle Ω_p is given by:

$$\Omega_p = \arctan\left(\frac{1}{2}\frac{v^2}{\sqrt{1-v^2}}\right) \qquad (4.40)$$

As is made evident in the above expressions, the states measured by Bob will carry *kinematic noise*, which makes them different from the states sent by Alice. Of course, because the Lorentz transformation is unitary, Bob could apply the inverse transformation to obtain the original state. However, this only can be done if Bob has an exact knowledge of the kinematics of the two inertial frames. In reality, of course, a sender and a receiver may not have a precise knowledge of their relative kinematic state. Thus, we can assume that Bob has no knowledge of the kinematics; therefore the presence of the unitary transformation is seen as 'noise' in the communications. We use the term 'kinematic' because it is produced exclusively by the kinematic relative state of motion between Alice and Bob. This kinematic noise is parameterized by the two real numbers a and b. But of course, there is only one free parameter, the Wigner angle Ω_p.

The probability that Bob measures '0' when Alice sends a '0' is:

$$P(0|0) = |\langle\tilde{0}|\tilde{0}\rangle_b|^2 = |\alpha^*(\alpha a - \beta b) + \beta^*(\alpha b + a\beta)|^2 = a^2 + 4b^2\delta^2 \qquad (4.41)$$

and the probability that Bob measures '1' when Alice sends a '1' is:

$$P(1|1) = |\langle\tilde{1}|\tilde{1}\rangle_b|^2 = |\beta(\beta^* a + \alpha^* b) - \alpha(\beta^* b - \alpha^* a)|^2 = a^2 + 4b^2\delta^2 \qquad (4.42)$$

where we have defined:

$$\delta \equiv \Im(\alpha^*\beta) \tag{4.43}$$

and $\Im$ represents the imaginary part of its argument.

Similarly, the error probabilities are given by:

$$P(0|1) = P(1|0) = 1 - a^2 - 4b^2\delta^2 = b^2 - 4b^2\delta^2 = b^2(1 - 4\delta^2) \tag{4.44}$$

As expected, this protocol forms a binary symmetric channel. Furthermore, ϵ, the error probability due to relativistic effects, can be expressed as:

$$\epsilon = b^2\left(1 - 4\delta^2\right) = \left(1 - 4\Im^2(\alpha^*\beta)\right)\left(\frac{1}{2} - \frac{\sqrt{1 - v^2}}{2 - v^2}\right) \tag{4.45}$$

Therefore, the total amount of error due to kinematic noise depends not only on the kinematic parameter v, but also on (α, β), the basis used to represent the classical information.

As an example, suppose that α and β are real numbers. Then, the probability of error is given by:

$$\epsilon = \frac{1}{2} - \frac{\sqrt{1 - v^2}}{2 - v^2} \tag{4.46}$$

which is independent of the basis. As can be seen in figure 4.1, the error probability is extremely small unless we consider the ultra-relativistic case where $v \approx 1$. For example, if the relative speed is $v \approx 0.9$, then we get an error probability of about 0.13. On the other hand, if the relative speed between Bob and Alice is about 100 times the speed of sound (3×10^2 m/s), then $v \approx 10^{-4}$, and the associated error probability would be $\approx 10^{-17}$.

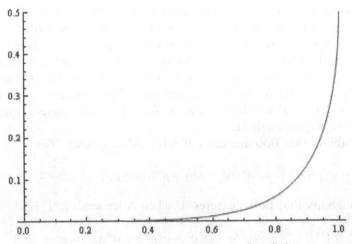

Figure 4.1. Plot of error probability ϵ versus the velocity v.

Therefore, in the case where α and β are real numbers and Alice and Bob are in the non-relativistic regime $v \ll 1$, $\epsilon \approx 0$ and the change to the channel capacity is negligible. On the other hand, in the ultra-relativistic regime, $v \approx 1$ and $\epsilon \approx 1/2$, so the channel capacity is reduced to zero.

4.3.2 Relativistic fixed points

Because the error probability depends on the basis parameters α and β, as well as the speed v, in this section we will consider the possibility of finding a basis that is resilient to these relativistic effects.

Looking at equation (4.44), it is clear that the probability of error is equal to zero when:

$$\delta = \Im(\alpha^*\beta) = \pm \frac{1}{2} \tag{4.47}$$

In particular, this clearly happens when:

$$(\alpha, \beta) = \frac{1}{\sqrt{2}}(1, i) \tag{4.48}$$

and as a consequence, the basis:

$$|\tilde{0}\rangle = \frac{|0\rangle + i|1\rangle}{\sqrt{2}} \qquad |\tilde{1}\rangle = \frac{i|0\rangle + |1\rangle}{\sqrt{2}} \tag{4.49}$$

does not feel the effects of the Lorentz transformation Λ.

The previous result can be better understood by studying the eigenvectors and eigenvalues of the Lorentz transformation. The two eigenvalues of $\hat{U}$ are given by:

$$\lambda = \begin{cases} \lambda_0 = a + ib \\ \lambda_1 = a - ib \end{cases} \tag{4.50}$$

with associated eigenvectors:

$$\begin{pmatrix} a & -b \\ b & a \end{pmatrix} \frac{1}{\sqrt{2}} \begin{pmatrix} 1 \\ i \end{pmatrix} = (a - ib) \frac{1}{\sqrt{2}} \begin{pmatrix} 1 \\ i \end{pmatrix} = \frac{\lambda_0}{\sqrt{2}} \begin{pmatrix} 1 \\ i \end{pmatrix}$$

$$\begin{pmatrix} a & -b \\ b & a \end{pmatrix} \frac{1}{\sqrt{2}} \begin{pmatrix} i \\ 1 \end{pmatrix} = (a + ib) \frac{1}{\sqrt{2}} \begin{pmatrix} i \\ 1 \end{pmatrix} = \frac{\lambda_1}{\sqrt{2}} \begin{pmatrix} i \\ 1 \end{pmatrix} \tag{4.51}$$

so the proposed basis states in equation (4.49) are eigenvectors of the Lorentz transformation $\hat{U}$. However:

$$|a \pm ib|^2 = a^2 + b^2 = 1 \tag{4.52}$$

so the eigenvalue is just a global phase of unit length which does not affect the calculation of probabilities. Indeed, if we use the basis defined by the eigenvectors of $\hat{U}$ we get:

$$P(0|0) = \left|\langle \tilde{0}|\tilde{0}\rangle_b\right|^2 = \left|\langle \tilde{0}|\hat{U}|\tilde{0}\rangle\right|^2 = |\lambda_0|^2 \left|\langle \tilde{0}|\tilde{0}\rangle\right|^2 = 1 \tag{4.53}$$

and similarly for $P(1|1)$ with eigenvalue λ_1, which renders $\epsilon = 0$ in this basis.

The physical interpretation of this fixed point is easy to understand. Indeed, the two Lorentz transformations involved are along the x and z axes. Under Lorentz transformations, physical variables such as length are contracted only across the axis parallel to the boost. Therefore, these Lorentz transformations affect the spin projection but only along the x and z axes, leaving the y component invariant. It is easy to confirm that this is the case:

$$\sigma_y|\tilde{0}\rangle = \frac{1}{\sqrt{2}}\begin{pmatrix} 0 & -i \\ i & 0 \end{pmatrix}\begin{pmatrix} 1 \\ i \end{pmatrix} = \frac{1}{\sqrt{2}}\begin{pmatrix} 1 & i \end{pmatrix} = +|\tilde{0}\rangle \qquad (4.54)$$

and:

$$\sigma_y|\tilde{1}\rangle = \frac{1}{\sqrt{2}}\begin{pmatrix} 0 & -i \\ i & 0 \end{pmatrix}\begin{pmatrix} i \\ 1 \end{pmatrix} = -\frac{1}{\sqrt{2}}\begin{pmatrix} i \\ 1 \end{pmatrix} = -|\tilde{1}\rangle \qquad (4.55)$$

Therefore, the Lorentz invariant basis corresponds to the σ_y basis.

4.4 The teleportation channel

Teleportation is a truly quantum mechanical communication protocol with no classical counterpart [7, 8]. Like the previous example, it is possible to teleport classical information steganographically by using an orthonormal basis to represent '0' and '1' [9]. As before, Alice and Bob select a basis to encode classical information:

$$|\tilde{0}\rangle = \alpha|0\rangle + \beta|1\rangle \qquad |\tilde{1}\rangle = -\beta^*|0\rangle + \alpha^*|1\rangle \qquad (4.56)$$

and access to a partly entangled state:

$$|\Phi\rangle = \eta_{00}|00\rangle + \eta_{01}|01\rangle + \eta_{10}|10\rangle + \eta_{11}|11\rangle \qquad (4.57)$$

The state does not need to be perfectly entangled, but it requires some amount of entanglement. Let us recall that the standard measure of entanglement for bipartite pure systems is given by the Shannon entropy of the Schmidt coefficients:

$$E(|\Phi\rangle) = -\sum_k |\lambda_k|^2 \log(|\lambda_k|^2) = -|\lambda_+|^2 \log(|\lambda_+|^2) - |\lambda_-|^2 \log(|\lambda_-|^2) \qquad (4.58)$$

where:

$$\lambda_\pm^2 = \frac{1}{2}\left(1 \pm \sqrt{1 - 4|\eta_{01}\eta_{10} - \eta_{00}\eta_{11}|^2}\right) \qquad (4.59)$$

are the Schmidt coefficients [7, 8]. For example, it is easy to check that, for an arbitrary Bell state Φ_B, the amount of entanglement is maximal and equal to:

$$E(|\Phi_B\rangle) = 1 \qquad (4.60)$$

where:

$$|\Phi_B\rangle = \begin{cases} |\Phi^+\rangle = \dfrac{|00\rangle + |11\rangle}{\sqrt{2}} \\[2ex] |\Phi^-\rangle = \dfrac{|00\rangle - |11\rangle}{\sqrt{2}} \\[2ex] |\Psi^+\rangle = \dfrac{|01\rangle + |10\rangle}{\sqrt{2}} \\[2ex] |\Psi^-\rangle = \dfrac{|01\rangle - |10\rangle}{\sqrt{2}} \end{cases} \qquad (4.61)$$

That is, all Bell states have maximal entanglement [7, 8].

It can be shown that this teleportation channel is binary symmetric and has the following probabilities for the successful and failed teleportation of '0' and '1':

$$P = |\eta_{00}|^2 + |\eta_{11}|^2 + 2\left(|\eta_{01} + \eta_{10}|^2 - |\eta_{00} - \eta_{11}|^2\right)|\alpha|^2|\beta|^2$$

$$\epsilon = 1 - |\eta_{00}|^2 - |\eta_{11}|^2 - 2\left(|\eta_{01} + \eta_{10}|^2 - |\eta_{00} - \eta_{11}|^2\right)|\alpha|^2|\beta|^2$$

$$(4.62)$$

A startling conclusion follows from these equations: the standard measure of entanglement for bipartite systems is not correlated with the amount of information that can be teleported using an entangled state [9]. That is, sometimes a perfectly entangled state does not correspond to maximum channel capacity (maximal probability of success). In what follows, we will show that relativistic teleportation is an instance where this happens.

4.4.1 Relativistic teleportation

Let us first consider the case where both qubits are at rest in Alice's frame of reference. She has a perfectly entangled state given by the Bell state:

$$|\Phi^+\rangle = \frac{1}{\sqrt{2}}(|00\rangle + |11\rangle) \qquad (4.63)$$

Because Alice and Bob's frames only differ from each other by a single boost, the Wigner rotation angle is $\Omega_p = 0$. In this situation, the Bell state remains invariant. In a similar manner, if Alice and Bob share entangled particles in their rest frames, their frames only differ by a single boost and once again the Wigner angle vanishes: $\Omega_p = 0$.

As a more meaningful example, let us consider the case where Alice produces an entangled pair in her inertial frame of reference and the second qubit is ejected at some speed $\mathbf{v} = v\hat{\mathbf{k}}$. According to her, the qubits are described as the Bell state expressed above. That is, Alice has a perfectly entangled state in the sense that $E(|\Phi^+\rangle) = 1$.

Let us also imagine that Bob is travelling on an orthogonal direction with a velocity $\tilde{v} = \tilde{v}\hat{i}$. Then, he will perceive the entangled state to be:

$$\left(\mathbb{I} \otimes \hat{U}\right)|\Phi^+\rangle = \frac{1}{\sqrt{2}}\left(a|00\rangle + b|01\rangle - b|10\rangle + a|11\rangle\right) \tag{4.64}$$

where a and b are the same coefficients that depend on the Wigner angle Ω_p that were defined before in equation (4.38).

So, for Bob, the entangled state does not look like a Bell state anymore. However, it remains an entangled state with the exact same degree of entanglement as the original Bell state. Indeed, the Schmidt coefficients calculated by Bob are:

$$\lambda_\pm^2 = \frac{1}{2}\left(1 \pm \sqrt{1 - 4\left|\frac{a^2 + b^2}{2}\right|^2}\right) = \frac{1}{2} \tag{4.65}$$

Clearly, these values also correspond to a maximally entangled state. Therefore, the net amount of entanglement remains the same after the Lorentz transformation:

$$E(|\Phi^+\rangle) = E\left(\mathbb{I} \otimes \hat{U}|\Phi^+\rangle\right) = 1 \tag{4.66}$$

That is, *the amount of entanglement is a relativistic invariant for momentum eigenvector states*[1].

Nonetheless, the probability of successful teleportation of classical information is not invariant:

$$P = |\eta_{00}|^2 + |\eta_{11}|^2 + 2\left(|\eta_{01} + \eta_{10}|^2 - |\eta_{00} - \eta_{11}|^2\right)|\alpha|^2|\beta|^2$$

$$= \left|\frac{a}{\sqrt{2}}\right|^2 + \left|\frac{a}{\sqrt{2}}\right|^2 + 2\left(\left|\frac{b-b}{\sqrt{2}}\right|^2 - \left|\frac{a-a}{\sqrt{2}}\right|^2\right)|\alpha|^2|\beta|^2$$

$$= a^2 \tag{4.67}$$

so:

$$\epsilon = 1 - a^2 = b^2 \tag{4.68}$$

Indeed, even though the entangled state remains with a maximal level of entanglement, the capacity of the teleportation channel decreases.

As expected, in the non-relativistic regime, $v \ll 1$, and as a consequence, $\epsilon \approx 0$, which means that the change to the channel capacity is negligible. On the other hand, in the ultra-relativistic regime, $v \approx 1$, and as a consequence, $\epsilon \approx 1/2$, which implies that the capacity of the teleportation channel is reduced to zero.

---

[1] It is worth mentioning that if, instead of using two momentum-spin eigenstates with entangled spin, we use states with a spread in momentum entangled in spin, the result of applying a boost is a state with some degree of spin entanglement, and some degree of entanglement in the momentum [3].

It is also important to note that, in this case, the probability of success depends exclusively on the kinematic state (v), and it is independent of the basis used to communicate the information (α and β). In other words, the relativistic effect does not depend on the basis used to represent the information.

4.4.2 Absence of relativistic fixed points

As noted in the previous example, the relativistic effect depends on the kinematic variable v, but it is independent of the basis (α, β) used to represent the information. Therefore, the notion of relativistic fixed points as described in section 4.2 does not apply to this implementation of the teleportation channel. However, we could explore the case where Alice and Bob use a perfectly entangled state in some other basis. Indeed, suppose we use the relativistic invariant base for the case in consideration:

$$|\bar{0}\rangle = \frac{1}{\sqrt{2}}\begin{pmatrix} 1 \\ i \end{pmatrix} \qquad |\bar{1}\rangle = \frac{1}{\sqrt{2}}\begin{pmatrix} i \\ 1 \end{pmatrix} \tag{4.69}$$

and Alice uses this basis to generate what she perceives to be a perfectly entangled Bell state:

$$|\Phi\rangle = \frac{1}{\sqrt{2}}\left(|\bar{0}\bar{0}\rangle + |\bar{1}\bar{1}\rangle\right) \tag{4.70}$$

Applying the Lorentz transformation leads to the following state being perceived by Bob:

$$(\mathbb{I} \otimes \hat{U})|\Phi\rangle = \frac{1}{\sqrt{2}}\left(|\bar{0}\rangle\hat{U}|\bar{0}\rangle + |\bar{1}\rangle\hat{U}|\bar{1}\rangle\right) = \frac{1}{\sqrt{2}}\left(\lambda_0|\bar{0}\bar{0}\rangle + \lambda_1|\bar{1}\bar{1}\rangle\right) \tag{4.71}$$

and the probability of successful transmission is given by:

$$\begin{aligned} P &= |\eta_{00}|^2 + |\eta_{11}|^2 + 2\left(|\eta_{01} + \eta_{10}|^2 - |\eta_{00} - \eta_{11}|^2\right)|\alpha|^2|\beta|^2 \\ &= \frac{|\lambda_0|^2}{2} + \frac{|\lambda_1|^2}{2} - |\lambda_0 - \lambda_1|^2|\alpha|^2|\beta|^2 \\ &= 1 - b^2|\alpha|^2|\beta|^2 \end{aligned} \tag{4.72}$$

while the error probability is:

$$\epsilon = b^2|\alpha|^2|\beta|^2 \tag{4.73}$$

Therefore, even with the use of the relativistic invariant basis, the teleportation protocol feels the relativistic effects. Furthermore, the error probability now depends on the basis (α, β) used to represent the information.

In the non-relativistic regime we have $v \ll 1$ and as a consequence $\epsilon \approx 0$. However, in the ultra-relativistic regime:

$$v \approx 1 \Rightarrow \epsilon \approx \frac{|\alpha|^2|\beta|^2}{2} \tag{4.74}$$

In such a case, the only way to reduce the relativistic effects is to move the basis (α, β) used to represent the information toward the computational basis:

$$\left\{|\bar{0}\rangle, |\bar{1}\rangle\right\} \rightarrow \left\{|0\rangle, |1\rangle\right\} \Rightarrow (\alpha, \beta) \rightarrow (1, 0) \Rightarrow \epsilon \rightarrow 0 \tag{4.75}$$

Unfortunately, in the limiting case, the use of quantum teleportation may become redundant from a perspective of steganographic communications. Indeed, in such a case the information is no longer 'hidden' in the choice of bases in the steganographic protocol.

We could also consider the case where Alice and Bob purposely use an imperfectly entangled state to try to overcome the effects of kinematic noise. The state Alice generates is of the form:

$$|\Phi\rangle = \kappa_{00}|00\rangle + \kappa_{11}|11\rangle \tag{4.76}$$

And for Bob, this partly entangled state looks like:

$$\begin{aligned}
\left(\mathbb{I} \otimes \hat{U}\right)|\Phi\rangle &= \kappa_{00}|0\rangle \otimes (a|0\rangle + b|1\rangle) + \kappa_{11}|1\rangle \otimes (-b|0\rangle + a|1\rangle) \\
&= a\,\kappa_{00}|00\rangle + b\,\kappa_{00}|01\rangle - b\,\kappa_{11}|10\rangle + a\,\kappa_{11}|11\rangle
\end{aligned} \tag{4.77}$$

so the probability of successful transmission of information is given by:

$$\begin{aligned}
P &= |\eta_{00}|^2 + |\eta_{11}|^2 + 2\left(|\eta_{01} + \eta_{10}|^2 - |\eta_{00} - \eta_{11}|^2\right)|\alpha|^2|\beta|^2 \\
&= a^2|\kappa_{00}|^2 + a^2|\kappa_{11}|^2 + 2\left(b^2|\kappa_{00} - \kappa_{11}|^2 - a^2|\kappa_{00} - \kappa_{11}|^2\right)|\alpha|^2|\beta|^2 \\
&= a^2 - 2(a^2 - b^2)|\kappa_{00} - \kappa_{11}|^2|\alpha|^2|\beta|^2
\end{aligned} \tag{4.78}$$

and because:

$$2\left(a^2 - b^2\right)|\kappa_{00} - \kappa_{11}|^2|\alpha|^2|\beta|^2 \geqslant 0 \tag{4.79}$$

we conclude that:

$$P \leqslant a^2 \Rightarrow \epsilon \geqslant b^2 \tag{4.80}$$

This is a bound that cannot be surpassed by any selection of the entanglement parameters κ_{00} and κ_{11}, nor for any selection of the encoding basis α and β.

Furthermore, in the ultra-relativistic regime, $v \approx 1$, and as a consequence, $\epsilon \approx 1/2$, which means that the channel capacity is reduced to zero without regard to the entanglement parameters κ_{00} and κ_{11} and the encoding basis α and β.

4.5 Spread momentum states

In the previous sections, we only considered the transformation of the spinor–momentum eigenstate $|p, \pm 1/2\rangle$. This state can be formally written as:

$$|p, \pm 1/2\rangle \rightarrow |p, \pm\rangle = |p\rangle \otimes |\pm\rangle \tag{4.81}$$

Indeed, the Dirac spinor eigenstate is the tensor product state of a momentum eigenstate and a spin eigenstate. This is a crucial observation because the Lorentz transformation acts in a different manner in the momentum and spin Hilbert spaces.

Indeed, the Lorentz transformation of the only two possible states with definite momentum p can be written as:

$$\hat{U}(\Lambda)|p,+\rangle = |\Lambda p\rangle \otimes \left(\cos\left(\frac{\Omega_p}{2}\right)|+\rangle + \sin\left(\frac{\Omega_p}{2}\right)|-\rangle \right)$$

$$\hat{U}(\Lambda)|p,-\rangle = |\Lambda p\rangle \otimes \left(-\sin\left(\frac{\Omega_p}{2}\right)|+\rangle + \cos\left(\frac{\Omega_p}{2}\right)|-\rangle \right)$$

(4.82)

Consequently, if the original state is a momentum eigenstate, then the Lorentz transformation leads to separable states in the spin and momentum degrees of freedom.

On the other hand, let us assume a particle with a non-definite momentum state, but definite helicity state, described by:

$$|\psi\rangle = \frac{1}{\sqrt{2}}(|p\rangle + |q\rangle) \otimes |+\rangle$$

(4.83)

which clearly is a separable state in the spin and momentum degrees of freedom. Then, the Lorentz transformation of this state leads to:

$$|\psi\rangle \rightarrow |\tilde{\psi}\rangle = \hat{U}(\Lambda)|\psi\rangle$$

$$= \frac{1}{\sqrt{2}}\hat{U}(\Lambda)|p\rangle \otimes |+\rangle + \frac{1}{\sqrt{2}}\hat{U}(\Lambda)|q\rangle \otimes |+\rangle$$

$$= \frac{1}{\sqrt{2}}|\Lambda p\rangle \otimes \left(\cos\left(\frac{\Omega_p}{2}\right)|+\rangle + \sin\left(\frac{\Omega_p}{2}\right)|-\rangle \right)$$

$$+ \frac{1}{\sqrt{2}}|\Lambda q\rangle \otimes \left(\cos\left(\frac{\Omega_q}{2}\right)|+\rangle + \sin\left(\frac{\Omega_q}{2}\right)|-\rangle \right)$$

$$= \frac{1}{\sqrt{2}}\left(\cos\left(\frac{\Omega_p}{2}\right)|\Lambda p\rangle + \cos\left(\frac{\Omega_q}{2}\right)|\Lambda q\rangle \right) \otimes |+\rangle$$

$$+ \frac{1}{\sqrt{2}}\left(\sin\left(\frac{\Omega_p}{2}\right)|\Lambda p\rangle + \sin\left(\frac{\Omega_q}{2}\right)|\Lambda q\rangle \right) \otimes |-\rangle$$

(4.84)

which in general is a non-separable state. The two Schmidt coefficients are given by:

$$\lambda_\pm^2 = \frac{1}{2}\left(1 \pm \sqrt{1 - \left(\cos\frac{\Omega_p}{2}\sin\frac{\Omega_q}{2} - \cos\frac{\Omega_q}{2}\sin\frac{\Omega_p}{2} \right)^2} \right)$$

$$= \frac{1}{2}\left(1 \pm \sqrt{1 - \sin^2\left(\frac{\Omega_q - \Omega_p}{2}\right)} \right)$$

$$= \frac{1}{2}\left(1 \pm \cos\left(\frac{\Omega_q - \Omega_p}{2}\right) \right)$$

(4.85)

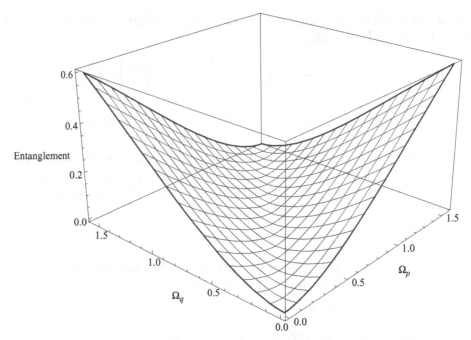

Figure 4.2. Relativistic entanglement due to the Lorentz transformation of a state with no definite momentum.

and therefore the two Schmidt coefficients can be simply expressed as:

$$\lambda_+^2 = \cos^2\left(\frac{\Omega_q - \Omega_p}{4}\right)$$

$$\lambda_-^2 = \sin^2\left(\frac{\Omega_q - \Omega_p}{4}\right)$$

(4.86)

which leads to a level of entanglement given by equation (4.58). Thus, for non-definite momentum states, a Lorentz transformation could generate entangled states out of non-entangled states.

Figure 4.2 shows the behaviour of the relativistic entanglement with respect to the two Wigner angles Ω_p and Ω_q:

$$0 \leqslant \Omega_p \leqslant \frac{\pi}{2} \qquad 0 \leqslant \Omega_q \leqslant \frac{\pi}{2}$$

(4.87)

It can be observed that if $\Omega_p = \Omega_q$, then the relativistic entanglement is zero. Indeed, in this case $|p\rangle = |q\rangle$ and consequently the state after a Lorentz transformation remains a separable state. On the other hand, if $\Omega_p \approx 0$ and $\Omega_q \approx \pi/2$, the entanglement reaches a maximal value of ≈ 0.6, which corresponds to the situation in which p is the non-relativistic limit, while q is the ultra-relativistic limit.

A similar rationale and conclusions can be obtained if we take a state with a spread in momentum, but definite position, of the form:

$$|\psi\rangle \propto \int \sum_\sigma |\mathbf{p}, \sigma\rangle \, \mathcal{G}(\mathbf{p}) \, \mathrm{d}^3\mathbf{p} \qquad (4.88)$$

where $\mathcal{G}(\mathbf{p})$ is a smooth and non-zero distribution over a wide range of energies. Furthermore, it can be shown that, for bipartite systems of definite position states, any initial entanglement in the spin can be transformed into entanglement in the momentum degrees of freedom [3].

4.6 Summary

In this chapter we have discussed relativistic quantum information in the context of inertial frames. In particular we have analysed the performance of a steganographic communications protocol and teleportation. Although most of our discussion was limited to the case of momentum eigenstates, we also briefly discussed the case of qubits in non-definite momentum states (e.g. the superposition of two momentum eigenstates).

The Lorentz transformation is unitary, and, as such, the inverse operation can always be applied. Therefore, all the effects of kinematic noise can be removed if we know exactly the kinematics of all the qubits and observers involved in the communication or teleportation protocols. However, this may not be the case in a realistic situation.

Finally, if the qubit is in a momentum eigenstate, then the Lorentz transformation will lead to a separable state. That is, the amount of entanglement is conserved after a Lorentz transformation of momentum eigenstates. On the other hand, for a super-position of two different momentum eigenstates, the Lorentz transformation leads to a non-local operation which changes the amount of entanglement in the system.

Bibliography

[1] Alsing P M and Milburn G J 2002 On entanglement and Lorentz invariance *Quantum Inf. Comput.* **2** 487

[2] Bergou A J, Gingrich R M and Adami C 2003 Entangled light in moving frames *Phys. Rev.* A **68** 042102

[3] Gingrich R M and Adami C 2002 Quantum entanglement of moving bodies *Phys. Rev. Lett* **89** 270402

[4] Peres A and Terno D R 2004 Quantum information and relativity theory *Rev. Mod. Phys.* **76** 93–123

[5] Soo C and Lin C C Y 2004 Wigner rotations, Bell States, and Lorentz invariance of entanglement and von Neumann entropy *Int. J. Quantum Inf.* **2** 183–200

[6] Martin K 2007 Steganographic communication with quantum information *Information Hiding: Lecture Notes in Computer Science* **4567** 32–49

[7] Nielsen M A and Chuang I L 2000 *Quantum Computation and Quantum Information* (Cambridge: Cambridge University Press)

[8] Vedral V 2006 *Introduction to Quantum Information Science* (Cambridge: Cambridge University Press)

[9] Lanzagorta M and Martin K 2012 Teleportation with an imperfect state *Theoretical Computer Science* **430** 117–25

IOP Concise Physics

Quantum Information in Gravitational Fields

Marco Lanzagorta

Chapter 5

Quantum fields in curved spacetimes

In this chapter we will discuss the dynamics of quantum fields in the presence of a classical gravitational field. As we will see, the absence of a global reference frame in the context of general relativity has a deep impact on our understanding of quantum phenomena. Once again, the reader is urged to read the more comprehensive literature available on the topic [1–7].

The approach for scalar and vector fields is quite straightforward, and will be limited to the application of the minimal substitution rule obtained from the principle of general covariance. We will also discuss the structure of the quantum vacuum for the case of a real scalar quantum field in the presence of a classical gravitational field.

Considerably more time will be devoted to the discussion of spin-$\frac{1}{2}$ particles in curved spacetimes. As we will see, it is necessary to introduce the tetrad field formalism to properly describe the dynamics of Dirac fields in the presence of a classical gravitational field. We will conclude our presentation of the topic with an analysis of the approximate solutions to the Dirac equation in classical gravitational fields, and use these results to analyse further the coupling between spin and curvature.

5.1 Scalar fields in curved spacetime

Let us consider the classical Klein–Gordon field equation:

$$\left(\eta^{\mu\nu}\partial_\mu\partial_\nu - m^2\right)\phi = 0 \tag{5.1}$$

in flat spacetime. By using the principle of general covariance, the generalization of the Klein–Gordon equation in the presence of a gravitational field is given by:

$$\left(g^{\mu\nu}\mathcal{D}_\mu\partial_\nu - m^2\right)\phi = 0 \tag{5.2}$$

doi:10.1088/978-1-627-05330-3ch5   5-1

where we have used the fact that, for scalar fields, the covariant derivative reduces to the standard derivative:

$$\mathcal{D}_\mu \phi = \partial_\mu \phi \qquad (5.3)$$

and, consequently, the gravitational effects are simply introduced by the covariant derivative and the metric tensor that contracts the two derivatives [1, 6].

The above expression for the Klein–Gordon equation in the presence of gravity is often referred as the *minimally coupled case* [1]. Indeed, it is possible to add a more general gravitational interaction term that couples the scalar field with the spacetime curvature. At the same time, it can be shown that the only geometric scalar that can be added to the Klein–Gordon equation is a term proportional to the spacetime curvature $R = R(x)$ [1, 6]. Then, in the most general case, the Klein–Gordon equation in curved spacetime looks like:

$$\left(g^{\mu\nu}\mathcal{D}_\mu \partial_\nu - m^2 + \xi R\right)\phi = 0 \qquad (5.4)$$

and the minimally coupled case is obtained when $\xi = 0$. This expression can be rewritten as:

$$\frac{1}{\sqrt{-g}}\partial_\mu\left(\sqrt{-g}\,g^{\mu\nu}\partial_\nu\phi\right) - m^2\phi + \xi R\phi = 0 \qquad (5.5)$$

Both expressions make clear that the Klein–Gordon equation in curved spacetime involves a standard derivative of the field followed by a covariant derivative. Indeed, the second derivative is not acting on a scalar, but on the four-vector $\partial_\mu\phi$. Also, notice that in the absence of gravitational fields, $R = 0$ and $g_{\mu\nu} = \eta_{\mu\nu}$, and the expressions reduce to the Klein–Gordon equation in flat spacetime.

It is important to notice that ξ is a dimensionless constant and it cannot be very large [6]. Indeed, the non-minimal coupling arises from an interaction term in the Lagrangian that looks like:

$$\mathcal{L}_\xi \propto \xi R \phi^2 \qquad (5.6)$$

and, consequently, a large ξ will produce an *effective gravitational constant* that may vary markedly over space and time because of the fluctuations in the scalar field ϕ.

Furthermore, let us notice that the interaction term mediated by ξ does not reduce to zero in a local inertial frame. Indeed, the curvature R does not vanish in a local inertial frame, even though the affine connection is zero at that point (recall that the curvature scalar R involves derivatives of the affine connection that not necessarily vanish at that point). Therefore the presence of the curvature term will affect the behaviour of the scalar fields that exhibit a non-minimal coupling to gravity [8–10]. Consequently, these non-minimal terms in the dynamics of the scalar fields will lead to a violation of the strong principle of equivalence, which states that all local effects of gravity must disappear in a local inertial frame [5].

5.2 Quantum dynamics in general relativity

The description of quantum particles in gravitational fields offers interesting conceptual challenges. On the one hand, general relativity is a theory based on the description of test particles moving in precise geodesic lines with determined four-momentum. On the other hand, particles in quantum theory obey Heisenberg's uncertainty principle; consequently, the position and momentum cannot have simultaneously precise values. Furthermore, the canonical quantization of fields requires the use of Fourier expansions that involve plane-wave approximations. And of course, a plane wave is a very ambiguous concept in the context of curved spacetime. In this section we will discuss these issues within the context of Klein–Gordon fields.

5.2.1 The plane wave approximation

Most of the results in relativistic and non-relativistic quantum mechanics use the *plane wave approximation* [5, 11]. That is, we assume that it is possible to decompose the quantum fields in terms of plane waves of the form:

$$e^{i\mathbf{k}\cdot\mathbf{x}-i\omega_k t} \tag{5.7}$$

that describe the dynamics of particles (i.e. quantum field excitations) with momentum $\mathbf{k}$ and energy ω_k. For example, in one dimension, a localized particle can be expressed as the wave packet $\Psi(x)$ given by:

$$\Psi(x) \propto \int dq \, \mathcal{G}(q) e^{iqx-i\omega_q t} \tag{5.8}$$

where $\mathcal{G}(q)$ is non-zero and smooth over a wide range of energies. For the case of a Gaussian wave packet of momentum centred in k, we have:

$$\mathcal{G}(q) \propto e^{-(q-k)^2 \Delta\lambda^2} \tag{5.9}$$

with a spatial spread of $\Delta\lambda$ and a momentum spread of $\Delta k \approx 1/\Delta\lambda$, which are consistent with Heisenberg's uncertainty principle:

$$\Delta k \, \Delta\lambda \approx 1 \tag{5.10}$$

Furthermore, the state of this particle is well defined as long as:

$$\Delta k \ll k \tag{5.11}$$

therefore, because of Heisenberg's uncertainty principle, the spatial spread of the wave packet satisfies:

$$\Delta\lambda \approx \frac{1}{\Delta k} \implies \Delta\lambda \gg \frac{1}{k} \tag{5.12}$$

That is, in the absence of gravity, a localized quantum state is well defined through a plane wave expansion for as long as the spatial spread of the wave packet is greater than the inverse of the momentum [5, 11].

However, the plane wave approximation is firmly grounded on the specific mathematical properties of Minkowski space. In general, curved spacetimes have different properties that may break down the plane wave approximation [12]. Consequently, there will be ambiguities in the definition of 'particle' and 'vacuum' when studying quantum fields in the presence of gravity [5, 11].

Indeed, if we now consider a particle in the presence of a gravitational field, then the spacetime curvature may vary from point to point. In such a case, the plane wave approximation may fail to describe the state of a quantum particle: it does not make much sense to talk about plane waves in a region of space with a highly variable curvature. Therefore, the plane wave approximation remains valid only when the spacetime curvature does not vary in a significant manner over a region of size $\Delta\lambda$. In other words, the plane wave approximation can be used when the spatial spread of the localized quantum particle is smaller than the region of space where the curved space-time is approximately Minkowskian. That is, the plane wave expansion is valid when:

$$R_s \gg \Delta\lambda \quad \Rightarrow \quad R_s \gg \frac{1}{k} \tag{5.13}$$

where R_s is the *curvature scale*: the maximum distance where the curved spacetime can be approximated by Minkowski spacetime.

In addition, within the formalism of quantum field theory, it is known that the localization spread $\Delta\lambda$ cannot be made arbitrarily small [13]. Indeed, the localization of massive particles is limited by the Compton wavelength:

$$\lambda_c \equiv \frac{2\pi}{m} \tag{5.14}$$

expressed in natural units ($\hbar = 1, c = 1$). In particular, a wave packet made of positive energy solutions will have a non-vanishing tail that decays with the distance r as:

$$e^{-r/\lambda_c} \tag{5.15}$$

Therefore:

$$\Delta\lambda > \lambda_c \quad \Rightarrow \quad R_s \gg \lambda_c \tag{5.16}$$

That is, we can only talk about localized particles when the curvature scale is much larger than the associated Compton wavelength. Thus, we can safely consider a highly localized particle with spread $\Delta\lambda$ in a gravitational field such that:

$$R_s \gg \Delta\lambda > \lambda_c \tag{5.17}$$

even though the exponential tail may fall outside the Minkowskian region.

Finally, it is important to recall from equation (3.259) that, in flat spacetime, the time derivative of the oscillating terms in the Fourier expansion of a quantum field are used to distinguish between positive and negative energy solutions. This idea can be generalized to curved spacetimes as follows. If there is a time-like Killing field ξ, so that the *Lie derivative* of the frequency mode u_i is such that:

$$\pounds_\xi u_j = -i\omega u_j \qquad \omega > 0 \tag{5.18}$$

then, we can distinguish between positive and negative energy modes [26]. As expressed in equation (2.180), for a scalar field S, as is the case for the mode functions u_j, the Lie derivative is given by:

$$\pounds_\xi S = \xi^\mu \mathcal{D}_\mu S \qquad (5.19)$$

where $\mathcal{D}_\mu$ represents the covariant derivative.

5.2.2 Hilbert spaces

The spacetime history of a localized state Ψ is a one-parameter sequence of states $\Psi(\lambda)$, each of these associated with a specific point $x^\mu(\lambda)$ in the particle's world line [14]. Formally, each member of this sequence needs to be considered as belonging to different Hilbert spaces $\mathcal{H}_\lambda$ attached to each point $x^\mu(\lambda)$ of the world line. Furthermore, as the localization cannot be made arbitrarily small, the particle traverses a superposition of paths over different world lines. In the absence of other external forces, these path superpositions can be represented by geodetic congruences.

Consequently, it is an ill-defined task to try to compare the quantum state of a particle at two different points of its world line, $\Psi(\lambda_1)$ and $\Psi(\lambda_2)$, as these states belong to different Hilbert spaces, $\mathcal{H}_{\lambda_1}$ and $\mathcal{H}_{\lambda_2}$, respectively. On the other hand, imagine we have two particles following two different world lines. One of them follows the world line $x^\mu(\lambda)$ while the other follows $x^\mu(\sigma)$. Then, if both have exactly the same initial and final conditions:

$$x^\mu(\lambda_0) = x^\mu(\sigma_0) \qquad x^\mu(\lambda_1) = x^\mu(\sigma_1) \qquad (5.20)$$

it is possible to compare the quantum state of the particles at both positions. Indeed, in such a case, the Hilbert spaces are identical at the initial and final positions:

$$\mathcal{H}_{\lambda_0} = \mathcal{H}_{\sigma_0} \qquad \mathcal{H}_{\lambda_1} = \mathcal{H}_{\sigma_1} \qquad (5.21)$$

Furthermore, in general the Hilbert space will not only depend on the position x^μ, but also on the four-momentum p^μ. That is: $\mathcal{H} = \mathcal{H}(x, p)$.

That said, the specific choice of the Hilbert space used to describe the dynamics of quantum particles in curved spacetime is not determined. Indeed, let us recall that the momentum $\hat{\mathbf{P}}$ and orbital angular momentum $\hat{\mathbf{L}}$ operators do not commute:

$$[\hat{\mathbf{P}}, \hat{\mathbf{L}}] \neq 0 \qquad (5.22)$$

where:

$$\hat{\mathbf{L}} \equiv \hat{\mathbf{R}} \times \hat{\mathbf{P}} \qquad (5.23)$$

and $\hat{\mathbf{R}}$ is the position operator. However, each one of these commutes with the Hamiltonian $\hat{H}$ (for spinless particles). Then, we have two possible choices of simultaneously commuting Hermitian operators that can be used to describe the behaviour of scalar quantum particles:

- Energy and orbital angular momentum: $\{\hat{H}, \hat{\mathbf{L}}^2, \hat{L}_z\}$
- Energy and momentum: $\{\hat{H}, \hat{\mathbf{P}}\}$

Depending on the problem at hand, each set offers its own advantages and disadvantages. In what follows we will present the solutions to the minimally coupled Klein–Gordon field in the presence of a gravitational field using both sets of operators. The method described can be easily generalized to other quantum fields[1].

5.2.3 Scalar orbital angular momentum eigenstates

Let us consider a scalar quantum particle in the presence of a spherically symmetric gravitational field described by the Schwarzschild metric. This problem has been extensively studied in the scientific literature [15–25] The method to follow is strikingly similar to that performed to determine the quantum structure of the hydrogen atom. The Schwarzschild metric tensor is given in chapter 6, but its specific components are not important for our current discussion. Instead, let us recall that the spatial part of a spherically symmetric problem is usually best described in terms of spherical coordinates made of one radial r and two angular variables (θ, φ).

Indeed, the mathematical structure of the angular part of a spherically symmetric problem is well known in quantum mechanics and classical electrodynamics [11, 26, 27]. As is well known, orbital angular momentum is conserved for particles of spin-0. In particular, we are looking for wave functions that satisfy the Klein–Gordon equation and are eigenstates of angular momentum:

$$\hat{L}\phi = l(l+1)\phi \qquad \hat{L}_z\phi = m\phi \tag{5.24}$$

Then, as the gravitational field is spherically symmetric and independent of the angular variables θ and φ, we can write the general solution to the Klein–Gordon equation as:

$$\phi(t, r, \theta, \varphi) = \frac{f(r, t)}{r} Y_{lm}(\theta, \varphi) \tag{5.25}$$

where $Y_{lm}(\theta, \varphi)$ are the spherical harmonic functions [28]. These functions satisfy the eigenstate equations, in the sense that:

$$\hat{L}Y_{lm}(\theta, \varphi) = l(l+1)Y_{lm}(\theta, \varphi) \qquad \hat{L}_z Y_{lm}(\theta, \varphi) = mY_{lm}(\theta, \varphi) \tag{5.26}$$

The structure of the spherical harmonics can be found in the mathematical physics literature [27, 29].

Therefore, in the case of a spherically symmetric gravitational field, the Klein–Gordon equation is reduced to a differential equation in r and t that can be written as:

$$\frac{\partial^2 f}{\partial t^2} - \frac{\partial^2 f}{\partial r_*^2} + \mathcal{V}^2 f = 0 \tag{5.27}$$

---

[1] Notice, however, that for spin-$\frac{1}{2}$ particles, the orbital angular momentum does not commute with the Dirac Hamiltonian. In such a case, we need to consider eigenstates related to the *total angular momentum* operator $\hat{\mathbf{J}} = \hat{\mathbf{L}} + \frac{1}{2}\hat{\boldsymbol{\Sigma}}$.

where the *effective scalar potential* is given by:

$$\mathcal{V}^2 = \left(1 - \frac{2M}{r}\right)\left[\frac{l(l+1)}{r^2} + \frac{2M}{r^3} + m^2\right] \tag{5.28}$$

and the *Regge–Wheeler tortoise coordinate* is defined as:

$$r_* \equiv r + 2M \ln\left(\frac{r}{2M} - 1\right) \tag{5.29}$$

Furthermore, if the field is considered stationary (i.e. the metric tensor is independent of the time coordinate, as is the case for the Schwarzschild metric), then:

$$f(r,t) = u(r)e^{-iEt} \tag{5.30}$$

and the radial component obeys the equation:

$$\left(E^2 - \mathcal{V}^2\right)u = -\frac{\partial^2 u}{\partial r_*^2} \tag{5.31}$$

where, as usual, E can be identified with the energy of the particle. Notice that the effect of using r_* is to shift the event horizon from $r = 2M$ to $-\infty$.

The behaviour of the oscillatory component f of the solution depends on the structure of the scalar effective potential. We observe the following limit behaviour:

$$\lim_{r \to 2M} \mathcal{V}^2 = 0$$
$$\lim_{r \to \infty} \mathcal{V}^2 = m^2 \tag{5.32}$$

The behaviour of $\mathcal{V}^2 \times M^2/100$ with respect to r_* for different values of l is illustrated in figure 5.1. From the lower to the upper curve, the values of l are 0, 30, 40, 50 and 70, respectively. In addition, we have assumed that M and m are such that:

$$\frac{GMm}{\hbar c} = 10 \tag{5.33}$$

where we have explicitly reintroduced the universal constants G, $\hbar$ and c.

It can be observed that after a certain value of the orbital angular momentum number l, the curves have a local minimum and a local maximum, which suggests the existence of bound states [30]. Figure 5.2 shows the large r_* region of figure 5.1, and the position of the local minimum is highlighted with a black dot. As shown in the figure, there are no local minima for the curves that correspond to l equal to 0 and 30. The condition for the existence of a local minimum can be found analytically and is given by:

$$\frac{GMm}{\hbar c} > \frac{\sqrt{108}}{1728}\sqrt{3 + 2l(l+1) + 3l^2(l+1)^2 + \sqrt{1 + l(l+1) - l^2(l+1)^2 - l^3(l+1)^3}} \tag{5.34}$$

where for clarity we have reintroduced the fundamental constants $\hbar$, c and G. For the same case shown in figure 5.1 and 5.2, this analytical expression gives $l \geqslant 35$, which can be easily verified using numerical computation tools.

The above discussion suggests that the use of angular momentum eigenstates in curved spacetimes is better suited to describe problems such as pair production in the

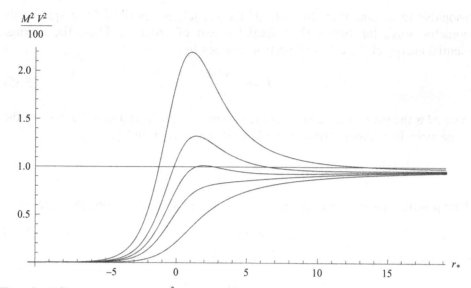

Figure 5.1. Effective scalar potential $\mathcal{V}^2$ for several values of l. From the lower to the upper curve: $l = 0, 30, 40, 50$ and 70.

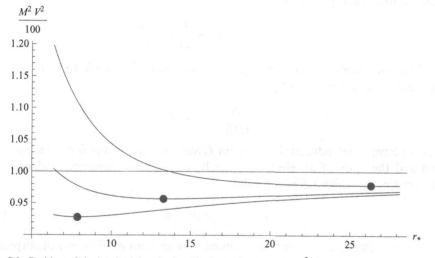

Figure 5.2. Position of the local minima in the effective scalar potential $\mathcal{V}^2$ for several values of l. From the lower to the upper curve: $l = 40, 50$ and 70.

neighbourhood of a black hole and other instances where specific paths or geodetic congruences are not relevant.

As a comparison, we can make a rough estimation of the smallest radius r_0 of a stationary bound state in Newtonian gravitation. To this end, we will follow the same rationale used for the hydrogen atom, invoking the uncertainty principle [26]. We will require the use of non-relativistic mechanics, as general relativity does not offer a clear distinction between kinetic and potential energy. Furthermore, it is reasonable to assume that the state of the particle is described by a spherically symmetric wave function with a spatial extent of order r_0. Thus, the average potential energy of the orbiting electron is given by:

$$\overline{V} \approx -\frac{Mm}{r_0} \tag{5.35}$$

where M is the mass of the source of the gravitational field and m is the mass of the test particle. In addition, the average kinetic energy is given by:

$$\overline{T} \approx \frac{\overline{p}^2}{2m} \tag{5.36}$$

where p is the momentum of the test particle. The uncertainty principle states that:

$$\Delta p \, r_0 \approx \hbar \tag{5.37}$$

and, consequently, the minimum kinetic energy is bounded by:

$$\overline{T} \gtrsim \overline{T}_{\min} \approx \frac{(\Delta p)^2}{2m} \approx \frac{\hbar^2}{2mr_0^2} \tag{5.38}$$

Then, the total energy is bounded by:

$$E_{\min} = \overline{T}_{\min} + \overline{V} \approx \frac{\hbar^2}{2mr_0^2} - \frac{Mm}{r_0} \tag{5.39}$$

Therefore, the minimal value of r_0, which is compatible with the principles of quantum mechanics, is given by:

$$a_g \approx \frac{\hbar^2}{GMm^2} \approx 10^{-23} m \tag{5.40}$$

where we have re-introduced the constant G and used the values for the mass of the Earth and the mass of an electron. Clearly, a_g is gravitational equivalent to the Bohr's radius a_e, which takes the value:

$$a_e \approx \frac{\hbar}{me^2} \approx 10^{-11} m \tag{5.41}$$

and as we might have expected, a_g is much smaller than a_0. This is not surprising: after all, gravity is much weaker than electromagnetic interactions [31]. In general relativity, however, the source has an event horizon at $r = r_s$, which for a black hole of the same mass as Earth corresponds to about $r_s \approx 0.44$ cm, whereas the circular orbit with the smallest radius compatible with general relativity is at about $r_o \approx 0.66$ cm. Both figures are much larger than a_g.

Figure 5.3 shows the conceptual behaviour of the effective scalar potential $\mathcal{V}^2$ in the presence of extreme points. Let us call $\mathcal{V}_m^2$ and $\mathcal{V}_M^2$ the local minimum and local maximum values of $\mathcal{V}^2$, respectively. Now, let us consider the case of a scalar particle dropped from infinite with energy E that moves toward the source of the gravitational field. Notice that the region near the local minimum of $\mathcal{V}_m^2$ acts as a

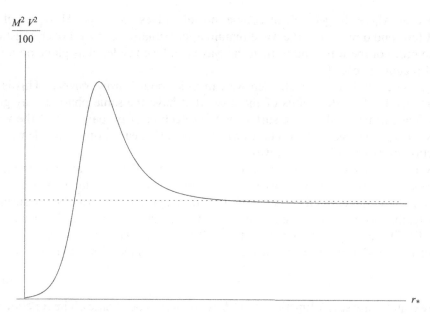

Figure 5.3. Conceptual behaviour of the effective scalar potential in the presence of extreme points. The dashed line corresponds to its asymptotic value at infinite: $\mathcal{V}^2 \to m^2$.

potential well, whereas the region near $\mathcal{V}_M^2$ acts as a potential barrier. Then, the following cases encompass the behaviour of this particle:

- $E^2 \gg \mathcal{V}_M^2$: the particle is mostly transmitted through the barrier.
- $m^2 \lesssim E^2 \lesssim \mathcal{V}_M^2$: the particle is mostly reflected by the barrier.
- $E^2 \approx \mathcal{V}_M^2$: the particle is partly reflected and partly transmitted.
- $\mathcal{V}_m^2 \lesssim E^2 \lesssim m^2$: the particle is in a bounded state and the energy may be quantized, taking a discrete spectrum (notice that this case is similar to the behaviour of an electron in the hydrogen atom [11, 26]).

Finally, if the potential does not have extreme points, then the particle will just accelerate until it reaches the event horizon.

5.2.4 Scalar four-momentum eigenstates

As mentioned before, the dynamics of a partly localized state $\Psi(x)$ has to be considered as a superposition of states traversing different world lines. Consideration of such path superpositions in the case of geodetic motion of quantum particles leads to the concept of geodesic congruences discussed in chapter 2. The geometrical optics approximation provides a good conceptual analogy that can be used to understand better the motion of quantum particles in curved spacetime.

The geometrical optics approximation

Let us assume a classical electromagnetic wave propagating in flat spacetime. If this is a plane wave, then the direction of propagation and amplitude of the wave are the same everywhere. In general, of course, not all waves are plane. However, if the amplitude and direction of the wave remain approximately constant over a distance of the order of the wavelength, then the wave can be considered as plane on a very small region of space [32].

If these conditions are met, then we can talk about a *wave surface*. This is the surface made of all the points of the wave that have the same phase at any given time. The normal to the wave surface is the direction of propagation of the wave. Thus, *rays* are defined as those curves whose tangents at each point coincide with the direction of propagation of the wave.

In a sense, if the above-stated conditions are met, then we can talk about the propagation of light without making a direct reference to its undulatory behaviour. In such a case, it is enough to understand the propagation of rays and wave surfaces. Clearly, these conditions are satisfied if we assume that the wavelength is very small ($\lambda \to 0$). This asymptotic regime is often referred as *geometrical optics*.

Let us consider an arbitrary wave, not necessarily plane, described by:

$$f = ae^{i\psi} \tag{5.42}$$

where a and ψ are scalar functions that depend on the coordinates. The function ψ is referred as the *eikonal*. As we already know, if f is a plane wave, then the amplitude a is a constant and ψ is given by:

$$\psi_p(\mathbf{r}, t) = \mathbf{k} \cdot \mathbf{r} - \omega t + \alpha \tag{5.43}$$

where α is a constant phase and $\mathbf{k}$ and ω correspond to the momentum and frequency of the wave, respectively. This expression implies that, for a plane wave, the wave surfaces are planes perpendicular to the direction of propagation $\mathbf{k}$, and the rays are straight lines parallel to $\mathbf{k}$. Similarly, the wave surfaces of a spherical wave are concentric spheres, whereas the rays are parallel to their radial vector.

For more general waves, the eikonal may not have such a simple mathematical expression. However, in the geometrical optics approximation, λ is very small, and consequently ψ is very large (recall that $|\mathbf{k}| \propto \lambda^{-1}$). Nonetheless, in a small region of space we can expand the eikonal as a series. At first order we find that:

$$\psi \approx \psi_0 + \mathbf{r} \cdot \frac{\partial \psi}{\partial \mathbf{r}} + t \frac{\partial \psi}{\partial t} \tag{5.44}$$

As we are working on the geometric optics approximation, in a very small region of space and in a small interval of time, the wave can be expressed as a plane wave of momentum $\mathbf{k}$ and frequency ω. Therefore, we can make the following identification:

$$\mathbf{k} = \frac{\partial \psi}{\partial \mathbf{r}} = \nabla \psi \qquad \omega = -\frac{\partial \psi}{\partial t} \tag{5.45}$$

which in terms of four-vectors can be written as:

$$k_\mu = \frac{\partial \psi}{\partial x^\mu} = \partial_\mu \psi \qquad k_\mu = -(\omega, \mathbf{k}) \tag{5.46}$$

For light, we know that the four-momentum is normalized as:

$$k_\mu k^\mu = 0 \qquad (5.47)$$

which leads to the *eikonal equation*:

$$\partial_\mu \psi \partial^\mu \psi = 0 \qquad (5.48)$$

If the wave has constant frequency ω, then the wave can be written as:

$$f = a e^{i\psi(x)} e^{-i\omega t} \qquad (5.49)$$

and the eikonal equation becomes:

$$(\nabla\psi(x))^2 = \omega^2 \qquad (5.50)$$

where $\psi(x)$ is a function that only depends on the spatial coordinates. Then, in this case we can see that the wave surfaces correspond to the surfaces of constant eikonal $\psi(x)$, and the direction of the rays at each point of space is given by the direction of the gradient of the eikonal $\nabla\psi(x)$.

The semi-classical approximation
Let us now solve the Klein–Gordon equation in the presence of gravity. To this end, we apply the *semi-classical approximation* to the Klein–Gordon equation in curved spacetime as follows. We recall that the minimally coupled Klein–Gordon equation in the presence of a gravitational field is given by:

$$\left(\hbar^2 g^{\mu\nu} \mathcal{D}_\mu \partial_\nu - m^2\right)\phi = 0 \qquad (5.51)$$

Now, we assume that the field can be written as:

$$\phi = A e^{i\Phi/\hbar} \qquad (5.52)$$

where A and Φ are real functions of the coordinates. With this expression for ϕ, the Klein–Gordon equation can be rewritten as:

$$\hbar^2 \mathcal{D}_\mu \partial^\mu A + i\hbar\left(2\partial_\mu A \partial^\mu \Phi + A\mathcal{D}_\mu \partial^\mu \Phi\right) - A\partial^\mu \Phi \partial_\mu \Phi - Am^2 = 0 \qquad (5.53)$$

Separating real and imaginary parts, we have:

$$\begin{aligned}
\hbar^2 \mathcal{D}_\mu \partial^\mu A - A\partial^\mu \Phi \partial_\mu \Phi - Am^2 &= 0 \\
2\partial_\mu A \partial^\mu \Phi + A\mathcal{D}_\mu \partial^\mu \Phi &= 0
\end{aligned} \qquad (5.54)$$

This system of two coupled second-order differential equations is *strictly equivalent* to the Klein–Gordon equation (5.51).

Notice that, multiplying both sides by A, the second equation can be written as:

$$\mathcal{D}_\mu\left(A^2 \partial^\mu \Phi\right) = 0 \qquad (5.55)$$

which can be interpreted as a continuity equation for the four-current:

$$j^\mu = A^2 \partial^\mu \Phi \qquad (5.56)$$

We can compare this current with the four-current for the Klein–Gordon field in flat spacetime given in equation (3.27):

$$\tilde{j}^\mu = i(\phi^* \partial^\mu \phi - \phi \partial^\mu \phi^*) = -2A^2 \partial^\mu \Phi = -2j^\mu \qquad (5.57)$$

which, except for a multiplicative constant, are the same.

Let us now consider the first equation of the system, which for convenience can be written as:

$$\partial^\mu \Phi \partial_\mu \Phi + m^2 = \hbar^2 \frac{\mathcal{D}_\mu \partial^\mu A}{A} \qquad (5.58)$$

The semi-classical approximation considers the limit $\hbar \to 0$ in the quantum dynamical equations [11, 26, 33]. Applying the semi-classical approximation to the Klein–Gordon equation leads to:

$$\partial^\mu \Phi \partial_\mu \Phi + m^2 \approx 0 \qquad (5.59)$$

Notice that this last equation is *not* the Klein–Gordon equation. Indeed, this is a relativistic classical equation that does *not* involve Planck's constant.

Furthermore, we observe that expression (5.59) corresponds to the Hamilton–Jacobi equation for a classical particle in a gravitational field (as discussed in section 2.6). Thus, Φ has to be identified with the classical action $\mathcal{S}$:

$$\Phi = \mathcal{S} = -m \int d\tau = \int p_\mu dx^\mu \qquad (5.60)$$

where $d\tau$ is the Lorentz interval and p^μ is the four-momentum of a classical particle in the gravitational field under consideration. Then, the Hamilton–Jacobi equations imply that:

$$p^\mu = \partial^\mu \Phi \qquad (5.61)$$

and the conserved four-current in the continuity equation becomes:

$$j^\mu = A^2 p^\mu \qquad (5.62)$$

which is equivalent to the expression found in the flat spacetime case in equation (3.29). Notice that in the present case, these equations define a family of *integral curves* $x^\mu(\tau)$ of u^μ, the four-velocity vector field[2]. That is, the mathematical object $p^\mu = p^\mu(x) = mu^\mu(x)$ does not correspond to a single geodesic world line, but it identifies an entire geodesic congruence. In other words, $x^\mu(\tau)$ is a set of integral curves of the vector field u^μ, in such a way that there is only one curve through each point of the manifold.

---

[2] Let us recall that, if **f** is a vector field and $\mathbf{x}(t)$ is a solution to the autonomous system of ordinary differential equations $\dot{\mathbf{x}}(t) = \mathbf{f}(\mathbf{x}(t))$, then $\mathbf{x}(t)$ is said to be an integral curve of **f** [34].

Clearly, the semi-classical approximation is valid as long as the $\hbar^2$ term in equation (5.58) is small. This equation can be rewritten as:

$$\partial^\mu \Phi \partial_\mu \Phi = m^2 \left(\frac{\hbar^2}{m^2} \frac{\mathcal{D}_\mu \partial^\mu A}{A} - 1 \right) \tag{5.63}$$

which makes clear that the semi-classical approximation requires the expression:

$$\left| \frac{\hbar^2}{m^2} \frac{\mathcal{D}_\mu \partial^\mu A}{A} \right| \ll 1 \tag{5.64}$$

to be valid everywhere. In terms of the *reduced Compton wavelength* $\lambda_c \equiv \hbar/m$, the requirement of validity of the semi-classical approximation becomes:

$$\lambda_c^2 \ll \left| \frac{A}{\mathcal{D}_\mu \partial^\mu A} \right| \tag{5.65}$$

The geometric meaning of this equation will depend on the specific metric in consideration. That said, we can make a 'simple semi-qualitative analysis'. If the range of variation of the field amplitude is of order d, then:

$$\partial_\mu A \approx \frac{A}{d} \tag{5.66}$$

therefore, the semi-classical condition implies that:

$$\lambda_c^2 \ll d^2 \tag{5.67}$$

We can notice the similarity between this equation and equation (5.16). That is, the semi-classical approximation will be valid for as long as the reduced Compton wavelength is much smaller than the curvature scale.

If the semi-classical condition holds, then the Klein–Gordon field can be written as:

$$\phi = A e^{\frac{i}{\hbar} \int p_\mu dx^\mu} \tag{5.68}$$

and the amplitude A can be obtained by integrating the continuity condition:

$$\mathcal{D}_\mu \left(A^2 p^\mu \right) = 0 \tag{5.69}$$

and p^μ is the four-momentum of a classical particle that can be found by solving the corresponding classical dynamical equations. Thus, in the absence of any other non-gravitational forces, p^μ describes a geodesic congruence corresponding to classical particles in the presence of a gravitational field.

Also, if the four-momentum has zero covariant divergence, then:

$$\mathcal{D}_\mu p^\mu = 0 \Rightarrow \mathcal{D}_\mu \left(A^2 p^\mu \right) = 2 A p^\mu \partial_\mu A = 0 \tag{5.70}$$

and consequently:

$$\frac{dA}{d\tau} = \frac{p^\mu}{m} \partial_\mu A = 0 \tag{5.71}$$

which means that the amplitude A remains constant.

It is interesting to note that the expression for the Klein–Gordon field in the semi-classical approximation can be written as:

$$\phi = Ae^{-\frac{im}{\hbar}\int d\tau} = Ae^{\frac{i}{\hbar}S} = Ae^{-\frac{im}{\hbar}\int\sqrt{-dS^2}} \tag{5.72}$$

where S represents the classical action of a classical particle in a gravitational field and dS represents the Lorentz interval.

Finally, let us notice that in the absence of gravitation, p^μ is constant, therefore A is also a constant, and consequently:

$$\phi = Ae^{\frac{i}{\hbar}\int p_\mu dx^\mu} = Ae^{\frac{i}{\hbar}p_\mu\int dx^\mu} = Ae^{\frac{i}{\hbar}p_\mu x^\mu} = Ae^{\frac{i}{\hbar}(\mathbf{p}\cdot\mathbf{x} - Et)} \tag{5.73}$$

which is the standard expression for the propagation of a plane wave. Furthermore, because A is a constant, the condition for the semi-classical approximation is valid everywhere.

The Wentzel–Kramers–Brillouin approximation
Let us now consider now the *Wentzel–Kramers–Brillouin method*, also known as the WKB approximation [26, 33, 35–36]. This is a powerful mathematical tool for finding approximate solutions to linear partial differential equations. The WKB approximation consists of assuming the following ansatz:

$$\phi = Ae^{i\Phi/\hbar} \tag{5.74}$$

and introducing a series expansion in powers of $\hbar$ for A and Φ:

$$A = A_0 + \hbar A_1 + \hbar^2 A_2 + \cdots$$
$$\Phi = \Phi_0 + \hbar\Phi_1 + \hbar^2\Phi_2 + \cdots \tag{5.75}$$

and then neglecting higher orders in $\hbar^n$ for some n. These expressions are introduced in the coupled system of equations:

$$\hbar^2\mathcal{D}_\mu\partial^\mu A - A\partial^\mu\Phi\partial_\mu\Phi - Am^2 = 0$$
$$2\partial_\mu A\partial^\mu\Phi + A\mathcal{D}_\mu\partial^\mu\Phi = 0 \tag{5.76}$$

that are formally equivalent to the Klein–Gordon equation.

If we group terms in $\hbar^0$ and $\hbar^1$, we get:

$$\hbar^0: \quad \partial_\mu\Phi_0\partial^\mu\Phi_0 + m^2 = 0$$
$$\mathcal{D}_\mu\left(A_0^2\partial^\mu\Phi_0\right) = 0$$
$$\hbar^1: \quad \partial_\mu\Phi_1\partial^\mu\Phi_0 = 0$$
$$\mathcal{D}_\mu\left(A_0^2\partial^\mu\Phi_1 + 2A_0A_1\partial^\mu\Phi_0\right) = 0 \tag{5.77}$$

We can observe that the two $\hbar^0$ equations correspond to the semi-classical approximation:

$$A_0 = A_{\text{semi-classical}}$$

$$\Phi_0 = \Phi_{\text{semi-classical}}$$

(5.78)

therefore, the $\hbar^1$ equations now look like:

$$\hbar^1: \quad p^\mu \, \partial_\mu \Phi_1 = 0 \Leftrightarrow \frac{d\Phi_1}{d\tau} = 0$$

$$\mathcal{D}_\mu \left(A_0^2 \, \partial^\mu \Phi_1 + 2A_0 A_1 p^\mu \right) = 0$$

(5.79)

which means Φ_1 is a constant over the path of the classical particle with four-momentum p^μ. The integration of these differential equations leads to expressions for Φ_1 and A_1.

At this point, we can notice that the WKB approximation is very similar to the semi-classical approximation. However, the WKB approximation is much more general, and it allows us to solve quantum equations in regions of space that are inaccessible to classical mechanics. For instance, the WKB solution can be used to study quantum tunnelling effects, which have no classical counterpart.

The validity of the WKB approximation is rather complicated and depends on the interactions in the wave equation. Furthermore, the expansion in powers of $\hbar$ not always converges. Even so, the method provides a good approximation of Φ and A if the series are broken at some power of $\hbar$ [33]. In general, the WKB approximation remains valid for slowly varying potentials and, in the case of gravity, for large curvature scales R_s (when compared with the Compton wavelength).

The optical analogy
In the semi-classical approximation, the Klein–Gordon field is given by:

$$\phi = A e^{\frac{i}{\hbar} S}$$

(5.80)

where S is the classical action that satisfies the Hamilton–Jacobi equation:

$$\partial_\mu S \partial^\mu S = -m^2$$

(5.81)

Also, let us assume that A is approximately constant (this means that the amplitude varies much more slowly than the phase in the context of the WKB approximation).

These two equations resemble the equations that describe the motion of an arbitrary wave of constant amplitude in the geometrical optics approximation. In this case, however, the eikonal equation has an m^2 term (because the Klein–Gordon field has mass m whereas photons in the electromagnetic waves are massless).

This means that we can interpret the solutions of the Klein–Gordon equation in the semi-classical approximation as the propagation of classical waves in the geometric optics approximation with an eikonal given by the classical action S. In this case, the wave surfaces are given by those surfaces where the phase S is a constant. Furthermore, the direction of the rays $\mathcal{R}^\mu$ correspond to the direction of the four-momentum defined by the gradient of S:

$$\mathcal{R}^\mu \| \partial^\mu S = p^\mu \tag{5.82}$$

That is, the trajectories are orthogonal to the surfaces of constant phase S.

In a sense, the geometrical optics approximation in classical electrodynamics is similar to the semi-classical approximation in quantum mechanics [11]. Indeed, if we take the limit $\hbar \to 0$ directly in the Klein–Gordon field given in equation (5.74), then the entire phase factor $\Phi/\hbar$ is very large. This is equivalent to taking the $\lambda \to 0$ limit in the classical field expression in equation (5.42), which leads to a large eikonal ψ.

To summarize, in a small region of spacetime, we can interpret the solution of the Klein–Gordon equation in the presence of gravitation as waves defined by the rays $\partial_\mu S$ and the surface waves characterized by constant S. Notice that the quantum problem is solved by using and re-interpreting the classical solutions embedded in the four-momentum p^μ.

Feynman's path integrals

The semi-classical approximation and the WKB method provide solutions to the Klein–Gordon equation that can be interpreted through the optical analogy as the propagation of matter surface waves and rays. Unfortunately, this analogy breaks down if the semi-classical or WKB conditions are not satisfied.

In the general case, the Klein–Gordon field can be expressed through Feynman's path integrals. A review of path integrals is completely outside the scope of this book, and the interested reader can consult the extensive literature on the topic [37–41]. That said, we will briefly describe the path integral formulation, as this helps with the understanding of how quantum particles propagate in spacetime.

For simplicity, let us consider a non-relativistic, one-dimensional quantum mechanical system described by the wave function $\psi(q_i t_i)$ at time t_i and position q_i. Then, the wave function at time t_f and position q_f is given by:

$$\psi(q_f t_f) = \int K(q_f t_f; q_i t_i)\psi(q_i, t_i)\mathrm{d}q_i \tag{5.83}$$

where K is the *propagator*, which can be expressed as the transition probability between $(q_i t_i)$ and $(q_f t_f)$:

$$K(q_f t_f; q_i t_i) = \langle q_f t_f | q_i t_i \rangle \tag{5.84}$$

The propagator can be expressed as a *path integral*:

$$\langle q_f t_f | q_i t_i \rangle \propto \int \mathcal{D}q \ e^{\frac{i}{\hbar}S(t_i, t_f, [q, \dot{q}])} \tag{5.85}$$

where $S(t_i, t_f, [q, \dot{q}])$ is the classical action evaluated at the path $q(t)$ and integrated between t_i and t_f. The path integral is taken over *all possible paths* $q(t)$ that connect $(q_i t_i)$ and $(q_f t_f)$.

The classical limit of the path integral expression, when $\hbar \to 0$, is of interest to our discussion. Indeed, in the classical limit, the integrand $iS/\hbar$ will be very large, and the exponential term will oscillate very rapidly as q varies. These rapidly oscillating modes will cancel each other upon integration of all paths q. The only terms that will contribute to the integral are those in which S remains approximately constant. This happens when S is at a stationary point $\delta S \approx 0$, which corresponds to the classical path $q_c(t)$:

$$\delta S(t_i, t_f, [q_c, \dot{q}_c]) \approx 0 \qquad (5.86)$$

Therefore, in the classical limit, the wave equation at $(q_f t_f)$ will look like:

$$\psi(q_f t_f) \propto \int e^{\frac{i}{\hbar} S(t_i, t_f, [q_c, \dot{q}_c])} \psi(q_i, t_i) dq_i \propto e^{\frac{i}{\hbar} S(t_i, t_f, [q_c, \dot{q}_c])} \qquad (5.87)$$

That is, we recover once again the semi-classical approximation and the path of the particle can be interpreted using the optical analogy.

Furthermore, the preceding observations suggest that in the most general case, quantum behaviour can be interpreted as fluctuations over the classical trajectory [38]. Therefore, the motion of a quantum particle in curved spacetime involves fluctuations around the geodesic curves. It is only in the semi-classical limit that we can apply the optical analogy: the particle behaves as a surface wave with associated rays that are tangent to the geodesics. In other words, in the most general case, we cannot interpret the motion of quantum particles as surface waves because of the substantial quantum fluctuations around the classical trajectory.

Spherically symmetric gravitational field
As an example, let us consider the case of a particle moving in a stationary, spherically symmetric spacetime described by the Schwarzschild metric. This example is equivalent to that previously discussed using orbital angular momentum eigenstates.

Except for a region close to the singularity, where the curvature scale R_s is very small, we can apply the WKB approximation. Then, as discussed previously, the quantum particle is described by the relativistic wave function:

$$\phi \propto e^{iS} \qquad (5.88)$$

and S represents the action of the classical particle. Then, using the optical analogy, we conclude that the quantum scalar particle is represented by surface waves of constant S and rays parallel to $\partial_\mu S$. We can easily solve these classical equations for general free-fall motion in Schwarzschild metric in the equatorial plane $\theta = \pi/2$. (Chapter 6 further details the Schwarzschild metric and its geodesics.) The result is:

$$\partial_\mu S = m u^\mu \qquad (5.89)$$

where the components of u^μ are given by:

$$u^t = \frac{K}{f}$$

$$u^r = \sqrt{K^2 - f\frac{J^2}{r^2} - f}$$ (5.90)

$$u^\theta = 0$$

$$u^\varphi = \frac{J}{r^2}$$

where:

$$f \equiv 1 - \frac{2M}{r}$$ (5.91)

and J and K are integration constants [9, 10]. Clearly, J is related to the orbital angular momentum and K to the total energy. Thus, we set:

$$K = \frac{E}{m}$$ (5.92)

so we have:

$$\lim_{r \to \infty} u^t = E$$ (5.93)

where E is the total energy and m is the mass. In such a case, the radial equation takes the form:

$$(u^r)^2 = \frac{E^2}{m^2} - f\frac{J^2}{r^2} - f$$ (5.94)

therefore:

$$m^2(u^r)^2 + \left(\mathcal{V}^2 - E^2\right) = 0$$ (5.95)

where the effective scalar potential is given by:

$$\mathcal{V}^2 \equiv f\left(\frac{m^2 J^2}{r^2} + m^2\right)$$ (5.96)

which can be rewritten as:

$$\mathcal{V}^2 = \frac{1}{4M^2}\left(1 - \frac{2M}{r}\right)\left(\left(\frac{2M}{r}\right)^2 m^2 J^2 + 4M^2 m^2\right)$$ (5.97)

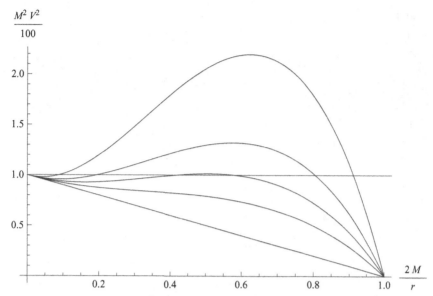

Figure 5.4. Effective scalar potential $\mathcal{V}^2$ for several values of mJ. From the lower to the upper curve: $mJ = 0$, 30, 40, 50 and 70.

The behaviour of $\mathcal{V}^2 \times M^2/100$ for several values of mJ is illustrated in figure 5.4. For this plot we have assumed that:

$$GMmc^2 = 10 \tag{5.98}$$

It can be observed that if J is larger than a certain number, then the curves will have two extreme points: a maximum and a minimum. This condition for J can be calculated analytically to give:

$$J > 2\sqrt{3}\,M \tag{5.99}$$

And for the specific example in consideration ($Mm = 10$) we have:

$$mJ \gtrsim 2\sqrt{3}\,Mm \approx 34.6 \tag{5.100}$$

Figure 5.5 shows the location of the minima of the scalar effective potential for large r for selected values of mJ.

Figure 5.6 shows the conceptual behaviour of the effective scalar potential in the presence of extreme points. Notice that the behaviour of a particle with such an effective scalar potential is the same as the one discussed before for the case of orbital angular momentum eigenstates. Indeed, even though the solutions are expressed in terms of different bases, the underlying physical description is the same. Thus, we have exactly the same results that determine if the particle is partly reflected or partly transmitted through the potential barrier, as well as the possibility of bounded states given by the minimum of the potential.

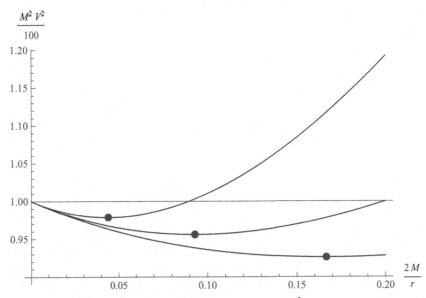

Figure 5.5. Position of the local minima in the effective scalar potential $\mathcal{V}^2$ for several values of mJ. From the lower to the upper curve: $mJ = 0, 30, 40, 50$ and 70.

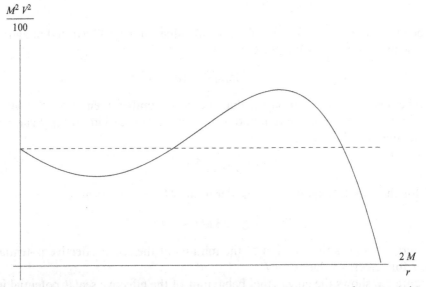

Figure 5.6. Conceptual behaviour of the effective scalar potential in the presence of extreme points. The dashed line corresponds to its value at infinite: $\mathcal{V}^2 \approx m^2$.

To summarize, we found the dynamics of a classical particle in a static, spherically symmetric gravitational field. The dynamics are given in terms of the classical action S and the four-momentum p^μ. These two functions are related by $p_\mu = \partial_\mu S$. Then, using the optical analogy and the WKB approximation, we found that the quantum particle can be interpreted as waves with wave fronts given by the surfaces of constant S and rays parallel to $\partial_\mu S$. Therefore, the analysis of the classical effective potential $\mathcal{V}^2$ gives rough information as to the behaviour of the quantum particle in the gravitational field. With this information we can see when the particle waves are reflected or transmitted through the potential wall, or when the particle may form an orbital bounded state. A more detailed description that includes, for instance, a formal mathematical description of tunnelling through the potential walls, would require the use of the next-order terms in the WKB approximation.

Therefore, the above discussion suggests that momentum eigenstates are more apt to describe the path superpositions of quantum particles traversing world lines that are far away from the singularity and the event horizon.

5.3 The quantum vacuum in a gravitational field

Let us consider the case of a real scalar quantum field in a *spatially flat Friedman spacetime* [5]. This is an important cosmological model as it is often used to represent the universe at large scales (>100 Mpc) [9, 10, 42].

Let us recall that the *Friedman–Lemaitre–Robertson–Walker metric* is defined through the line element given by:

$$\mathrm{d}S^2 = -\mathrm{d}t^2 + a^2\delta_{ik}\mathrm{d}x^i\mathrm{d}x^j \qquad a \equiv a(t) \tag{5.101}$$

which describes a spatially homogenous, isotropic and expanding universe. The time-dependent spatial components of the metric are gauged by the scale factor $a(t)$, which explicitly depends on time, but not on the spatial variables.

It is convenient to introduce the time-like parameter η defined as:

$$\eta \equiv \int^t \frac{\mathrm{d}\tilde{t}}{a(\tilde{t})} \quad \Rightarrow \quad \mathrm{d}\eta = \frac{\mathrm{d}t}{a(t)} \tag{5.102}$$

so we can rewrite the line element of the Friedman spacetime as:

$$\mathrm{d}S^2 = a^2\left(-\mathrm{d}\eta^2 + \delta_{ij}\mathrm{d}x^i\mathrm{d}x^j\right)$$
$$= a^2\left(\eta_{\mu\nu}\mathrm{d}x^\mu\mathrm{d}x^\nu\right) \tag{5.103}$$

where, for notational convenience, we have renamed the time-like variable η as t:

$$\eta \to t \tag{5.104}$$

with the understanding that this new t is different from that shown in equation (5.101) but it is given by η as expressed in equation (5.102). Then, we can express the metric tensor as a function of the Minkowski tensor:

$$g_{\mu\nu} = a^2\eta_{\mu\nu} \qquad g^{\mu\nu} = \frac{1}{a^2}\eta^{\mu\nu} \tag{5.105}$$

Then, the action for a scalar quantum field is given by:

$$S = -\frac{1}{2}\int \sqrt{-g}\left(g^{\alpha\beta}\partial_\alpha\phi\partial_\beta\phi + m^2\phi^2\right)\mathrm{d}^4x \tag{5.106}$$

where:

$$g \equiv \det g_{\mu\nu} = -a^8 \tag{5.107}$$

Using the Euler–Lagrange equations we conclude that the equation that describes the dynamics of a scalar quantum field in Friedman spacetime is given by:

$$\frac{1}{a^4} \partial_\mu \left(a^2 \partial^\mu \phi \right) - m^2 \phi = 0 \tag{5.108}$$

For notational convenience we can define the χ scalar quantum field as:

$$\chi \equiv a \phi \tag{5.109}$$

which satisfies the field equation:

$$\eta^{\mu\nu} \partial_\mu \partial_\nu \chi - \left(m^2 a^2 - \frac{\ddot{a}}{a} \right) \chi = 0 \tag{5.110}$$

Therefore, the field equation looks like the Klein–Gordon equation in flat spacetime, but with an effective mass $\tilde{m}$ given by:

$$\tilde{m}^2 = m^2 a^2 - \frac{\ddot{a}}{a} \qquad \tilde{m} \equiv \tilde{m}(t) \tag{5.111}$$

which depends on the time parameter t [5]. Clearly:

$$a = 1 \Rightarrow \tilde{m} = m \tag{5.112}$$

reduces the equation to its flat spacetime form, as expected.

As in the case of flat spacetime, the explicit quantization of the scalar field $\hat{\chi}$ looks like:

$$\hat{\chi} = \int \frac{d^3 k}{(2\pi)^{3/2} \sqrt{2}} \left(e^{i\mathbf{k}\cdot\mathbf{x}} v_{\mathbf{k}}^* \hat{a}_{\mathbf{k}}^\dagger + e^{-i\mathbf{k}\cdot\mathbf{x}} v_{\mathbf{k}} \hat{a}_{\mathbf{k}} \right) \tag{5.113}$$

but now the coefficients need to satisfy the equation:

$$\frac{d^2 v_{\mathbf{k}}}{dt^2} + \left(k^2 + \tilde{m}^2 \right) v_{\mathbf{k}} = 0 \tag{5.114}$$

which implies a time-dependent frequency given by:

$$\tilde{\omega}_k^2(t) = k^2 + \tilde{m}^2(t) \tag{5.115}$$

and the number operator and vacuum state are defined such that:

$$\hat{N}_a \equiv \hat{a}^\dagger \hat{a} \qquad \hat{N}_a |0\rangle = 0 \tag{5.116}$$

However, as we previously discussed within the context of flat spacetimes in chapter 3, other vacuum states can be generated by means of the Bogolyubov transformations:

$$\hat{a}_{\mathbf{k}} = \alpha_{\mathbf{k}}^* \hat{b}_{\mathbf{k}} + \beta_{\mathbf{k}} \hat{b}_{-\mathbf{k}}^\dagger$$

$$\hat{a}_{-\mathbf{k}}^\dagger = \beta_{\mathbf{k}}^* \hat{b}_{\mathbf{k}} + \alpha_{\mathbf{k}} \hat{b}_{-\mathbf{k}}^\dagger \qquad (5.117)$$

$$u_{\mathbf{k}} = \alpha_{\mathbf{k}} v_{\mathbf{k}} + \beta_{\mathbf{k}} v_{\mathbf{k}}^*$$

that satisfy the normalization condition:

$$|\alpha_{\mathbf{k}}|^2 - |\beta_{\mathbf{k}}|^2 = 1 \qquad (5.118)$$

Let us recall that, in the case of flat spacetime, the ambiguity of choosing the right vacuum was eliminated by finding the minimum of the expectation value of the vacuum energy of the quantum field as given by the field Hamiltonian. Thus, we can proceed in a similar manner to find the minimum of:

$$_a\langle 0|\hat{H}|0\rangle_a = \frac{1}{4} \int \mathrm{d}^3k \left({}_a\langle 0|\hat{a}_{\mathbf{k}}\hat{a}_{-\mathbf{k}}|0\rangle_a F_{\mathbf{k}}^* + {}_a\langle 0|\hat{a}_{\mathbf{k}}^\dagger \hat{a}_{-\mathbf{k}}^\dagger|0\rangle_a F_{\mathbf{k}} \right)$$

$$+ \frac{1}{2} \int \mathrm{d}^3k\, {}_a\langle 0|\hat{a}_{\mathbf{k}}^\dagger \hat{a}_{\mathbf{k}}|0\rangle_a E_{\mathbf{k}} + \frac{1}{4} \int \mathrm{d}^3k \delta^3(0) E_{\mathbf{k}} \qquad (5.119)$$

where:

$$F_{\mathbf{k}}(t) = \dot{v}_{\mathbf{k}}^2 + \tilde{\omega}_{\mathbf{k}}^2(t) v_{\mathbf{k}}^2$$

$$E_{\mathbf{k}}(t) = |\dot{v}_{\mathbf{k}}|^2 + \tilde{\omega}_{\mathbf{k}}^2(t) |v_{\mathbf{k}}|^2 \qquad (5.120)$$

$$\tilde{\omega}_{\mathbf{k}}^2(t) = k^2 + \tilde{m}^2(t)$$

Notice that the difference between flat spacetime and curved spacetime comes down to the fact that now the frequency depends on time $\tilde{\omega}_{\mathbf{k}}(t)$.

We could fix the time to some fixed value t_0, and then minimize the energy as we did before. In this case, the solution to the minimization problem is given by:

$$v_{\mathbf{k}} = \frac{e^{i\tilde{\omega}_{\mathbf{k}}t_0}}{\sqrt{\tilde{\omega}_{\mathbf{k}}}} \qquad \tilde{\omega}_{\mathbf{k}} \equiv \tilde{\omega}_{\mathbf{k}}(t_0) \qquad (5.121)$$

However, because the frequency depends on time, this solution will be different at each point of time. Therefore, the number of particles in the vacuum state will not always be the same. Indeed, if we fix the vacuum state with the values obtained at $t = t_0$, then the expectation value of the number of particles in the vacuum at time t_0 is given by:

$$_{t_0}\langle 0|\hat{N}_{\mathbf{k}}(t_0)|0\rangle_{t_0} = 0 \qquad (5.122)$$

as expected. On the other hand, if we calculate the expectation value of the number of particles in the vacuum at some different time t_1, we obtain:

$$_{t_1}\langle 0|\hat{N}_{\mathbf{k}}(t_0)|0\rangle_{t_1} = |\beta_{\mathbf{k}}|^2 \neq 0 \qquad (5.123)$$

where the creation and annihilation operators at time t_0 and at time t_1 are connected through a Bogolyubov transformation characterized by the $\beta_{\mathbf{k}}$ coefficient. In other words, the quantum vacuum is in a constant state of change as a consequence of the constantly expanding Friedman universe. As such, the quantum vacuum in curved spacetime can only be defined as the *instantaneous minimum energy state* [1, 5, 6].

Furthermore, the expectation value of vacuum energy density at time t_0 is given by:

$$_{t_0}\langle 0|\hat{H}(t_0)|0\rangle_{t_0} = \frac{1}{4}\int d^3\mathbf{k}\,\delta^3(0)E_k(t_0) = \frac{1}{2}\int d^3\mathbf{k}\,\delta^3(0)\omega_k(t_0) \qquad (5.124)$$

which is a divergent quantity that explicitly depends on time. Thus, removing this time-dependent divergency is not a trivial undertaking as it was for the case of flat spacetime. Consequently, we are required to use more sophisticated renormalization procedures [1, 5, 6].

Let us also point out that, if we require the vacuum to be constant over a period of time (i.e. the vacuum is an eigenstate of the Hamiltonian at all times), then we have to require that:

$$F_{\mathbf{k}}(t) = 0 \qquad \forall t \in [t_0, t_1] \qquad (5.125)$$

This means that:

$$F_{\mathbf{k}}(t) = \dot{v}_{\mathbf{k}}^2 + \tilde{\omega}_{\mathbf{k}}^2(t)\, v_{\mathbf{k}}^2 = 0 \qquad (5.126)$$

which has the solution:

$$v_{\mathbf{k}}(t) = C e^{\pm i \int \omega_{\mathbf{k}}(\eta)d\eta} \qquad (5.127)$$

However, this solution does not satisfy the field equation, except when $\omega_{\mathbf{k}}(t)$ is a constant, which corresponds to the case of a flat spacetime.

At this point it is important to recall that the quantum vacuum state is invariant under Poincare transformations. That is, all inertial observers see exactly the same vacuum state. Consequently, all inertial observers can agree on the quantum field mode expansion. This is not true, however, in the case of curved spacetime. Indeed, as we have seen, the vacuum state is not invariant under general coordinate transformations, and different observers will identify different vacuum states.

However, it can be shown that, because of the principle of equivalence, the vacuum state ambiguity disappears if we restrict our attention to localized quantum states within a small region $\Delta\lambda \ll R_s$ [14]. Furthermore, as discussed in section 5.1, if the spatial spread of the quantum state is much smaller than the curvature scale, then it is possible to use the plane wave approximation. In such a case, it is also feasible to talk about a specific vacuum state.

5.4 The spin-statistics connection

The *spin-statistics theorem* states that bosonic quantum fields (i.e. particles with integral spin) obey commutation relations and the Bose–Einstein statistics, whereas fermionic quantum fields (i.e. particles with half-odd integer spin) obey anti-commutation relations and the Fermi–Dirac statistics [43–46]. Quite surprisingly, curved spacetimes offer a new way to derive the spin-statistics connection theorem [6]. To show how this is the case, let us assume that in the asymptotic far past a scalar quantum field is described by:

$$\hat{\phi}(t \to -\infty) = \int \frac{d^3k}{(2\pi)^{3/2}\sqrt{2}} \left(e^{i\mathbf{k}\cdot\mathbf{x}} v_{\mathbf{k}}^* \hat{a}_{\mathbf{k}}^\dagger + e^{-i\mathbf{k}\cdot\mathbf{x}} v_{\mathbf{k}} \hat{a}_{\mathbf{k}} \right) \qquad (5.128)$$

whereas in the asymptotic distant future it is given by:

$$\hat{\phi}(t \to +\infty) = \int \frac{d^3k}{(2\pi)^{3/2}\sqrt{2}} \left(e^{i\mathbf{k}\cdot\mathbf{x}} u_{\mathbf{k}}^* \hat{b}_{\mathbf{k}}^\dagger + e^{-i\mathbf{k}\cdot\mathbf{x}} u_{\mathbf{k}} \hat{b}_{\mathbf{k}} \right) \qquad (5.129)$$

where both expressions are written in such a way that the instantaneous quantum vacuum state is well defined. Then, the field operators $\hat{a}$ and $\hat{b}$ are related by a Bogolyubov transformation:

$$\hat{a}_{\mathbf{k}} = \alpha_{\mathbf{k}}^* \hat{b}_{\mathbf{k}} + \beta_{\mathbf{k}} \hat{b}_{-\mathbf{k}}^\dagger$$
$$\hat{a}_{-\mathbf{k}}^\dagger = \beta_{\mathbf{k}}^* \hat{b}_{\mathbf{k}} + \alpha_{\mathbf{k}} \hat{b}_{-\mathbf{k}}^\dagger \qquad (5.130)$$

that satisfies the normalization condition:

$$|\alpha_{\mathbf{k}}|^2 - |\beta_{\mathbf{k}}|^2 = 1 \qquad (5.131)$$

Now, let us assume that we do not know if the scalar field operators $\hat{a}$ and $\hat{b}$ obey commutation or anti-commutation relations. So, we write the anti-commutation or commutation relations for the $\hat{b}$ operators, which are valid in the far future, as follows:

$$\left[\hat{b}_{\mathbf{k}}, \hat{b}_{\mathbf{q}}^\dagger \right]_{\pm} = \delta_{\mathbf{k},\mathbf{q}}$$
$$\left[\hat{b}_{\mathbf{k}}, \hat{b}_{\mathbf{q}} \right]_{\pm} = 0 \qquad (5.132)$$
$$\left[\hat{b}_{\mathbf{k}}^\dagger, \hat{b}_{\mathbf{q}}^\dagger \right]_{\pm} = 0$$

where a single choice of sign is taken to determine the type of relation satisfied by the operators. The commutation relations for the field operators $\hat{a}$, that are valid in the distant past, can be found by applying a Bogolyubov transformation to the commutation relations for the $\hat{b}$ operators. That is:

$$\left[\hat{a}_{\mathbf{k}}, \hat{a}_{\mathbf{q}}^{\dagger}\right]_{\pm} = \left[\alpha_{\mathbf{k}}^{*}\hat{b}_{\mathbf{k}} + \beta_{\mathbf{k}}\hat{b}_{-\mathbf{k}}^{\dagger}, \beta_{\mathbf{q}}^{*}\hat{b}_{\mathbf{q}} + \alpha_{\mathbf{q}}\hat{b}_{-\mathbf{q}}^{\dagger}\right] = \left(|\alpha_{\mathbf{k}}|^{2} \pm |\beta_{\mathbf{k}}|^{2}\right)\delta_{\mathbf{k},\mathbf{q}}$$

$$\left[\hat{a}_{\mathbf{k}}, \hat{a}_{\mathbf{q}}\right]_{\pm} = \left[\alpha_{\mathbf{k}}^{*}\hat{b}_{\mathbf{k}} + \beta_{\mathbf{k}}\hat{b}_{-\mathbf{k}}^{\dagger}, \alpha_{\mathbf{q}}^{*}\hat{b}_{\mathbf{q}} + \beta_{\mathbf{q}}\hat{b}_{-\mathbf{q}}^{\dagger}\right] = \left(\alpha_{\mathbf{k}}^{*}\beta_{\mathbf{k}} \pm \alpha_{\mathbf{k}}^{*}\beta_{\mathbf{k}}\right)\delta_{\mathbf{k},\mathbf{q}} \qquad (5.133)$$

$$\left[\hat{a}_{\mathbf{k}}^{\dagger}, \hat{a}_{\mathbf{q}}^{\dagger}\right]_{\pm} = \left[\beta_{\mathbf{k}}^{*}\hat{b}_{\mathbf{k}} + \alpha_{\mathbf{k}}\hat{b}_{-\mathbf{k}}^{\dagger}, \beta_{\mathbf{q}}^{*}\hat{b}_{\mathbf{q}} + \alpha_{\mathbf{q}}\hat{b}_{-\mathbf{q}}^{\dagger}\right] = \left(\alpha_{\mathbf{k}}\beta_{\mathbf{k}}^{*} \pm \alpha_{\mathbf{k}}\beta_{\mathbf{k}}^{*}\right)\delta_{\mathbf{k},\mathbf{q}}$$

Now, it is reasonable to ask for the field operators to have the same type of commutation or anti-commutation relations in the far past as in the distant future [6]. Clearly, this happens when:

$$|\alpha_{\mathbf{k}}|^{2} \pm |\beta_{\mathbf{k}}|^{2} = 1$$

$$\alpha_{\mathbf{k}}\beta_{\mathbf{k}}^{*} \pm \alpha_{\mathbf{k}}\beta_{\mathbf{k}}^{*} = 0 \qquad (5.134)$$

$$\alpha_{\mathbf{k}}^{*}\beta_{\mathbf{k}} \pm \alpha_{\mathbf{k}}^{*}\beta_{\mathbf{k}} = 0$$

If $\beta_{\mathbf{k}} \neq 0$, then these equations are only satisfied when we take the '$-$' sign (in such a case, the first equation reduces to the normalization condition of the Bogolyubov transformation). Thus, this result becomes a consistency requirement that requires the field operators to obey commutation relations:

$$\left[\hat{a}_{\mathbf{k}}, \hat{a}_{\mathbf{q}}^{\dagger}\right]_{\pm} \rightarrow \left[\hat{a}_{\mathbf{k}}, \hat{a}_{\mathbf{q}}^{\dagger}\right]_{-}$$

$$\left[\hat{b}_{\mathbf{k}}, \hat{b}_{\mathbf{q}}^{\dagger}\right]_{\pm} \rightarrow \left[\hat{b}_{\mathbf{k}}, \hat{b}_{\mathbf{q}}^{\dagger}\right]_{-} \qquad (5.135)$$

Therefore, the quantum field $\hat{\phi}$ describes particles of spin zero that obey Bose–Einstein statistics. That is, Bose–Einstein statistics are the only ones consistent with the dynamics of spin-0 particles in curved spacetime. Following a similar argument, it is possible to arrive at the equivalent result that Pauli–Dirac statistics are the only ones consistent with the dynamics of spin-$\frac{1}{2}$ particles in curved spacetime [47, 48].

Notice that this consistency connection between spin and statistics can only be obtained because we have assumed that $\beta_{\mathbf{k}} \neq 0$, which results from assuming a curved spacetime. Indeed, in Minkowski space we always have $\beta_{\mathbf{k}} = 0$ because the instantaneous vacuum state is a Poincare invariant, and therefore the quantum field operators do not change with time.

5.5 Quantum vector fields in curved spacetime

Although this book is mostly concerned with spin-$\frac{1}{2}$ particles, it is useful to consider the case of a photon in a gravitational field. The photon is a spin-1 vector field that obeys the explicitly covariant Maxwell's equations:

$$\partial_{\mu}F^{\mu\nu} = \partial_{\mu}\partial^{\mu}A^{\nu} - \partial_{\mu}\partial^{\nu}A^{\mu} = 0 \qquad (5.136)$$

where $F^{\mu\nu}$ is the electromagnetic tensor field and A^{μ} is the electromagnetic four-vector potential [10, 40]. If we impose the *Lorentz gauge condition*:

$$\partial_{\mu}A^{\mu} = 0 \qquad (5.137)$$

the dynamical equation for the vector field in the absence of gravitational fields is simply given by:

$$\eta_{\mu\nu}\partial^{\mu}\partial^{\nu}A^{\alpha} = 0 \qquad (5.138)$$

At this point, one needs to be careful and properly understand to what equation one should apply the principle of equivalence. Indeed, the use of the Lorentz gauge condition to obtain equation (5.138) implies that it is possible to permute the order of the two derivatives. Although this can be done in flat spacetime, in general this cannot be done in curved spacetime.

For example, the second covariant derivative of the vector field A^{α} is given by:

$$\mathcal{D}_{\nu}\mathcal{D}_{\mu}A^{\alpha} = \mathcal{D}_{\nu}\left(\partial_{\mu}A^{\alpha} + \Gamma^{\alpha}_{\mu\beta}A^{\beta}\right)$$

$$= \partial_{\nu}\partial_{\mu}A^{\alpha} + A^{\beta}\partial_{\nu}\Gamma^{\alpha}_{\mu\beta} + \Gamma^{\alpha}_{\mu\beta}\partial_{\nu}A^{\beta} + \Gamma^{\alpha}_{\beta\nu}\partial_{\mu}A^{\beta} - \Gamma^{\beta}_{\mu\nu}\partial_{\beta}A^{\alpha}$$

$$+ \Gamma^{\alpha}_{\lambda\nu}\Gamma^{\lambda}_{\mu\beta}A^{\beta} - \Gamma^{\lambda}_{\mu\nu}\Gamma^{\alpha}_{\lambda\beta}A^{\beta} \qquad (5.139)$$

therefore one can show that the commutation of two covariant derivatives involves curvature terms:

$$[\mathcal{D}_{\nu},\mathcal{D}_{\mu}]A^{\alpha} = \mathcal{D}_{\nu}\mathcal{D}_{\mu}A^{\alpha} - \mathcal{D}_{\mu}\mathcal{D}_{\nu}A^{\alpha} = -A^{\beta}R^{\alpha}_{\beta\nu\mu} \qquad (5.140)$$

or equivalently:

$$[\mathcal{D}_{\nu},\mathcal{D}_{\mu}]A_{\alpha} = \mathcal{D}_{\nu}\mathcal{D}_{\mu}A_{\alpha} - \mathcal{D}_{\mu}\mathcal{D}_{\nu}A_{\alpha} = A_{\beta}R^{\beta}_{\alpha\nu\mu} \qquad (5.141)$$

where $R^{\alpha}_{\beta\nu\mu}$ is the curvature tensor [10].

If we take these expressions into account, then the covariant derivative of the electromagnetic tensor can be rewritten as:

$$\mathcal{D}_{\mu}F^{\mu\nu} = \mathcal{D}_{\mu}\mathcal{D}^{\mu}A^{\nu} - \mathcal{D}_{\mu}\mathcal{D}^{\nu}A^{\mu}$$

$$= \mathcal{D}_{\mu}\mathcal{D}^{\mu}A^{\nu} - g^{\nu\alpha}\mathcal{D}_{\mu}\mathcal{D}_{\alpha}A^{\mu}$$

$$= \mathcal{D}_{\mu}\mathcal{D}^{\mu}A^{\nu} - g^{\nu\alpha}\left(\mathcal{D}_{\alpha}\mathcal{D}_{\mu}A^{\mu} + A^{\beta}R^{\mu}_{\beta\alpha\mu}\right)$$

$$= \mathcal{D}_{\mu}\mathcal{D}^{\mu}A^{\nu} - \mathcal{D}^{\nu}\mathcal{D}_{\mu}A^{\mu} - g^{\nu\alpha}g^{\mu\delta}A^{\beta}R_{\delta\beta\alpha\mu}$$

$$= \mathcal{D}_{\mu}\mathcal{D}^{\mu}A^{\nu} - \mathcal{D}^{\nu}\mathcal{D}_{\mu}A^{\mu} + g^{\nu\alpha}A^{\beta}R_{\beta\alpha} \qquad (5.142)$$

which means that the most general description of a photon interacting with a gravitational field is given by:

$$\mathcal{D}_{\mu}\mathcal{D}^{\mu}A^{\nu} - \mathcal{D}^{\nu}\mathcal{D}_{\mu}A^{\mu} + g^{\nu\alpha}A^{\beta}R_{\beta\alpha} = 0 \qquad (5.143)$$

which explicitly couples the photon tensor field with the Ricci tensor [1].

If we now impose the Lorentz gauge condition in terms of the covariant derivative:

$$\mathcal{D}_\mu A^\mu = 0 \tag{5.144}$$

we obtain a simplified version of Maxwell's equations for a photon field in curved spacetime:

$$\mathcal{D}_\mu \mathcal{D}^\mu A^\nu + A^\beta R_\beta^{\ \nu} = 0 \tag{5.145}$$

which is different to what we would have obtained by wrongly applying the principle of equivalence to equation (5.137).

5.6 Spinors in curved spacetimes

In this section we will discuss the case of spinors in gravitational fields. Let us first consider the Dirac equation in flat Minkowski spacetime:

$$\left(i\eta_{\mu\nu}\gamma^\mu\partial^\nu - m\right)\psi = 0 \tag{5.146}$$

If we invoke the principle of general covariance, one would *wrongly* conclude that the Dirac equation in curved spacetime is given by:

$$\left(ig_{\mu\nu}\gamma^\mu\mathcal{D}^\nu - m\right)\psi = 0 \tag{5.147}$$

This equation is wrong because the principle of general covariance requires the flat spacetime equation to be written in terms of objects that behave as tensor fields under Lorentz transformations. Unfortunately, ψ is a spinor field, not a tensor field.

Indeed, it is not possible to directly express spinor fields within the context of general relativity. The origin of this impediment resides on the fact that general coordinate transformations in general relativity are described through the *general linear group* GLR(4) (also denoted by GL(4, $\mathbb{R}$) or GL$_4(\mathbb{R})$) made of all real regular (invertible) 4×4 matrices [49]. However, it is known that GLR(4) does not have a spinor representation [49, 50].

Consequently, the covariant derivative of the spinor field $\mathcal{D}_\mu\psi$ becomes an ill-defined term, as covariant differentiation was originally conceived for objects that transform as covariant or contravariant tensors under a general coordinate transformation. In addition, in the most general case, the Dirac matrices γ^μ may depend on the spacetime coordinates:

$$\gamma^\mu = \gamma^\mu(x) \tag{5.148}$$

The solution to these problems is to introduce the tetrad formalism described in chapter 2. The goal is to construct an appropriate tetrad field that covers the entire spacetime. Then, we can analyse the dynamics of the spinors with reference to the local inertial frames described by the tetrad fields. As such, we only need to work with spinors as observed by the inertial observers at each point of the tetrad field. Consequently, there is no need to express these spinors in the general coordinate system. Indeed, within the context of general relativity, we can only talk about spinors in reference to some local inertial frame described by a tetrad field.

Let us recall that an arbitrary tensor field Θ_μ expressed in the general coordinate system of some curved spacetime can be written in terms of Θ_a, its projection into a local inertial frame of reference and the associated tetrad field $e^a{}_\mu(x)$:

$$\Theta_\mu(x) = e^a{}_\mu(x)\Theta_a(x) \tag{5.149}$$

and vice versa:

$$\Theta_a(x) = e_a{}^\mu(x)\Theta_\mu(x) \tag{5.150}$$

Bearing this in mind, the Dirac matrices in curved spacetime γ^μ can be expressed in terms of γ^a, the Dirac matrices defined in the Minkowski space of a local inertial frame, as follows:

$$\gamma^\mu(x) = e_a{}^\mu(x)\gamma^a \tag{5.151}$$

where we know that the γ^a matrices satisfy the anti-commutation rule:

$$\gamma^a\gamma^b + \gamma^b\gamma^a = -2\eta^{ab}(x) \tag{5.152}$$

Therefore, the generalization of the anti-commutation relation that defines the Dirac matrices in curved spacetime $\gamma^\mu(x)$ is given by:

$$\gamma^\mu(x)\gamma^\nu(x) + \gamma^\nu(x)\gamma^\mu(x) = -2g^{\mu\nu}(x) \tag{5.153}$$

In addition, we can define the action of a covariant derivative on a spinor field as:

$$\mathcal{D}_\mu\psi \equiv \left(\partial_\mu - \Gamma_\mu\right)\psi \tag{5.154}$$

where Γ_μ is the *spinorial affine connection*, and the covariant derivative of a Dirac matrix is given by:

$$\mathcal{D}_\mu\gamma_\nu(x) \equiv \partial_\mu\gamma_\nu(x) - \Gamma^\lambda_{\mu\nu}\gamma_\lambda - \Gamma_\mu\gamma_\nu(x) + \gamma_\nu(x)\Gamma_\mu \tag{5.155}$$

which, as we will show in a subsequent section devoted to the detailed discussion of covariant derivatives, is required to satisfy:

$$\mathcal{D}_\mu\gamma_\nu(x) = 0 \tag{5.156}$$

Furthermore, it is possible to show that the spinorial affine connection that satisfies the above condition and reduces to zero in Minkowski spacetime is given by:

$$\Gamma_\mu(x) = -\frac{1}{4}\gamma_a\gamma_b e^{a\alpha}(x)\nabla_\mu e^b{}_\alpha(x) \tag{5.157}$$

where the *coordinate covariant derivative* of the tetrad field has the standard tensor form:

$$\nabla_\mu e^b{}_\alpha(x) \equiv \partial_\mu e^b{}_\alpha(x) - \Gamma^\nu_{\mu\alpha}e^b{}_\nu(x) \tag{5.158}$$

It is convenient to define the *connection one-form* as:

$$\omega_\mu{}^{ab}(x) \equiv e^{a\alpha}(x)\,\nabla_\mu e^b{}_\alpha(x) \tag{5.159}$$

which can be proved to be anti-symmetric in the $(a,\,b)$ indices:

$$\omega_\mu{}^{ab}(x) = -\omega_\mu{}^{ba}(x) \tag{5.160}$$

Then, the spinorial affine connection can be written as:

$$\Gamma_\mu(x) = -\frac{1}{4}\gamma_a\gamma_b\omega_\mu{}^{ab}(x)$$

$$= -\frac{1}{8}[\gamma_a,\gamma_b]\omega_\mu{}^{ab}(x)$$

$$= -i\,\Sigma_{ab}\,\omega_\mu{}^{ab}(x) \tag{5.161}$$

where:

$$\Sigma^{ab} \equiv -\frac{i}{8}\left[\gamma^a,\gamma^b\right] \tag{5.162}$$

is independent of the coordinates and provides a matrix representation for the Lie algebra of the Lorentz group. Then, the covariant derivative of a spinor field takes the form:

$$\mathcal{D}_\mu\psi = \left(\partial_\mu + i\,\Sigma_{ab}\,\omega_\mu{}^{ab}(x)\right)\psi \tag{5.163}$$

Therefore, the Dirac equation in curved spacetime is written as:

$$\left(i\gamma^\mu(x)\mathcal{D}_\mu - m\right)\psi = 0 \tag{5.164}$$

or equivalently:

$$\left(ie_a{}^\mu\gamma^a\partial_\mu - ie_a{}^\mu\gamma^a\Gamma_\mu - m\right)\psi = 0 \tag{5.165}$$

and consequently:

$$\left(i\gamma^a\partial_a - i\gamma^a\Gamma_a - m\right)\psi = 0 \tag{5.166}$$

or:

$$\left(i\gamma^a\partial_a - m\right)\psi = i\gamma^a\Gamma_a\psi \tag{5.167}$$

where the left-hand side of the equation is the Dirac equation in Minkowski space, the right-hand side describes the interaction of the field ψ with the gravitational field with respect to the local inertial frame associated with the tetrad field $e_a^\mu(x)$, and:

$$\Gamma_a(x) = -\frac{1}{8}[\gamma_b,\gamma_c]\,e_a^\mu(x)\omega_\mu{}^{bc}(x) \tag{5.168}$$

is the spinorial affine connection projected into the local inertial frame. Once more, let us emphasize that the description of spinors in general relativity can only be done with respect to some local inertial frame defined by a tetrad field.

Let us now discuss the meaning of the dynamics of the Dirac spinor ψ in the presence of gravity. Formally, a momentum eigenstate of the Dirac equation in curved spacetime should be written with all its functional dependencies in an explicit manner. That is:

$$\psi \leftrightarrow \left| p^a(x),\, j,\, \sigma;\, x^\mu,\, e_a^{\,\mu}(x),\, g_{\mu\nu}(x) \right\rangle \qquad (5.169)$$

which represents an *extended state* of spin j, helicity σ and definite momentum $p^a(x)$ as observed from the position $x^a = e_{\,\mu}^a(x)x^\mu$ of the local inertial frame defined by the tetrad field $e_{\,\mu}^a(x)$ in a curved spacetime described by the metric tensor $g_{\mu\nu}(x)$ [51]. Indeed, the description of a Dirac state can only be given with respect to the tetrad field and the local inertial frame that it describes.

At this point, we should try to understand the meaning of the tricky concept of a Dirac spinor moving in curved spacetime. Indeed, it does not make any sense to try to find a meaning to ψ in the context of the general coordinate system. This is clear because spinors, such as ψ, are ill-defined in the general coordinate system. Consequently, it is important to notice that ψ, as given by equation (5.169), does not represent a quantum particle of momentum p^μ localized at x^μ.

If we need to represent a localized state as observed in the local inertial frame, then we need to build a wave packet such as:

$$\Psi(x) = \int d^3\mathbf{p}\, \mathcal{G}(\mathbf{p}) \left| p^a(x),\, j,\, \sigma;\, x,\, e_a^{\,\mu}(x),\, g_{\mu\nu}(x) \right\rangle \qquad (5.170)$$

where $\mathcal{G}(\mathbf{p})$ is non-zero and smooth over a finite range of energies. For the case of a Gaussian wave packet:

$$\mathcal{G}(\mathbf{p}) \propto e^{-(\mathbf{p}-\mathbf{q})^2 \lambda^2} \qquad (5.171)$$

the state $\Psi(x)$ represents a state localized around $x \approx 0$ with a position spread of $\Delta x = \lambda$ and a momentum distributed around $\mathbf{q}$ with a spread of $\Delta p = |\mathbf{p} - \mathbf{q}| = 1/\lambda$, which is consistent with Heisenberg's uncertainty principle [5].

In the subsequent sections of this book we will mostly restrict our discussions to spin-$\frac{1}{2}$ momentum eigenstates defined in local inertial frames described by the tetrad field and *transported* along parametrized world lines $x^\mu(\lambda)$ that form a geodetic congruence. That is, we will use a fully covariant spinor transport theory that represents the motion of Dirac particles over path superpositions traversing different world lines. As we will concentrate on the regions of spacetime where R_s is large, we will invoke the WKB approximation and rely on the geometrical optics approximation to visualize the quantum dynamics of spinors in the presence of gravitational fields. As such, the world lines traversed by the quantum particle are described by the geodetic congruences that result from solving the Hamilton–Jacobi equations for a classical particle.

The astute reader may have noticed a potential problem with this approach. Indeed, although the tetrad formalism allows us to recover local Lorentz invariance, the translational symmetry of the Poincare group is absent. In other words, in an arbitrary curved spacetime $\mathcal{M}$, the Lorentz group, not the Poincare group, is the structure group that acts on orthonormal frames in the tangent spaces of $\mathcal{M}$ [52]. As such, quantum states that are representations of the Poincare group become ill-defined (recall that the Poincare group is the Lorentz group extended by trans-lations). Nonetheless, it is possible to justify the use of Poincare representations in this setting [14].

Furthermore, it is worth mentioning that it has been shown that relating space-time torsion to intrinsic spin, Einstein–Cartan theories restore the relevance of the Poincare group to describe relativistic quantum states [53]. That is, imposing the relevance of the Poincare group in local inertial frames represented by a tetrad field will naturally lead to spacetime torsion as described in Einstein–Cartan theories.

5.7 Covariant derivative for fields of arbitrary spin

So far, we have offered *ad hoc* constructions for the covariant derivative of scalar, vector and spinor fields. It is possible, however, to unify the construction of covariant derivatives for fields of arbitrary spin by using the tetrad formalism along with the group theoretical arguments discussed in chapter 3.

As before, we can use a tetrad field to generalize the ideas of quantum field theory in curved spacetime by taking the projection of the field components in the general coordinates frame into a local inertial frame. There are two invariance conditions that need to be met for such a construction.

First, the action S of the quantum field given by:

$$S = \int \mathcal{L}\sqrt{-g}\, \mathrm{d}^4 x \qquad (5.172)$$

must be general covariant (where $\mathcal{L}$ is the field Lagrangian). That is, *the action S must be a coordinate scalar* [10]. This is an important requirement, as it can be shown that the energy–momentum tensor $T^{\mu\nu}$ is conserved if, and only if, the action is a scalar:

$$S \to \tilde{S} = S \Leftrightarrow \mathcal{D}_\mu T^\mu_{\ \nu} = 0 \qquad (5.173)$$

Consequently, this is equivalent to demanding that, except for the tetrad field, all other fields need to be treated as coordinate scalars (so they can produce a coor-dinate scalar action). Indeed, let us recall that the tetrad field is used to replace the four-vector A^μ for the set of four coordinate scalars A^a, $a = 0, 1, 2, 3$.

Second, the principle of equivalence needs to remain valid in all the local inertial frames. That is, the laws of physics are invariant with respect to the choice of the locally inertial frame defined at each point in spacetime by the tetrad field. In other words, the components of the fields in the locally inertial frame have to be Lorentz scalars, Lorentz tensors or Lorentz spinors. Furthermore, the field equations have to be invariant under these local Lorentz transformations. That is, *the action S must be a Lorentz scalar* [10].

Let us consider the case of an arbitrary vector field A^μ. Using a tetrad field we determine the four relevant coordinate scalar components to be:

$$A^a = e^a_{\ \mu}(x)\, A^\mu \tag{5.174}$$

where A^a forms a Lorentz vector and a set of four coordinate scalars. Now, if the vector field Lagrangian is made of a single quadratic term in the field variables:

$$\begin{aligned}
\mathcal{L} &= A^\mu A_\mu \\
&= e^{\ \mu}_b(x) A^b e^a_{\ \mu}(x) A_a \\
&= A^b A_a \delta^a_b \\
&= A^a A_a \tag{5.175}
\end{aligned}$$

which is completely independent of the tetrad field. Thus, as desired, the action S will be a coordinate scalar and a Lorentz scalar. In general, a Lorentz invariant Lagrangian that only depends on the fields, but not on their derivatives or the tetrad field, will automatically produce an action that is a coordinate scalar and a Lorentz scalar. However, this Lagrangian does not introduce gravitational effects. Indeed, equation (5.175) does not introduce any information about the gravitational field.

In field theory, gravitational effects are introduced by the *derivatives* of the fields. Thus, in general, *covariant derivatives* are introduced in the theory to guarantee that the action remains a coordinate scalar and a Lorentz scalar, despite the presence of field derivative terms. Consequently, covariant derivatives need to involve the tetrad field so that they can procure the right transformation rule for the action S.

For example, if the Lagrangian has a vector field derivative term such as:

$$\begin{aligned}
\mathcal{L} &= \partial_\mu A^\mu \\
&= e^a_{\ \mu}(x)\partial_a\!\left(e^{\ \mu}_b(x) A^b\right) \\
&= e^a_{\ \mu}(x) e^{\ \mu}_b(x)\partial_a A^b + e^a_{\ \mu}(x) A^b\,\partial_a e^{\ \mu}_b(x) \\
&= \delta^a_b \partial_a A^b + e^a_{\ \mu}(x) A^b \partial_a e^{\ \mu}_b(x) \\
&= \partial_a A^a + e^a_{\ \mu}(x) A^b\,\partial_a e^{\ \mu}_b(x) \tag{5.176}
\end{aligned}$$

then the action S will not transform as a coordinate scalar or as a Lorentz scalar. Indeed, let us consider the standard derivative in the general coordinate system, which clearly transforms as a covariant vector under a general coordinate transformation:

$$\partial_\mu \to \frac{\partial x^\nu}{\partial \tilde{x}^\mu}\partial_\nu \tag{5.177}$$

Such a differential operator can only be included in the Lagrangian if it transforms as a coordinate scalar. To this end, the differential operator in the Lagrangian should be of the form:

$$e_a^{\ \mu}(x)\,\partial_\mu \qquad\qquad (5.178)$$

which clearly transforms as a coordinate scalar. This way, the standard derivative of a Lorentz scalar field ϕ:

$$e_a^{\ \mu}(x)\,\partial_\mu\phi \qquad\qquad (5.179)$$

transforms as a coordinate scalar and as a Lorentz vector.

However, the standard derivative of an arbitrary field Ψ does not transform as a Lorentz scalar, or as a Lorentz vector, but it has the following transformation rule under a local Lorentz transformation $\Lambda(x)$:

$$e_a^{\ \mu}(x)\partial_\mu\Psi \rightarrow \Lambda_a^{\ b}e_b^{\ \mu}(x)\,\partial_\mu(D(\Lambda)\Psi)$$
$$= \Lambda_a^{\ b}e_b^{\ \mu}(x)\,D(\Lambda)\partial_\mu\Psi + \Lambda_a^{\ b}e_b^{\ \mu}(x)\,\left(\partial_\mu D(\Lambda)\right)\Psi$$
$$= \Lambda_a^{\ b}e_b^{\ \mu}(x)\,\left\{D(\Lambda)\partial_\mu\Psi + \left(\partial_\mu D(\Lambda)\right)\Psi\right\} \qquad (5.180)$$

where $D(\Lambda)$ is a matrix representation of the (infinitesimal) Lorentz transformation $\Lambda_a^{\ b}$, and the field transforms as:

$$\Psi \rightarrow \tilde{\Psi} = D(\Lambda)\Psi \qquad\qquad (5.181)$$

under a given representation of the Lorentz group.

Therefore, we need to define a *covariant derivative* $\mathcal{D}_a$ in such a way that the covariant derivative of a general field transforms as a Lorentz vector and as a coordinate scalar. That is, we require:

$$\mathcal{D}_a\Psi \rightarrow \Lambda_a^{\ b}D(\Lambda)\mathcal{D}_b\Psi \qquad\qquad (5.182)$$

which explicitly transforms as a Lorentz vector, and by virtue of:

$$\mathcal{D}_\mu = e^a_{\ \mu}(x)\mathcal{D}_a \qquad\qquad (5.183)$$

then $\mathcal{D}_a$ transforms as a coordinate scalar.

At this point, one could get the wrong impression that the action and the field equations will depend on the specific choice of tetrad field. After all, even though different tetrad fields may lead to exactly the same metric tensor, the tetrad field explicitly appears in the equations. Needless to say, because the metric tensor is the quantity with physical meaning, it is crucial for the action and field equations to be independent of the specific choice of the tetrad field. It can be shown that such is the case if the action is a Lorentz scalar [10].

Indeed, let us assume two different tetrad fields $e_a^{\ \mu}(x)$ and $\tilde{e}_a^{\ \mu}(x)$ that lead to the same metric tensor $g_{\mu\nu}$. Then, as we showed in chapter 2, both tetrad fields have to be related by a local Lorentz transformation:

$$\tilde{e}_a^{\ \mu}(x) = \Lambda_a^{\ b}(x)e_b^{\ \mu}(x) \qquad\qquad (5.184)$$

In other words, the metric tensor is invariant under such local Lorentz transformations. Thus, if the action S is a Lorentz scalar, then it is invariant under local Lorentz transformations. Therefore, the action and related field equations will be independent of the specific choice of tetrad field.

To summarize, a suitable action has to be constructed with a Lagrangian that depends on the field Ψ and its covariant derivatives $\mathcal{D}_a\Psi$ in such a way that they form Lorentz invariant terms.

5.7.1 General form

Let us consider an arbitrary field Ψ that transforms as a scalar under general coordinate transformations and as:

$$\Psi \to \tilde{\Psi} = D(\Lambda)\Psi \tag{5.185}$$

under a local Lorentz transformation $\Lambda_b{}^a(x)$ (for notational convenience, this Lorentz transformation will be denoted as Λ for as long the indices and explicit dependence on spacetime coordinates are not required). In other words, the field Ψ transforms under some representation of the Lorentz group. The matrix $D(\Lambda)$ denotes an adequate representation of the Lorentz group and satisfies the group multiplication rule:

$$D(\Lambda_1)D(\Lambda_2) = D(\Lambda_1\Lambda_2) \tag{5.186}$$

Because Λ is a local Lorentz transformation, the standard derivative will not transform as a Lorentz vector. To have a differential operator that transforms as a coordinate scalar and as a Lorentz vector, we define the covariant derivative as:

$$\mathcal{D}_a \equiv e_a^{\mu}(x)\big(\partial_\mu - \Gamma_\mu\big) \tag{5.187}$$

where the matrix Γ_μ is known as the *connection*. For the covariant derivative to be a Lorentz vector and a coordinate scalar, the connection needs to transform as:

$$\Gamma_\mu \to \tilde{\Gamma}_\mu = D(\Lambda)\Gamma_\mu D^{-1}(\Lambda) - \big(\partial_\mu D(\Lambda)\big)D^{-1}(\Lambda) \tag{5.188}$$

under a local Lorentz transformation [10]. Notice that the connection does not transform as a covariant tensor.

To determine the structure of the connection, let us consider the infinitesimal local Lorentz transformation given by:

$$\Lambda^a{}_b(x) = \delta^a_b + \xi^a{}_b \tag{5.189}$$

where:

$$\xi_{ab} = -\xi_{ba}$$
$$\left|\xi^a{}_b\right|^2 \ll 1 \tag{5.190}$$

Then, the matrix representation of this local Lorentz transformation looks like:

$$D(\Lambda) = D(1 + \xi) = 1 + i\xi^{ab}\Sigma_{ab} \tag{5.191}$$

where the generators of the Lorentz group Σ_{ab} are antisymmetric in a and b:

$$\Sigma_{ab} = -\Sigma_{ba} \tag{5.192}$$

and satisfy:

$$[\Sigma_{ab}, \Sigma_{cd}] = \eta_{cb}\Sigma_{ad} - \eta_{ac}\Sigma_{bd} + \eta_{db}\Sigma_{ca} - \eta_{da}\Sigma_{cb} \tag{5.193}$$

which defines the Lie algebra of the Lorentz group. Then, under this local Lorentz transformation, it can be shown that the connection transforms as:

$$\Gamma_\mu \rightarrow \tilde{\Gamma}_\mu = \Gamma_\mu + i\xi^{ab}[\Sigma_{ab}, \Gamma_\mu] - i\Sigma_{ab}\partial_\mu\xi^{ab} \tag{5.194}$$

while the tetrad field transforms as:

$$e^a_{\ \mu}(x) \rightarrow \tilde{e}^a_{\ \mu}(x) = e^a_{\ \mu}(x) + \xi^a_{\ b}(x)e^b_{\ \mu}(x) \tag{5.195}$$

From these equations we can find that the connection Γ_μ that satisfies these transformation rules is determined by:

$$\Gamma_\mu(x) = -i\,\Sigma^{ab}e^{\ \nu}_a(x)\left(\partial_\mu e_{b\nu}(x) - \Gamma^\alpha_{\mu\nu}e_{b\alpha}(x)\right) \tag{5.196}$$

in the case of a torsion-free metric:

$$\Gamma^\alpha_{\mu\nu} = \Gamma^\alpha_{\nu\mu} \tag{5.197}$$

Furthermore, because the connection takes its value in the Lie algebra of the Lorentz group, we can write it as:

$$\Gamma_\mu(x) = -i\omega_\mu^{\ ab}(x)\Sigma_{ab} \tag{5.198}$$

where $\omega_\mu^{\ ab}$ is the connection one-form defined through the expression:

$$\omega_{\mu ab} \equiv e^{\ \nu}_a(x)\left(\partial_\mu e_{b\nu}(x) - \Gamma^\alpha_{\mu\nu}e_{b\alpha}(x)\right) \tag{5.199}$$

Therefore, the covariant derivative of the coordinate scalar field component Ψ is given by:

$$\mathcal{D}_\mu\Psi = \partial_\mu\Psi + i\omega_\mu^{\ ab}(x)\Sigma_{ab}\Psi \tag{5.200}$$

and it can be shown that the commutation of two covariant derivatives obeys:

$$[\mathcal{D}_\mu, \mathcal{D}_\nu]\Psi = -iR_{\mu\nu}^{\ \ ab}\Sigma_{ab}\Psi \tag{5.201}$$

where:

$$R_{\mu\nu}^{\ \ ab} = e^a_{\ \alpha}(x)e^{b\beta}(x)R^\alpha_{\ \beta\mu\nu} \tag{5.202}$$

and $R^\alpha_{\ \beta\mu\nu}$ is the Riemann tensor [6]. Furthermore, the Riemann tensor can be defined solely in terms of the connection one-forms and its derivatives as follows:

$$R_{\mu\nu}^{\ \ ab} = \partial_\nu\omega_\mu^{\ ab} - \partial_\mu\omega_\nu^{\ ab} + \omega_\nu^{\ a}_{\ c}\omega_\mu^{\ cb} - \omega_\mu^{\ a}_{\ c}\omega_\nu^{\ cb} \tag{5.203}$$

Let us now consider the explicit form of the covariant derivative for the four specific cases of interest: scalar fields, vector fields, tetrad fields and Dirac fields.

5.7.2 Scalar fields

Scalar fields transform as:

$$\phi \to \tilde{\phi}(\tilde{x}) = \phi(\tilde{x}(x)) = \phi(x) \tag{5.204}$$

under a local Lorentz transformation. In section 3.9 we determined that, in the case of scalar fields, the matrix representation for the Lie algebra of the Lorentz group is given by:

$$\Sigma_{ab} = 0 \tag{5.205}$$

Therefore, the covariant derivative of a scalar field is simply given by the standard derivative:

$$\mathcal{D}_a \phi = \partial_a \phi \tag{5.206}$$

as we already knew.

5.7.3 Vector fields

Under a local Lorentz transformation, a coordinate scalar and Lorentz vector field A^a transforms as:

$$A^a \to \tilde{A}^a = \Lambda^a{}_b(x) A^b \tag{5.207}$$

The appropriate matrix representation of the Lie algebra of the Lorentz group for the case of a vector field was determined in section 3.9 to be:

$$[\Sigma_{cd}]^a{}_b = -\frac{i}{2} \left(\delta^a_c \eta_{db} - \delta^a_d \eta_{cb} \right) \tag{5.208}$$

Therefore, the covariant derivative of the vector field A^a is given by:

$$\mathcal{D}_\mu A^a = \partial_\mu A^a + \omega_\mu{}^a{}_b A^b \tag{5.209}$$

which, after some algebra, may be transformed into the well-known standard expression:

$$\mathcal{D}_\mu A^\alpha = \partial_\mu A^\alpha + \Gamma^\alpha_{\mu\nu} A^\nu \tag{5.210}$$

presented in chapter 2, where we identify:

$$\Gamma^\alpha_{\mu\nu} = e^b{}_\nu(x) e_a{}^\alpha(x) \omega_\mu{}^a{}_b \tag{5.211}$$

for the expression of the affine connection in terms of the connection one-form.

5.7.4 Tetrad fields

Tetrad fields are coordinate vectors and Lorentz vectors. Thus, it is not a trivial task to establish their covariant derivatives. However, let us recall that the covariant derivative of a contravariant vector in the general coordinate system is given by:

$$\mathcal{D}_\mu A^\nu = \partial_\mu A^\nu + \Gamma^\nu_{\mu\beta} A^\beta \tag{5.212}$$

We can use now the tetrad fields to project the contravariant vector to the local inertial frame. If we demand that the covariant derivative follows the Leibniz rule, we have:

$$
\begin{aligned}
\mathcal{D}_\mu A^\nu &= \mathcal{D}_\mu \left(e_a^{\ \nu}(x) A^a \right) \\
&= A^a \mathcal{D}_\mu e_a^{\ \nu}(x) + e_a^{\ \nu}(x) \mathcal{D}_\mu A^a \\
&= A^a \mathcal{D}_\mu e_a^{\ \nu}(x) + e_a^{\ \nu}(x) \left(\partial_\mu A^a + \omega_{\mu\ b}^{\ a} A^b \right) \\
&= A^a \mathcal{D}_\mu e_a^{\ \nu}(x) + \partial_\mu \left(e_a^{\ \nu}(x) A^a \right) - A^a \partial_\mu e_a^{\ \nu}(x) + e_a^{\ \nu}(x) \omega_{\mu\ b}^{\ a} A^b \\
&= A^a \mathcal{D}_\mu e_a^{\ \nu}(x) + \partial_\mu A^\nu - A^a \partial_\mu e_a^{\ \nu}(x) + e_a^{\ \nu}(x) \omega_{\mu\ b}^{\ a} A^b \\
&= A^a \mathcal{D}_\mu e_a^{\ \nu}(x) + \mathcal{D}_\mu A^\nu - \Gamma^\nu_{\mu\beta} A^\beta - A^a \partial_\mu e_a^{\ \nu}(x) + e_a^{\ \nu}(x) \omega_{\mu\ b}^{\ a} A^b
\end{aligned}
\tag{5.213}
$$

and consequently:

$$
\begin{aligned}
0 &= A^a \mathcal{D}_\mu e_a^{\ \nu}(x) - \Gamma^\nu_{\mu\beta} A^\beta - A^a \partial_\mu e_a^{\ \nu}(x) + e_a^{\ \nu}(x) \omega_{\mu\ b}^{\ a} A^b \\
&= A^a \left(\mathcal{D}_\mu e_a^{\ \nu}(x) - \Gamma^\nu_{\mu\beta} e_a^{\ \beta}(x) - \partial_\mu e_a^{\ \nu}(x) + e_b^{\ \nu}(x) \omega_{\mu\ a}^{\ b} \right)
\end{aligned}
\tag{5.214}
$$

which leads to the expression for the covariant derivative of the tetrad field:

$$\mathcal{D}_\mu e_a^{\ \nu}(x) = \partial_\mu e_a^{\ \nu}(x) + \Gamma^\nu_{\mu\beta} e_a^{\ \beta}(x) - e_b^{\ \nu}(x) \omega_{\mu\ a}^{\ b} \tag{5.215}$$

Inspection of this equation reveals that we could have guessed the right expression for the covariant derivative of the tetrad field. Indeed, the first term is the standard derivative which always shows up. The second term involves the affine connection $\Gamma^\nu_{\mu\beta}$ because the tetrad field is a coordinate vector. Finally, the third term involves the connection one-form $\omega_{\mu\ a}^{\ b}$ because the tetrad field is a Lorentz vector.

Furthermore, let us recall from equation (5.199) that for a torsion-free metric such that:

$$\Gamma^\nu_{\mu\beta} = \Gamma^\nu_{\beta\mu} \tag{5.216}$$

the connection one-form is given by:

$$\omega_{\mu ab} = e_a^{\ \nu}(x) \left(\partial_\mu e_{b\nu}(x) - \Gamma^\alpha_{\mu\nu} e_{b\alpha}(x) \right) \tag{5.217}$$

and consequently the covariant derivative of the tetrad field reduces to:

$$\mathcal{D}_\mu e_a^{\ \mu}(x) = 0 \tag{5.218}$$

That is, the covariant derivative is an operator that commutes with the conversion of tensors from the general coordinate system to the local inertial frame. For example:

$$\mathcal{D}_\mu A^\nu = \mathcal{D}_\mu \big(e_a^\nu(x) A^a\big) = e_a^\nu(x) \mathcal{D}_\mu A^a \tag{5.219}$$

For notational convenience, we can define:

$$\nabla_\mu e_{b\nu}(x) \equiv \partial_\mu e_{b\nu}(x) - \Gamma_{\mu\nu}^\alpha e_{b\alpha}(x) \tag{5.220}$$

as the *coordinate covariant derivative* of the tetrad field $e_{b\nu}(x)$.

5.7.5 Dirac fields

The Dirac field ψ is a coordinate scalar and a Lorentz spinor that transforms as:

$$\psi \to \tilde{\psi} = U(\Lambda)\psi \tag{5.221}$$

under a local Lorentz transformation Λ and where $U(\Lambda)$ is a spinor representation of the Lorentz group. Let us recall that, because the group of general coordinate transformations from general relativity does not have a spinor representation, it does not make much sense to talk about spinors outside the context of a local inertial frame. As discussed in section 3.9, the appropriate matrix representation of the Lie algebra of the Lorentz group is given by:

$$\Sigma_{ab} = -\frac{i}{8}[\gamma_a, \gamma_b] \tag{5.222}$$

where γ_i are an arbitrary representation of the Dirac matrices defined in Minkowski space.

Consequently, the covariant derivative of the Dirac field ψ is given by:

$$\mathcal{D}_\mu \psi = \partial_\mu \psi + \frac{1}{8}\omega_{\mu ab}[\gamma^a, \gamma^b]\psi \tag{5.223}$$

in agreement with equation (5.163). Then, the commutation of two covariant derivates acting on a Dirac field is given by:

$$[\mathcal{D}_\mu, \mathcal{D}_\nu]\psi = -\frac{1}{8}R_{\mu\nu}{}^{ab}[\gamma_a, \gamma_b]\psi \tag{5.224}$$

where $R_{\mu\nu\alpha\beta}$ is the curvature tensor [6].

Let us now consider the transformation properties of the Dirac matrices. Let us recall that in curved spacetime the Dirac equation takes the form:

$$\big(i\gamma^\mu(x)\mathcal{D}_\mu - m\big)\psi = 0 \tag{5.225}$$

where:

$$\gamma^\mu(x) = e_a^\mu(x)\gamma^a \tag{5.226}$$

Under a combination of local Lorentz and general coordinate transformations such as:

$$\psi \to \tilde{\psi} = D(\Lambda)\psi$$

$$e_a^{\ \mu}(x) \to \tilde{e}_a^{\ \mu}(\tilde{x}) = \frac{\partial \tilde{x}^\mu}{\partial x^\nu} \Lambda_a^{\ b}(x) e_b^{\ \nu}(x) \tag{5.227}$$

the covariance of the Dirac equation requires the gamma matrices to satisfy the condition:

$$D(\Lambda)\gamma^a D^{-1}(\Lambda) \Lambda^b_{\ a} = \gamma^b \tag{5.228}$$

therefore, the Dirac matrices defined in the general coordinate system transform as:

$$\gamma^\mu(x) \to \tilde{\gamma}^\mu(\tilde{x}) = \tilde{e}_a^{\ \mu}(\tilde{x})\gamma^a$$

$$= \frac{\partial \tilde{x}^\mu}{\partial x^\nu} D(\Lambda)\gamma^\nu(x)D^{-1}(\Lambda) \tag{5.229}$$

under a general coordinate transformation [6]. If we require that the Dirac matrices and their covariant derivatives transform in the same way:

$$\mathcal{D}_\nu\gamma^\mu(x) \to \mathcal{D}_\nu\tilde{\gamma}^\mu(\tilde{x}) = \frac{\partial \tilde{x}^\mu}{\partial x^\alpha}\frac{\partial x^\beta}{\partial \tilde{x}^\nu} D(\Lambda)\mathcal{D}_\beta\gamma^\alpha(x)D^{-1}(\Lambda) \tag{5.230}$$

then, the covariant derivative of the Dirac matrices has to be given by:

$$\mathcal{D}_\mu\gamma^\nu(x) = \partial_\mu\gamma_\nu(x) + \Gamma^\nu_{\mu\alpha}\gamma^\alpha(x) - \left[\Gamma_\mu, \gamma^\nu(x)\right] \tag{5.231}$$

which differs from the covariant derivative of a vector field by the presence of the commutator term [6]. This makes sense, the first term corresponds to the standard derivative, the second term is due to the vector nature of the Dirac matrices and, finally, the third term is due to the spinor structure of the gamma matrices.

Furthermore, using the connection Γ_μ found for the Dirac field:

$$\Gamma_\mu = -\frac{1}{8}\omega_\mu^{\ ab}[\gamma_a, \gamma_b] \tag{5.232}$$

in a torsion-free metric such that:

$$\mathcal{D}_\mu e_a^{\ \mu} = 0 \tag{5.233}$$

leads to:

$$\mathcal{D}_\mu\gamma^\nu(x) = 0 \tag{5.234}$$

and:

$$\mathcal{D}_\mu\gamma^a = 0 \tag{5.235}$$

Consequently, after some algebra, we obtain the following expression:

$$\left(\gamma^\mu \mathcal{D}_\mu\right)^2 \psi = \left(\mathcal{D}^\mu \mathcal{D}_\mu + \frac{1}{4}R\right)\psi \tag{5.236}$$

where R is the curvature scalar [6].

Finally, it is important to mention that the commutation relations for the covariant derivative of spinor fields are given by:

$$[\mathcal{D}_\mu, \mathcal{D}_\nu]\psi = \frac{i}{4} R_{\mu\nu\alpha\beta}\, \sigma^{\alpha\beta}\psi$$

$$[\mathcal{D}_\mu, \mathcal{D}_\nu]\overline{\psi} = -\frac{i}{4} R_{\mu\nu\alpha\beta}\, \overline{\psi}\sigma^{\alpha\beta} \tag{5.237}$$

where:

$$\sigma^{\alpha\beta} = \frac{i}{2}\left[\gamma^\alpha, \gamma^\beta\right] = e_a^{\;\alpha}(x)e_b^{\;\beta}(x)\left[\gamma^a, \gamma^b\right] \tag{5.238}$$

is proportional to the commutator of the Dirac matrices.

5.8 Spinor dynamics with tetrad fields

As we mentioned before, the best way to study the dynamics of spin-$\frac{1}{2}$ particles in gravitational fields is with local inertial frames defined at each point of spacetime [51]. These local inertial frames are defined through a tetrad field $e^a_{\;\mu}(x)$, and the momentum in the local inertial frame is related to the momentum in the general coordinate system by:

$$p^a(x) = p^\mu(x)e^a_{\;\mu}(x) \tag{5.239}$$

Then, any change in the momentum in the local inertial frame can be expressed as a combination of changes due to (1) external forces other than gravity $\delta p^\mu(x)$ and (2) spacetime geometry effects (gravity) $\delta e^a_{\;\mu}(x)$:

$$\delta p^a(x) = \delta p^\mu(x)e^a_{\;\mu}(x) + p^\mu(x)\delta e^a_{\;\mu}(x) \tag{5.240}$$

where the variation of the four-momentum is given by:

$$\delta p^\mu(x) = ma^\mu(x)\,d\tau \tag{5.241}$$

where $a^\mu(x)$ is the four-acceleration due to external forces other than gravity. The variation of the tetrad field is given by:

$$\begin{aligned}
\delta e^a_{\;\mu}(x) &= \left(\nabla_\nu e^a_{\;\mu}(x)\right)u^\nu\,d\tau \\
&= -u^\nu e^b_{\;\mu} e^a_{\;\alpha} \nabla_\nu e_b^{\;\alpha} d\tau \\
&= -u^\nu(x)\omega^a_{\;\nu\,b}(x)e^b_{\;\mu}(x)\,d\tau
\end{aligned} \tag{5.242}$$

where $u^\mu(x)$ is the four-velocity of the particle and the connection one-forms is defined as:

$$\omega^a{}_b(x) = e^a{}_\nu(x)\nabla_\mu e_b{}^\nu(x) \qquad (5.243)$$

and ∇_μ is the coordinate covariant derivative operating on a tetrad field as defined in the previous section.

Notice that, to derive the variation of the tetrad field, we used the fact that:

$$0 = \nabla_\mu \delta_b^a = \nabla_\mu\left(e^a{}_\alpha e_b{}^\alpha\right) = e^a{}_\alpha \nabla_\mu e_b{}^\alpha + e_b{}^\alpha \nabla_\mu e^a{}_\alpha \qquad (5.244)$$

which implies that:

$$e_b{}^\alpha \nabla_\mu e^a{}_\alpha = -e^a{}_\alpha \nabla_\mu e_b{}^\alpha \qquad (5.245)$$

and consequently:

$$\nabla_\mu e^a{}_\beta = -e^b{}_\beta e^a{}_\alpha \nabla_\mu e_b{}^\alpha \qquad (5.246)$$

It can be shown that, as the particle moves in spacetime, its momentum in the local inertial frame will transform under an infinitesimal local Lorentz transformation:

$$p^a(x) = \Lambda^a{}_b(x)p^b(x) \qquad (5.247)$$

where:

$$\Lambda^a{}_b(x) = \delta^a{}_b + \lambda^a{}_b(x)\,d\tau \qquad (5.248)$$

is the infinitesimal Lorentz transformation as the particle moves across an infinitesimal proper time $d\tau$ [51]. After considerable algebra, it can be shown that the Lorentz transformation in the local inertial frame is given by:

$$\lambda^a{}_b(x) = -\frac{1}{m}\left(a^a(x)p_b(x) - p^a(x)a_b(x)\right) + \chi^a{}_b(x) \qquad (5.249)$$

with:

$$\chi^a{}_b(x) = -u^\nu(x)\omega_\nu{}^a{}_b(x) \qquad (5.250)$$

Details of how to produce these expressions can be found in the literature [51, 54]. In the absence of external forces, $a^\mu(x) = 0$ and the Lorentz transformation in the local inertial frame is simply given by:

$$\lambda^a{}_b(x) = -u^\nu(x)\omega_\nu{}^a{}_b(x) \qquad (5.251)$$

Such is the case for free-falling particles moving in orbital paths around a gravitational source.

It can also be shown that the associated infinitesimal Wigner rotation that affects the particle's spin is given by:

$$W^a{}_b(x) = \delta^a{}_b + \vartheta^a{}_b(x)\,d\tau \qquad (5.252)$$

where:

$$\vartheta^a_{\ b}(x) = \lambda^a_{\ b}(x) + \frac{\lambda^a_{\ 0}(x)u_b(x) - \lambda_{b0}(x)u^a(x)}{u^0(x) + 1} \tag{5.253}$$

and its two-spinor representation is:

$$D^{(1/2)}_{\sigma'\sigma}(W(x)) = I + \frac{i}{2}\left(\vartheta_{23}(x)\sigma_x + \vartheta_{31}(x)\sigma_y + \vartheta_{12}(x)\sigma_z\right) d\tau \tag{5.254}$$

where $\sigma_{x,y,z}$ are the Pauli matrices [51].

Therefore, the Wigner rotation for a particle that moves over a finite proper time interval is:

$$W^a_{\ b}(x_f; x_i) = \mathcal{T}\exp\left(\int_{\tau_i}^{\tau_f}\vartheta^a_{\ b}(x(\tau))\,d\tau\right) \tag{5.255}$$

where $\mathcal{T}$ indicates *a time-ordered product* and the integral is taken along the path of the particle. In chapters 6–8 we will discuss in detail the Wigner rotation for a spin-$\frac{1}{2}$ particle in the Schwarzschild and Kerr metrics.

5.9 Dirac spinors in curved spacetime

Let us now use a tetrad field to obtain solutions to the Dirac equation in curved spacetime that explicitly describe the spin–curvature coupling [55–58]. To this end, we will use the WKB approximation described in section 5.2.

For the case at hand, the WKB expansion assumes that a four-spinor $\psi(x)$ that satisfies the Dirac equation is given by a power series in $\hbar$ and a local phase factor $\mathcal{S}(x)$:

$$\psi(x) = e^{i\mathcal{S}(x)/\hbar}\sum_{n=0}^{\infty}(-i\hbar)^n\psi_n(x) \tag{5.256}$$

where all the ψ_i are four-spinors. Inserting this expression in the Dirac equation:

$$\left(i\gamma^\mu(x)\mathcal{D}_\mu - m\right)\psi = 0 \tag{5.257}$$

we obtain equations for each power of $\hbar$. In particular, at zero and first order we get:

$$\begin{aligned}
\hbar^0 &: (\gamma^\alpha\partial_\alpha\mathcal{S}(x) + m)\psi_0(x) = 0 \\
\hbar^1 &: (\gamma^\alpha\partial_\alpha\mathcal{S}(x) + m)\psi_1(x) = -\gamma^\alpha\mathcal{D}_\alpha\psi_0(x)
\end{aligned} \tag{5.258}$$

It is important to note that the above WKB approximation assumes that the phase $\mathcal{S}(x)$ is the slowest varying component of the field, whereas the spinorial components ψ_n are the fastest varying component of the field. As such, we do not need to perform a WKB expansion on the phase $\mathcal{S}(x)$. However, the wavelength of the spinorial oscillations is required to be much smaller than the spacetime curvature scale.

5.9.1 Solution at order $\hbar^0$

We observe that the $\hbar^0$-order equation is *not* a differential equation but a homogeneous system of four algebraic equations (one for each component of the spinor field ψ_0). The existence of a non-trivial solution for this system of algebraic equations is guaranteed by the condition:

$$\det(\gamma^\alpha \partial_\alpha S(x) + m) = 0 \qquad (5.259)$$

After some algebra, this condition leads to the Hamilton–Jacobi equation for a relativistic spinless particle described in chapter 2:

$$\partial_\alpha S(x) \partial^\alpha S(x) = -m^2 \qquad (5.260)$$

Thus, it is convenient to recast the four-momentum and four-velocity of the Dirac particle as:

$$p_\alpha(x) \equiv \partial_\alpha S(x)$$
$$u_\alpha(x) \equiv \frac{1}{m} \partial_\alpha S(x) \qquad (5.261)$$

which are properly normalized as:

$$p^\alpha(x) p_\alpha(x) = -m^2$$
$$u^\alpha(x) u_\alpha(x) = -1 \qquad (5.262)$$

and related by:

$$p^\alpha(x) = m u^\alpha(x) \qquad (5.263)$$

Therefore, at zero order in $\hbar$ we identify $S(x)$ with the classical action for a spinless particle in a gravitational field:

$$S(x) = \int p_\alpha dx^\alpha \qquad (5.264)$$

which also corresponds to the phase of a quantum mechanical particle in curved spacetime [59, 60]. Let us recall that these equations define a family of *integral curves* $x^\mu(\tau)$ of u^μ, the four-velocity vector field. In addition, it can be shown that the integral curves of $u^\mu(x)$ are *geodesic*:

$$\frac{Du_\alpha}{D\tau} = u^\beta D_\beta u_\alpha = 0 \qquad (5.265)$$

and *rotation-free*:

$$\omega_{\alpha\beta} = \frac{1}{2} \left(D_\alpha u_\beta - D_\beta u_\alpha \right) = 0 \qquad (5.266)$$

That is, the integral curves of u^μ form a geodesic congruence and they are orthogonal to the system of hyper-surfaces $S = $ constant [55].

Therefore, the computation of the quantum mechanical phase $S(x)$ at this approximation order can be accomplished by assuming that the Dirac particle is a spinless classical particle moving with momentum p^μ in curved spacetime. The associated four-velocity field u^μ can be found, for example, by solving the geodesic equations for a classical test particle in the specific metric under consideration. However, as the extended state of the quantum particle takes on a path super-position, it should be understood that u^α actually represents a geodesic congruence (because in a sense, the initial conditions are not given by the equations).

Now we need to solve the order $\hbar^0$ algebraic equation for the four-spinor ψ_0 associated with the geodesic congruence $u^\mu(x)$. Once more, let us remark that the solution to the Dirac equation is given *with respect to a specific tetrad field* $e^a_{\ \mu}(x)$ that relates the four-momentum defined in the general coordinate system p^μ with the four-momentum observed by a locally inertial observer k^a:

$$k^a = e^a_{\ \mu}(x)p^\mu \qquad (5.267)$$

where, for clarity of notation in what follows, we have decided to rename the momentum p^a observed in the local inertial frame as k^a. Then, the $\hbar^0$-order equation can be written as:

$$
\begin{aligned}
(\gamma^\alpha \partial_\alpha S(x) + m)\psi_0(x) &= (\gamma^\alpha p_\alpha + m)\psi_0(x) \\
&= \left[\left(e_a^{\ \alpha}(x)\gamma^a\right)\left(e^b_{\ \alpha}(x)k_b\right) + m \right]\psi_0(x) \\
&= \left(e_a^{\ \alpha}(x)e^b_{\ \alpha}(x)\gamma^a k_b + m \right)\psi_0(x) \\
&= \left(\delta^b_a \gamma^a k_b + m \right)\psi_0(x) \\
&= (\gamma^a k_a + m)\psi_0(x) \\
&= 0 \qquad (5.268)
\end{aligned}
$$

Notice that this is the free particle Dirac equation in flat spacetime for a particle with four-momentum k^a (cf. equation (3.181)). Therefore, the two positive energy solutions to the Dirac equation expressed in terms of the four-momentum observed at the locally inertial frame are given by the equations derived in section 3.6:

$$\psi_0^{(1)}(x) = \sqrt{\frac{k^0 + m}{2m}} \begin{pmatrix} 1 \\ 0 \\ \dfrac{k^3}{k^0 + m} \\ \dfrac{k^1 + ik^2}{k^0 + m} \end{pmatrix} \qquad (5.269)$$

and:

$$\psi_0^{(2)}(x) = \sqrt{\frac{k^0 + m}{2m}} \begin{pmatrix} 0 \\ 1 \\ \dfrac{k^1 - ik^2}{k^0 + m} \\ \dfrac{-k^3}{k^0 + m} \end{pmatrix} \tag{5.270}$$

normalized as:

$$\overline{\psi}_0^{(i)}\psi_0^{(j)} = \delta_j^i \tag{5.271}$$

And expressed in terms of the four-momentum measured in the general coordinate system:

$$\tilde{\psi}_0^{(1)}(x) = \sqrt{\frac{e^0_{\ \mu}(x)p^\mu + m}{2m}} \begin{pmatrix} 1 \\ 0 \\ \dfrac{e^3_{\ \mu}(x)p^\mu}{e^0_{\ \mu}(x)p^\mu + m} \\ \dfrac{e^1_{\ \mu}(x)p^\mu + ie^2_{\ \mu}(x)p^\mu}{e^0_{\ \mu}(x)p^\mu + m} \end{pmatrix} \tag{5.272}$$

and:

$$\tilde{\psi}_0^{(2)}(x) = \sqrt{\frac{e^0_{\ \mu}(x)p^\mu + m}{2m}} \begin{pmatrix} 0 \\ 1 \\ \dfrac{e^1_{\ \mu}(x)p^\mu - ie^2_{\ \mu}(x)p^\mu}{e^0_{\ \mu}(x)p^\mu + m} \\ \dfrac{-e^3_{\ \mu}(x)p^\mu}{e^0_{\ \mu}(x)p^\mu + m} \end{pmatrix} \tag{5.273}$$

Consequently, at order $\hbar^0$ the general form of the spinor that satisfies the WKB approximation to the Dirac equation in curved spacetime takes the form:

$$\psi_0(x) = \beta_1(x)\psi_0^{(1)}(x) + \beta_2(x)\psi_0^{(2)}(x) \tag{5.274}$$

where $\beta_1(x)$ and $\beta_2(x)$ are complex scalars that may depend on the coordinates (recall that, at $\hbar^0$-order, $\psi_0(x)$ is described by an algebraic equation), and the spinor ψ_0 is normalized as:

$$\overline{\psi}_0 \psi_0 = \beta_1^* \beta_1 + \beta_2^* \beta_2 \tag{5.275}$$

Therefore, the general solution ψ_0 can be said to describe a spin-$\frac{1}{2}$ particle of momentum p^a as observed from a point x^a in the local inertial frame defined by the tetrad field. Furthermore, because of the optical analogy, the quantum particle moves with surface waves of constant S and rays parallel to the geodesic congruence $u^a(x)$.

To order $\hbar^0$, the algebraic equation (5.268) uniquely defines the dynamics of the particle. However, the condition for the existence of a non-trivial solution at order $\hbar^1$ restricts the solution at order $\hbar^0$. Indeed, the $\hbar^1$-order equation for $\psi_1(x)$ is an inhomogenous linear algebraic equation. Therefore, for a Hermitian system, all solutions to the homogeneous equation have to be orthogonal to the inhomogeneity. That is:

$$\overline{\psi}_0^{(i)}(x) \gamma^a D_a \psi_0(x) = 0 \tag{5.276}$$

for $i = 1, 2$.

It can be shown that this condition implies:

$$u^a D_a \psi_0(x) = -\frac{\theta}{2} \psi_0(x) \tag{5.277}$$

where we have defined the *expansion* as:

$$\theta \equiv D_a u^a \tag{5.278}$$

In a sense, equation (5.277) describes the propagation of the spinor ψ_0 along the geodesic congruence $u^a(x)$, with the understanding that spinors can only be defined with respect to a tetrad field. Similarly, the complex amplitudes propagate across the geodesic congruence as:

$$u^a \partial_a \beta_i(x) = -\frac{\theta}{2} \beta_i(x) \tag{5.279}$$

for $i = 1, 2$ [55].

Furthermore, notice that the $\hbar^0$-order solution that emerges from the WKB process reduces to the well-known solution of the Dirac equation in the flat spacetime limit:

$$\psi = e^{iS(x)/\hbar} \psi_0(x) + \mathcal{O}(\hbar)$$

$$\approx e^{\frac{i}{\hbar} \int p_\mu dx^\mu} \psi_0(x)$$

$$\approx e^{\frac{i}{\hbar} p_\mu x^\mu} \sum_{s=1}^{2} \beta_s u^{(s)}(\mathbf{p}) \tag{5.280}$$

Notice also that in the computation of the phase, we have used the fact that, in flat spacetime, u^μ is a constant. Consequently, p^μ can be pulled outside the integral, and the constancy of u^μ makes the expansion to be zero: $\theta = 0$. Furthermore, $\theta = 0$ implies that β_1 and β_2 are complex constants that no longer depend on the coordinates along the path (their derivative with respect to the proper time τ is zero).

At this point it is convenient to define a new spinor $\xi_0(x)$ proportional to $\psi_0(x)$:

$$\psi_0(x) = f(x)\xi_0(x) \tag{5.281}$$

and normalized as:

$$\bar{\xi}_0(x)\xi_0(x) = 1 \tag{5.282}$$

which implies that:

$$f^2 = \bar{\psi}_0\psi_0 = \beta_1^*\beta_1 + \beta_2^*\beta_2 \tag{5.283}$$

Then, the propagation equations are given by:

$$u^\alpha \partial_\alpha f(x) = -\frac{\theta}{2}f(x) \tag{5.284}$$

and:

$$u^\alpha \mathcal{D}_\alpha \xi_0(x) = 0 \tag{5.285}$$

which explicitly shows that the spinor ξ_0 is parallelly transported along the u^α geodesic congruence [55].

The function f can be understood as an 'envelope' that describes the shape of the spinor. If the congruence of lines has zero divergence, then:

$$\mathcal{D}_\mu u^\mu = 0 \tag{5.286}$$

and the expansion is zero $\theta = 0$. Then, the propagation equation for f reduces to:

$$u^\alpha \partial_\alpha f(x) = 0 \tag{5.287}$$

which is equivalent to:

$$\frac{df}{d\tau} = \frac{\partial f}{\partial x^\alpha}\frac{dx^\alpha}{d\tau} = 0 \tag{5.288}$$

That is, the shape of the envelope in the particle's rest frame remains constant. In reality, of course, Heisenberg's uncertainty principle impedes the divergence of the velocity to be exactly zero. However, if the expectation value of the divergence of the velocity is small enough:

$$\langle \mathcal{D}_\alpha u^\alpha \rangle \ll \frac{1}{\Gamma} \tag{5.289}$$

then we can assume that the envelope of the Dirac spinor is rigidly transported along the lines that make the u^μ congruence [14].

5.9.2 Geodesic deviation at order $\hbar^1$

As we discussed in section 2.8, spin and curvature are coupled in a non-trivial way. Consequently, spinning particles falling in a gravitational field do not follow geodesics. Although our previous discussion in section 2.8 referred to classical extended particles, the same is true for elementary particles with intrinsic spin [55]. In the case of a Dirac particle, the deviation from geodetic motion is given as a correction of order $\hbar$, as we will see in what follows.

Let us consider the $\hbar^1$-order correction to the spin dynamics using the *Gordon decomposition* of the probability current [38, 61]. Let us recall that the covariant current for the Dirac field is given by:

$$j^\alpha = \overline{\psi}\gamma^\alpha\psi \tag{5.290}$$

which satisfies the covariant continuity equation:

$$\mathcal{D}_\alpha j^\alpha = 0 \tag{5.291}$$

If we define:

$$\sigma^{\mu\nu} \equiv \frac{i}{2}[\gamma^\mu, \gamma^\nu] = \frac{i}{2}(\gamma^\mu\gamma^\nu - \gamma^\nu\gamma^\mu) \tag{5.292}$$

then we can write the covariant current as :

$$j^\alpha = j^\alpha_c + j^\alpha_M \tag{5.293}$$

where:

$$j^\alpha_c = \frac{\hbar}{2mi}\left(\mathcal{D}^\alpha\overline{\psi}\psi - \overline{\psi}\mathcal{D}^\alpha\psi\right) \tag{5.294}$$

is the *convection current* and:

$$j^\alpha_M = \frac{\hbar}{2m}\mathcal{D}_\beta\left(\overline{\psi}\sigma^{\alpha\beta}\psi\right) \tag{5.295}$$

is the *magnetization current*. Each of these currents satisfies a covariant continuity equation:

$$\mathcal{D}_\alpha j^\alpha_c = 0 \qquad \mathcal{D}_\alpha j^\alpha_M = 0 \tag{5.296}$$

Using the expression for ψ obtained from the WKB expansion, we get:

$$j^\alpha_c = \left(f^2 + \frac{\hbar}{i}\left(\overline{\psi}_0\psi_1 - \overline{\psi}_1\psi_0\right)\right)u^\alpha + \frac{\hbar}{2mi}f^2\left(\mathcal{D}^\alpha\overline{\xi}_0\xi_0 - \overline{\xi}_0\mathcal{D}^\alpha\xi_0\right) + \mathcal{O}(\hbar^2) \tag{5.297}$$

and:

$$j^\alpha_M = \frac{\hbar}{2m}\mathcal{D}_\beta\left(\overline{\psi}_0\sigma^{\alpha\beta}\psi_0\right) + \mathcal{O}(\hbar^2) \tag{5.298}$$

It can be shown that the convection current is the relativistic extension to the three-vector current density that emerges from the Schrödinger equation [61]. That is, the convection current describes the probability flow of a particle moving with a four-velocity v^α:

$$j_c^\alpha \propto v^\alpha \qquad (5.299)$$

Therefore, the correction to the geodesic four-velocity u^μ due to the spin–curvature interaction can be found by normalizing the convection current to obtain:

$$v^\alpha = u^\alpha + \frac{\hbar}{2mi}\left(\mathcal{D}^\alpha \bar{\xi}_0 \xi_0 - \bar{\xi}_0 \mathcal{D}_\alpha \xi_0\right) + \mathcal{O}(\hbar^2) \qquad (5.300)$$

Notice that we have dropped the $\bar{\psi}_0\psi_1$ and $\bar{\psi}_1\psi_0$ terms because they combine $\hbar^0$ and $\hbar^1$ terms and are multiplied by an $\hbar$ factor. Therefore, the deviation from the geodesic at order $\hbar$ is given by:

$$\delta u^\alpha = \frac{\hbar}{2mi}\left(\mathcal{D}^\alpha \bar{\xi}_0 \xi_0 - \bar{\xi}_0 \mathcal{D}^\alpha \xi_0\right) \qquad (5.301)$$

which is orthogonal to u^α by virtue of equation (5.285):

$$u^\alpha \delta u_\alpha = \frac{\hbar}{2mi}\left(u^\alpha \mathcal{D}_\alpha \bar{\xi}_0 \xi_0 - \bar{\xi}_0 u^\alpha \mathcal{D}_\alpha \xi_0\right) = 0 \qquad (5.302)$$

Furthermore, because of the normalization of the ξ_0 spinor, we have:

$$\partial_\alpha\left(\bar{\xi}_0 \xi_0\right) = 0 \Rightarrow \left(\partial_\alpha \bar{\xi}_0\right)\xi_0 = -\bar{\xi}_0 \partial_\alpha \xi_0 \qquad (5.303)$$

and:

$$\begin{aligned}
\mathcal{D}_\alpha \xi_0 &= \partial_\alpha \xi_0 - \Gamma_\alpha \xi_0 \\
\mathcal{D}_\alpha \xi_0^\dagger &= \partial_\alpha \xi_0^\dagger - \xi_0^\dagger \Gamma_\alpha^\dagger
\end{aligned} \qquad (5.304)$$

therefore:

$$\begin{aligned}
\mathcal{D}_\alpha \bar{\xi}_0 \xi_0 &= \left(\partial_\alpha \xi_0^\dagger \gamma^0 - \xi_0^\dagger \Gamma_\alpha^\dagger \gamma^0\right)\xi_0 \\
&= -\bar{\xi}_0 \partial_\alpha \xi_0 - \xi_0^\dagger \gamma^0 \gamma^0 \Gamma_\alpha^\dagger \gamma^0 \xi_0 \\
&= -\bar{\xi}_0 \partial_\alpha \xi_0 + \bar{\xi}_0 \Gamma_\alpha \xi_0 \\
&= -\bar{\xi}_0 (\partial_\alpha \xi_0 - \Gamma_\alpha \xi_0) \\
&= -\bar{\xi}_0 \mathcal{D}_\alpha \xi_0
\end{aligned} \qquad (5.305)$$

where we have used the relation:

$$\Gamma_\alpha = -\gamma^0 \Gamma_\alpha^\dagger \gamma^0 \qquad (5.306)$$

and the properties of the Dirac matrices. Then, the expression for the $\hbar$-order correction to the geodesic four-velocity reduces to:

$$\delta u^\alpha = -\frac{\hbar}{mi}\bar{\xi}_0 \mathcal{D}^\alpha \xi_0 \tag{5.307}$$

Let us now assume the case that we can find a tetrad field $e^a_{\ \mu}(x)$ such that the spinor is at rest in the local reference frame:

$$p^a = e^a_{\ \mu}p^\mu = (m,0,0,0) \tag{5.308}$$

In such a case, the correction to the four-velocity reduces to:

$$\delta u^\alpha = -\frac{\hbar}{mi}g^{\alpha\beta}\bar{\chi}_0 \mathcal{D}_\beta \chi_0 \tag{5.309}$$

where:

$$\chi_0 = \frac{1}{\sqrt{\beta_1^*\beta_1 + \beta_2^*\beta_2}}\begin{pmatrix} \beta_1 \\ \beta_2 \\ 0 \\ 0 \end{pmatrix} \tag{5.310}$$

is the general expression for a spinor normalized to 1 in the rest frame in the appropriate coordinate system.

The deviation from the geodesic behaviour is due to a force produced by the spin–curvature coupling [55]. This force is given by:

$$f^\alpha = m\frac{Dv^\alpha}{D\tau} = mv^\beta \mathcal{D}_\beta v^\alpha \tag{5.311}$$

and using the fact that:

$$v^\alpha v_\alpha = -1 \quad \Rightarrow \quad v^\alpha \mathcal{D}_\mu v_\alpha = 0 \tag{5.312}$$

we get:

$$\frac{f^\alpha}{m} = v^\beta \mathcal{D}_\beta v^\alpha - g^{\alpha\mu}v^\beta \mathcal{D}_\mu v_\beta$$

$$= (u^\beta + \delta u^\beta)\mathcal{D}_\beta(u^\alpha + \delta u^\alpha) - g^{\alpha\mu}(u^\beta + \delta u^\beta)\mathcal{D}_\mu(u_\beta + \delta u_\beta) + \mathcal{O}(\hbar^2)$$

$$= u^\beta(\mathcal{D}_\beta\delta u^\alpha - g^{\alpha\mu}\mathcal{D}_\mu\delta u_\beta) + \delta u^\beta(\mathcal{D}_\beta u^\alpha - g^{\alpha\mu}\mathcal{D}_\mu u_\beta) + \mathcal{O}(\hbar^2) \tag{5.313}$$

The second term of the last expression can be written as:

$$\mathcal{D}_\beta u^\alpha - g^{\alpha\mu}\mathcal{D}_\mu u_\beta = g^{\alpha\mu}(\mathcal{D}_\beta u_\mu - \mathcal{D}_\mu u_\beta)$$

$$= g^{\alpha\mu}(\partial_\beta u_\mu - \partial_\mu u_\beta)$$

$$= -\frac{1}{m}g^{\alpha\mu}(\partial_\beta\partial_\mu - \partial_\mu\partial_\beta)S$$

$$= 0 \tag{5.314}$$

where we have used the expression that relates the geodesic four-velocity and the phase factor S in the WKB expansion. Therefore, ignoring $\mathcal{O}(\hbar^2)$ terms we get:

$$
\begin{aligned}
f^\alpha &= mu^\beta \left(\mathcal{D}_\beta \delta u^\alpha - g^{\alpha\mu} \mathcal{D}_\mu \delta u_\beta \right) \\
&= mg^{\alpha\mu} u^\beta \left(\mathcal{D}_\beta \delta u_\mu - \mathcal{D}_\mu \delta u_\beta \right) \\
&= \frac{\hbar}{2i} g^{\alpha\mu} u^\beta \left\{ 2\mathcal{D}_\alpha \bar{\xi}_0 \mathcal{D}_\beta \xi_0 - 2\mathcal{D}_\beta \bar{\xi}_0 \mathcal{D}_\alpha \xi_0 + (\mathcal{D}_\beta \mathcal{D}_\alpha - \mathcal{D}_\alpha \mathcal{D}_\beta) \bar{\xi}_0 \xi_0 \right. \\
&\qquad \left. - \bar{\xi}_0 (\mathcal{D}_\beta \mathcal{D}_\alpha - \mathcal{D}_\alpha \mathcal{D}_\beta) \xi_0 \right\} \\
&= \frac{\hbar}{2i} g^{\alpha\mu} u^\beta \left([\mathcal{D}_\beta, \mathcal{D}_\alpha] \bar{\xi}_0 \xi_0 - \bar{\xi}_0 [\mathcal{D}_\beta, \mathcal{D}_\alpha] \xi_0 \right) \\
&= \frac{\hbar}{4} g^{\alpha\mu} u^\beta R_{\mu\beta\gamma\delta} \bar{\xi}_0 \sigma^{\gamma\delta} \xi_0
\end{aligned}
\tag{5.315}
$$

where we used equation (5.224) that describes the commutator of a covariant derivative of a spinor field, as well as the propagation equation for the spinor moving along a geodesic (cf. equation 5.285). This expression makes explicit that the force that deviates the spinor off its geodesic path is an interaction that couples the curvature of spacetime (as given by the Riemann curvature tensor $R_{\mu\beta\gamma\delta}$) and the spin of the particle (given in terms of the spinors $\bar{\xi}_0$ and ξ_0) [55].

In addition, notice that the spin–curvature force does not depend on the mass of the test particle, which can be interpreted as a type of gravitational charge. Therefore, this force does not couple gravitational charges as in the case of the Newtonian gravitational force. That said, it is also known that the deviation from geodetic motion is very small, except for the case of supermassive compact objects and/or ultra-relativistic test particles [62–65].

On the other hand, the magnetization current is related to the curl of the spin density [61]. In particular, the spin vector for a free particle moving along the v^α curve given by the WKB expansion:

$$
\begin{aligned}
S^\alpha &= \frac{1}{2} \epsilon^{\alpha\beta\gamma\delta} v_\beta \frac{\bar{\psi} \sigma_{\gamma\delta} \psi}{\bar{\psi} \psi} \\
&= S_0^\alpha + \hbar S_1^\alpha + \mathcal{O}(\hbar^2)
\end{aligned}
\tag{5.316}
$$

Then, the spin at order $\hbar^0$ is found to be:

$$
S_0^\alpha = \frac{1}{2} \epsilon^{\alpha\beta\gamma\delta} u_\beta \bar{\xi}_0 \sigma_{\gamma\delta} \xi_0
\tag{5.317}
$$

which satisfies the equations:

$$
\begin{aligned}
D_\alpha S_0^\beta u^\alpha &= 0 \\
D_\alpha S_0^\beta v^\alpha &= \mathcal{O}(\hbar)
\end{aligned}
\tag{5.318}
$$

That is, at order $\hbar^0$ the spin is parallelly propagated along the geodesic u_α. However, the spin is not parallelly propagated along the $\hbar$-order corrected trajectory v^α. At this point, the reader should compare the quantum case equations (5.315) and (5.318) with the classical case equations (2.206). In chapter 7 we will apply these results to the simple example of a Dirac spinor in Schwarzschild spacetime.

5.10 The spin–curvature coupling

As we have shown, spin and curvature are coupled in a non-trivial way. Such a coupling will lead to new effects on the spin dynamics of falling particles that were not considered in the discussion presented in section 5.8.

As we showed in the previous section, the correction to the four-velocity of a spin-$\frac{1}{2}$ particle due to the spin–curvature coupling is given by:

$$u_\mu(x) \rightarrow u_\mu(x) + \hbar \delta u_\mu(x) + \mathcal{O}(\hbar^2) \tag{5.319}$$

where:

$$\delta u_\mu(x) = -\frac{1}{mi}\bar{\chi}_0(x)\mathcal{D}_\mu(x)\chi_0(x) \tag{5.320}$$

and $\chi_0(x)$ is the most general expression for a Dirac spinor in the local frame, and $u_\mu(x)$ is the geodesic four-velocity. As such, at the lowest order, the deviation from the geodesic path is $\mathcal{O}(\hbar)$, which could be a meaningful correction only in the case of relativistic particles and/or strong gravitational fields produced by supermassive objects.

It is possible to study the effect of spin–curvature coupling in quantum information. In particular, we can analyse the effect of this coupling on the Wigner rotation of the quantum state [54]. In this case, the correction to the Lorentz transformation is given by:

$$\lambda^a{}_b(x) \rightarrow \lambda^a{}_b(x) + \hbar \delta \lambda^a{}_b(x) + \mathcal{O}(\hbar^2) \tag{5.321}$$

where:

$$\delta \lambda^a{}_b(x) = \delta a^a(x)u_b(x) - u^a(x)\delta a_b(x) - e^a{}_\mu(x)\delta u^\beta(x)\nabla_\beta e_b{}^\mu(x) \tag{5.322}$$

and:

$$\delta a^\mu = \frac{f^\mu}{m\hbar} \tag{5.323}$$

where f^μ is the force due to the spin–curvature coupling. Therefore, the correction to the Wigner rotation is given by:

$$\vartheta^a{}_b(x) \rightarrow \vartheta^a{}_b(x) + \hbar \delta \vartheta^a{}_b(x) + \mathcal{O}(\hbar^2) \tag{5.324}$$

with:

$$\delta\vartheta^a_{\ b}(x) = \delta\lambda^a_{\ b}(x) + \frac{\delta\lambda^a_{\ 0}(x)u_b(x) - \delta\lambda_{b0}(x)u^a(x)}{u^0(x) + 1} \qquad (5.325)$$

In chapter 7 we will apply these results to the simple example of a Dirac spinor in Schwarzschild spacetime.

5.11 Summary

This chapter provided a brief overview of quantum fields in curved spacetimes. Our emphasis was the development of the equations that describe the dynamics of spin-$\frac{1}{2}$ particles. In particular we used the tetrad field formalism to incorporate the interaction between Dirac spinors and gravitational fields. However, our discussion of the covariant derivative was general and can be directly applied to quantum fields of arbitrary spin. Finally, we analysed the spinor solutions to the Dirac equations in the presence of a gravitational field. We used these expressions to derive expressions to estimate the deviation from geodesic motion due to the non-trivial spin–curvature interaction.

Bibliography

[1] Birrell N D and Davis P C W 1982 *Quantum Fields in Curved Space* (Cambridge: Cambridge University Press)

[2] Fronsdal C 1965 Elementary particles in a curved space *Rev. Mod. Phys.* **37** 221

[3] Fronsdal C 1974 Elementary particles in a curved space II *Phys. Rev. D* **10** 589

[4] Fulling S A 1989 *Aspects of Quantum Field Theory in Curved Spacetime* (Cambridge: Cambridge University Press)

[5] Mukhanov V F and Winitzki S 2007 *Introduction to Quantum Effects in Gravity* (Cambridge: Cambridge University Press)

[6] Parker L and Toms D 2009 *Quantum Field Theory in Curved Spacetime* (Cambridge: Cambridge University Press)

[7] Wald R 1994 *Quantum Field Theory in Curved Spacetime and Black Hole Thermodynamics* (Chicago, IL: Chicago University Press)

[8] Carroll S M 2004 *Spacetime and Geometry: An Introduction to General Relativity* (Reading, MA: Addison-Wesley)

[9] Hobson M P, Efstathiou G and Lasenby A N 2006 *General Relativity: An Introduction for Physicists* (Cambridge: Cambridge University Press)

[10] Weinberg S 1972 *Gravitation and Cosmology: Principles and Applications of The General Theory of Relativity* (New York: Wiley)

[11] Landau L D and Lifshitz E M 1981 *Quantum Mechanics: Non-Relativistic Theory, Course of Theoretical Physics* vol 3 3rd edn (Amsterdam: Elsevier)

[12] Friedlander F G 1975 *The Wave Equation on a Curved Space-Time* (Cambridge: Cambridge University Press)

[13] Newton T D and Wigner E P 1949 Localized states for elementary systems *Rev. Mod. Phys.* **21** 400–6

[14] Palmer M C, Takahashi M and Westman H F 2012 Localized qubits in curved spacetimes *Ann. Phys.* **237** 1078–131

[15] Beem J K and Parker P E 1983 Klein–Gordon solvability and the geometry of geodesics *Pac. J. Math.* **107** 1–14

[16] Clayton M A, Demopoulos L and Legare J 1998 The dynamical stability of the static real scalar field solutions to the Einstein–Klein–Gordon equations revisited *Phys. Lett.* A **248** 131–8

[17] Deruelle N and Ruffini R 1974 Quantum and classical relativistic energy states in stationary geometries *Phys. Lett.* B **52** 437–41

[18] Elizalde E 1988 Exact solutions of the massive Klein–Gordon–Schwarzschild equation *Phys. Rev.* D **37** 2127–31

[19] Madsen M S 1988 Scalar fields in curved spacetimes *Class. Quantum Grav.* **5** 627–39

[20] Rowan D J and Stephenson G 1977 The Klein–Gordon equation in a Kerr–Newman background space *J. Phys. A: Math. Gen.* **10** 15–23

[21] Rowan D J and Stephenson G 1976 Solutions of the time-dependent Klein–Gordon equation in Schwarzschild background space *J. Phys. A: Math. Gen.* **9** 1631–5

[22] Rowan D J and Stephenson G 1976 The massive scalar meson field in Schwarzschild background space *J. Phys. A: Math. Gen.* **9** 1261–5

[23] Schmoltzi K and Schücker Th 1991 The energy spectrum of the static, spherically symmetric solutions to the Einstein-Klein-Gordon equations *Phys. Lett.* **161** 212–6

[24] Teixeira Filho R M and Bezerra V B 2004 Scalar particles in weak gravitational fields *Class. Quantum Grav.* **21** 307–15

[25] Zhang Z Y 1993 Gravitation of the Klein–Gordon scalar field *Int. J. Theor. Phys.* **32** 2015–21

[26] Cohen-Tannoudji C, Diu B and Laloë F 2005 *Quantum Mechanics* (New York: Wiley)

[27] Jackson J D 1975 *Classical Electrodynamics* 2nd edn (New York: Wiley)

[28] Wald R 1984 *General Relativity* (Chicago, IL: University of Chicago Press)

[29] Arfken G B 2000 *Mathematical Methods for Physicists* 5th edn (New York: Academic Press)

[30] Dolan S R 2008 Black holes and wave mechanics, Lecture Notes, http://www2.ufpa.br/ppgf/ASQTA/2008_arquivos/C4.pdf

[31] Ross G G 1984 *Grand Unified Theories* (New York: Benjamin-Cummings)

[32] Landau L D and Lifshitz E M 1975 *The Classical Theory of Fields, Course of Theoretical Physics* vol 2 (Amsterdam: Elsevier)

[33] Messiah A 1999 *Quantum Mechanics* (New York: Dover)

[34] Lang S 1972 *Differential Manifolds* (Reading, MA: Addison-Wesley)

[35] Bender C M and Orszag S A 2010 *Advanced Mathematical Methods for Scientists and Engineers: Asymptotic Methods and Perturbation Theory* (Berlin: Springer)

[36] Cartier P and DeWitt-Morette C 2006 *Functional Integration: Action and Symmetries* (Cambridge: Cambridge University Press)

[37] Feynman R P and Hibbs A R 1965 *Quantum Mechanics and Integrals* (New York: McGraw-Hill)

[38] Itzykson C and Zuber J B 1980 *Quantum Field Theory* (New York: McGraw-Hill)

[39] Ramond P 1990 *Field Theory: A Modern Primer* 2nd edn (Reading, MA: Addison-Wesley)

[40] Ryder L H 1996 *Quantum Field Theory* 2nd edn (Cambridge: Cambridge University Press)

[41] Weinberg S 1995 *The Quantum Theory of Fields* vol 1 (Cambridge: Cambridge University Press)

[42] Weinberg S 2008 *Cosmology* (Cambridge: Cambridge University Press)

[43] Araki H 1999 *Mathematical Theory of Quantum Fields* (Oxford: Oxford University Press)

[44] Bogolubov N N *et al* 1990 *General Principles of Quantum Field Theory* (Dordrecht: Kluwer)

[45] Haag R 1996 *Local Quantum Physics: Fields, Particles, Algebras* 2nd edn (Berlin: Springer)

[46] Streater R F and Wightman A S 1964 *PCT, Spin and Statistics, and All That* (Princeton, NJ: Princeton University Press)

[47] Parker L 1969 Quantized fields and particle creation in expanding universes. I *Phys. Rev.* **183** 1057–68

[48] Parker L 1971 Quantized fields and particle creation in expanding universes. II *Phys. Rev.* D **3** 346–56

[49] Carmeli M 1977 *Group Theory General Relativity: Representations of the Lorentz Group and Their Applications to the Gravitational Field* (London: Imperial College Press)

[50] Cartan E 1981 *The Theory of Spinors* (New York: Dover)

[51] Terashima H and Ueda M 2004 Einstein–Rosen correlation in gravitational field *Phys. Rev.* A **69** 032113

[52] Trautman A 2006 Einstein–Cartan theory *Encyclopedia of Mathematical Physics* ed J-P Francoise, G L Naber and S T Tsou vol **1** (Oxford: Elsevier) pp 189–95

[53] Hehl F W, von der Heyde P and Kerlick G D 1976 General relativity with spin and torsion: foundations and prospects *Rev. Mod. Phys.* **48** 393–416

[54] Alsing P M, Stephenson G J and Kilian P 2009 Spin-induced non-geodesic motion, gyroscopic precession, Wigner rotation and EPR correlations of massive spin-$\frac{1}{2}$ particles in a gravitational field arXiv:0902.1396v1 [quant-ph]

[55] Audretsch J 1981 Trajectories and spin motion of massive spin-$\frac{1}{2}$ particles in gravitational fields *J. Phys. A: Math. Gen.* **14** 411–22

[56] Cianfrani F and Montani G 2008 Curvature-spin coupling from the semi-classical limit to the Dirac equation *Int. J. Mod. Phys.* A **23** 1274–77

[57] Cianfrani F and Montani G 2008 Dirac equations in curved space-time vs. Papapetrou spinning particles *Europhys. Lett.* **84** 3008

[58] Hafner D and Nicolas J P 2004 Scattering of massless Dirac fields by a Kerr Black Hole *Rev. Math. Phys.* **16** 29–123

[59] Alsing P M, Evans J C and Nandi K K 2001 The phase of a quantum mechanical particle in curved spacetime *Gen. Rel. Grav.* **33** 1459–87

[60] Stodolsky L 1979 Matter and light wave interferometry in gravitational fields *Gen. Rel. Grav.* **11** 391–405

[61] Bjorken J D and Drell S D 1964 *Relativistic Quantum Mechanics* (New York: McGraw-Hill)

[62] Plyatsko R 2005 Ultrarelativistic circular orbits of spinning particles in a Schwarzschild field *Class. Quantum Grav.* **22** 1545–51

[63] Silenko A J 2008 Classical and quantum spins in curved spacetimes *Acta Phys. Polonica B Proc. Suppl.* **1** 87–107

[64] Plyatsko R 1998 Gravitational ultrarelativistic spin-orbit interaction and the weak equivalence principle *Phys. Rev.* D **58** 084031

[65] Singh D 2008 Perturbation method for classical spinning particle motion: I. Kerr space-time *Phys. Rev.* D **78** 104028

[66] Chandraserkhar S 1992 *The Mathematical Theory of Black Holes* (Oxford: Oxford University Press)

[67] Misner C W, Thorne K S and Wheeler J A 1973 *Gravitation* (San Francisco, CA: W H Freeman)

IOP Concise Physics

Quantum Information in Gravitational Fields

Marco Lanzagorta

Chapter 6

Qubits in Schwarzschild spacetime

The Schwarzschild metric is the most widely known exact solution to Einstein's field equations. This solution represents a static and isotropic gravitational field [1–3]. This is an empty space or vacuum solution to the Einstein field equations. That is, the gravitational field is described *outside* the body that produces it, in the region of spacetime where the energy density tensor vanishes identically.

Because of its symmetries, the Schwarzschild metric describes a spherically symmetric source of gravity. Therefore, it is a fair approximation to the gravitational field produced by a planet or a star. It is certainly the simplest spacetime that can be used to study the effects of gravitation on quantum information on Earth's surface or transmitted between a ground base and a space-borne satellite. In this chapter we will discuss the dynamics of spin-$\frac{1}{2}$ qubits falling in a classical gravitational field described by the Schwarzschild metric.

6.1 Metric tensor

The Schwarzschild solution is *static*, in the sense that (1) all the components of the metric tensor are independent of the time-like coordinate t and (2) the line element dS^2 is invariant under a reflection transformation of the time-like coordinate. In addition, the Schwarzschild solution is *isotropic*, which implies that the line element dS^2 only depends on the rotational invariants of the space-like coordinates and their derivatives.

In spherical coordinates (r, θ, φ), the line element of the Schwarzschild metric that describes the static and isotropic gravitational field produced by a spherically symmetric source of mass M is given by:

$$dS^2 = -f\,dt^2 + \frac{1}{f}\,dr^2 + r^2\,(d\theta^2 + \sin^2\theta\,d\varphi^2) \qquad (6.1)$$

doi:10.1088/978-1-627-05330-3ch6

where:

$$f \equiv 1 - \frac{r_s}{r}, \qquad r_s \equiv 2M \qquad (6.2)$$

all given in natural units $(G = 1, c = 1)$.

6.2 Structure of Schwarzschild spacetime

Because the Schwarzschild metric is an empty space solution, it is only valid outside the spherically symmetric source of the gravitational field. That is, the Schwarzschild metric is valid in a radial range that goes from infinity to r_M, the radius of the spherical body that produces the gravitational field.

In addition, it can be observed that the metric diverges in two distinct points, at $r = r_s$ and $r = 0$. The distance r_s is known as the *Schwarzschild radius*. If $r_M < r_s$, then the source of the gravitational field is a *black hole* and the Schwarzschild radius determines the extension of its *event horizon*.

It is important to distinguish the different nature of both singularities [1, 4–6]. To this end, let us recall that coordinates are mere labels that we use to identify events in spacetime. In this regard, coordinates do not have much physical meaning. In addition, the specific choice of coordinates may inadvertently introduce singularities in the equations that do not necessarily reflect the underlying physical structure of spacetime.

Furthermore, all the information about the physics and the spacetime curvature is contained in the curvature tensor $R_{\alpha\beta\gamma\delta}$, which depends on the coordinates. Therefore, to determine when a metric has a singularity with some physical meaning, it is necessary to look at the structure of the non-trivial scalars that can be constructed from the curvature tensor [5–7]. Indeed, as these are scalars formed by the contraction of tensors, they are invariants (in the sense that their value does not change under general coordinate transformations). Furthermore, it is easy to show that there are 14 curvature scalars in four dimensions, which for convenience we can denote by R_i, $i = 1, \ldots, 14$ [3].

In particular, the Schwarzschild metric is obtained by solving the Einstein field equations in vacuum. This means that the Ricci tensor has the trivial value of zero in the entire exterior of the source of the gravitational field:

$$R_{\mu\nu} = 0 \qquad (6.3)$$

and consequently the curvature scalar also has the trivial value of zero:

$$R_1 \equiv R = g^{\mu\nu} R_{\mu\nu} = 0 \qquad (6.4)$$

The curvature scalar does not provide any meaningful information about the singularities in Schwarzschild spacetime. However, a more meaningful non-trivial coordinate scalar can be constructed from the curvature tensor as follows:

$$R_2 \equiv R^{\alpha\beta\gamma\delta} R_{\alpha\beta\gamma\delta} = \frac{12\, r_s^2}{r^6} \qquad (6.5)$$

Clearly, this curvature-related scalar is singular at $r = 0$, but remains finite at $r = r_s$. This means that the singularity at $r = 0$ is actually related to the divergent value of the R_2 curvature scalar. As such, $r = 0$ is called an *intrinsic singularity* [1, 4]. In general, a sufficient condition for a point to be an intrinsic singularity is to have a divergent curvature scalar R_i for *any* value of i.

On the other hand, it can be shown that, for the Schwarzschild metric, all curvature scalars are finite at $r = r_s$. This suggests that the singularity at $r = r_s$ is *not* an intrinsic singularity, but it may be due to our specific choice of coordinates. Let us remark that, in general, it is much more difficult to show that a given point is not singular [4–6]. For the case of the Schwarzschild metric, however, it can be proved that the point at $r = r_s$ is not an intrinsic singularity, but it is indeed related to the specific choice of coordinates. Consequently, the point $r = r_s$ is known as a *coordinate singularity* [1, 4].

Furthermore, the coordinate singularity can be removed from the metric by an appropriate change of coordinates. In particular, the *Eddington–Finkelstein coordinates* make the metric finite at $r = r_s$ [1, 8]. However, in this book we will limit our attention to the gravitational field in the region of space outside the Schwarzschild radius.

6.3 Tetrad fields and connection one-forms

In this section we will present the values of the affine connections $\Gamma^\alpha_{\mu\nu}$ and the connection one-forms $\omega^a_{\mu\ b}$ for the Schwarzschild metric. We will also choose a convenient tetrad field $e^\mu_a(x)$ that can be used to study free-falling, orbiting test particles. However, as we discussed before, the selection of the tetrad field is not unique.

6.3.1 Affine connections

Using the metric tensor for the Schwarzschild solution, one can compute the elements of the affine connection [3]. Let us recall that, because the Schwarzschild metric has no torsion, the affine connection is related to the metric tensor by:

$$\Gamma^\alpha_{\mu\nu} = \frac{1}{2} g^{\sigma\alpha} \left(\frac{\partial g_{\nu\sigma}}{\partial x^\mu} + \frac{\partial g_{\mu\sigma}}{\partial x^\nu} - \frac{\partial g_{\nu\mu}}{\partial x^\sigma} \right) \tag{6.6}$$

Therefore, the only non-zero affine connections of the Schwarzschild metric are given by:

$$\Gamma^r_{rr} = -\frac{\dot{f}}{2f}$$

$$\Gamma^r_{\theta\theta} = -rf$$

$$\Gamma^r_{\varphi\varphi} = -rf \sin^2 \theta$$

$$\Gamma^r_{tt} = \frac{f\dot{f}}{2}$$

$$\Gamma^{\theta}_{r\theta} = \Gamma^{\theta}_{\theta r} = \frac{1}{r}$$

$$\Gamma^{\varphi}_{\varphi r} = \Gamma^{\varphi}_{r\varphi} = \frac{1}{r}$$

$$\Gamma^{t}_{rt} = \Gamma^{t}_{tr} = \frac{\dot{f}}{2f} \tag{6.7}$$

$$\Gamma^{\theta}_{\varphi\varphi} = -\sin\theta\cos\theta$$

$$\Gamma^{\varphi}_{\varphi\theta} = \Gamma^{\varphi}_{\theta\varphi} = \frac{\cos\theta}{\sin\theta}$$

where:

$$\dot{f} = \frac{r_s}{r^2} \tag{6.8}$$

and the 'dot' operation on the function f implies differentiation with respect to the radial variable r (because the Schwarzschild solution is static, the function f does not depend on time).

6.3.2 Curvature tensor

We will require the explicit values of the curvature tensor when we analyse the spin–curvature coupling (in the following chapter). To simplify the problem, we will restrict our attention to test particles moving in the equatorial plane ($\theta = \pi/2$). Therefore, the non-trivial components of the curvature tensor in the equatorial plane are given by:

$$R_{r\theta r\theta} = \frac{r_s}{2f}$$

$$R_{r\varphi r\varphi} = \frac{r_s}{2f}$$

$$R_{\theta\varphi\theta\varphi} = -r_s r^2$$

$$R_{rtrt} = \frac{r_s}{r^2} \tag{6.9}$$

$$R_{\theta t\theta t} = -\frac{r_s}{2}f$$

$$R_{\varphi t\varphi t} = -\frac{r_s}{2}f$$

as well as those found by the symmetry and anti-symmetry properties of $R_{\mu\nu\alpha\beta}$:

$$R_{\mu\nu\alpha\beta} = R_{\alpha\beta\mu\nu}$$

$$R_{\mu\nu\alpha\beta} = -R_{\nu\mu\alpha\beta} = R_{\nu\mu\beta\alpha} = -R_{\mu\nu\beta\alpha} \tag{6.10}$$

6.3.3 Tetrad fields

It is convenient to use the tetrad field with the following non-vanishing components:

$$e_0{}^t = \frac{1}{\sqrt{f}}$$

$$e_1{}^r = \sqrt{f}$$

$$e_2{}^\theta = \frac{1}{r} \tag{6.11}$$

$$e_3{}^\varphi = \frac{1}{r\sin\theta}$$

and the inverse components:

$$e^0{}_t = \sqrt{f}$$

$$e^1{}_r = \frac{1}{\sqrt{f}}$$

$$e^2{}_\theta = r \tag{6.12}$$

$$e^3{}_\varphi = r\sin\theta$$

This tetrad field is used to represent a 'hovering' observer in the associated local inertial frame [9]. Indeed, let us assume an observer at rest in the local inertial frame:

$$u^a = (E, 0, 0, 0) \tag{6.13}$$

Then, in the general coordinate system, the four-velocity of the observer looks like:

$$u^\mu = e_a{}^\mu u^a = \left(\frac{E}{\sqrt{f}}, 0, 0, 0\right) \tag{6.14}$$

which corresponds to a 'hovering' observer that is not falling towards the source of the gravitational field. Clearly, for an observer to hover over a gravitational body, there has to be an external force acting against the force of gravity. Furthermore, as expected:

$$\lim_{r\to\infty} u^\mu = E \tag{6.15}$$

It is also important to notice how our choice of the tetrad field defines a local inertial frame with spatial axes $(\hat{\mathbf{i}}, \hat{\mathbf{j}}, \hat{\mathbf{k}})$ perpendicular to the spherical coordinate axes $(\hat{\mathbf{r}}, \hat{\theta}, \hat{\varphi})$ [9]. That is:

$$\hat{\mathbf{i}}\|\hat{\mathbf{r}} \quad \hat{\mathbf{j}}\|\hat{\theta} \quad \hat{\mathbf{k}}\|\hat{\varphi} \tag{6.16}$$

in all points of spacetime.

As we have mentioned before, there are many other choices of tetrad fields that we could use. The only requirement is that they produce the Schwarzschild metric tensor $g_{\mu\nu}(x)$ as:

$$g^{\mu\nu}(x) = e_a^{\mu}(x)\, e_b^{\nu}(x)\, \eta_{ab} \tag{6.17}$$

In all cases we would get equivalent results. In the next chapter, for instance, we will use two different tetrad fields that are much more convenient to analyse the spin–curvature coupling of a Dirac spinor in the Schwarzschild metric.

6.3.4 Connection one-forms

The connection one-form is given by:

$$\omega_{\mu ab} = e_a^{\nu}(x)\left(\partial_\mu e_{b\nu}(x) - \Gamma_{\mu\nu}^{\alpha}\, e_{ba}(x)\right) \tag{6.18}$$

Therefore, the non-vanishing connection one-forms for our choice of tetrad field are found to be:

$$\omega_{t\,1}^{\,0} = \omega_{t\,0}^{\,1} = \frac{r_s}{2r^2}$$

$$\omega_{\theta\,2}^{\,1} = -\omega_{\theta\,1}^{\,2} = -\sqrt{f}$$

$$\omega_{\varphi\,3}^{\,1} = -\omega_{\varphi\,1}^{\,3} = -\sqrt{f}\,\sin\theta \tag{6.19}$$

$$\omega_{\varphi\,3}^{\,2} = -\omega_{\varphi\,2}^{\,3} = -\cos\theta$$

6.4 Geodesics

We can calculate the Schwarzschild geodesics from the generally covariant equation of motion for a free-falling spinless particle:

$$\frac{Du^{\mu}}{D\tau} = \frac{d^2 x^{\mu}}{d\tau^2} + \Gamma_{\alpha\beta}^{\mu}\frac{dx^{\alpha}}{d\tau}\frac{dx^{\beta}}{d\tau} = 0 \tag{6.20}$$

In particular, for orbits in the equatorial plane ($\theta = \pi/2$), the system of four differential equations simplifies to:

$$\frac{du^t}{d\tau} + 2\Gamma_{tr}^{t} u^t u^r = 0$$

$$\frac{du^r}{d\tau} + \Gamma_{tt}^{r} u^t u^t + \Gamma_{\varphi\varphi}^{r} u^\varphi u^\varphi + \Gamma_{rr}^{r} u^r u^r = 0$$

$$\frac{du^\theta}{d\tau} = 0 \tag{6.21}$$

$$\frac{du^\varphi}{d\tau} + 2\Gamma_{\varphi r}^{\varphi} u^\varphi u^r = 0$$

and the solutions to these geodesic equations are found to be:

$$u^t = \frac{K}{f}$$

$$u^r = \pm\sqrt{K^2 - f\frac{J^2}{r^2} - Af}$$

$$u^\theta = 0 \tag{6.22}$$

$$u^\varphi = \frac{J}{r^2}$$

where A, J and K are integration constants [1, 3].

For massive particles, we can adopt the following normalization:

$$u^\mu u_\mu = -1 \tag{6.23}$$

which implies:

$$g_{tt}(u^t)^2 + g_{rr}(u^r)^2 + g_{\theta\theta}(u^\theta)^2 + g_{\varphi\varphi}(u^\varphi)^2 = -1 \tag{6.24}$$

and consequently:

$$A = 1 \tag{6.25}$$

The meaning of K can be understood by looking at the energy of the test particle at large distances:

$$\lim_{r\to\infty} u^\mu = \left(K, \pm\sqrt{K^2 - 1}, 0, 0\right) \tag{6.26}$$

Thus, if the test particle falls to the source of the gravitational field from a state of rest at infinite radius, then $K = 1$. In particular, if the particle is dropped from rest at some distance R:

$$u^r(R) = -\sqrt{K^2 - f(R)} = 0 \quad\Rightarrow\quad K^2 = f(R) \tag{6.27}$$

Furthermore, we also have that, in general:

$$\frac{E}{m} = \lim_{r\to\infty} u^t = K \tag{6.28}$$

where E is the energy of the particle. Because the metric is independent of the time component t, we have:

$$p_t = g_{tt}p^t = mg_{tt}u^t = -mK \tag{6.29}$$

is a conserved quantity:

$$\frac{dp_t}{d\tau} = 0 \quad\Rightarrow\quad \frac{dK}{d\tau} = 0 \tag{6.30}$$

which merely expresses the conservation of the energy of the test particle. Let us stress the fact that E is the total energy of the particle, which includes the potential energy due to gravity. However, within the context of general relativity, the notion of gravitational potential energy cannot be defined in a formal manner. On the other hand, the concept of total energy is well defined as long as there is a time-like Killing vector.

Finally, the expression for u^φ makes clear that the integration constant J is associated with the angular momentum of the test particle.

6.5 Quantum dynamics

As discussed in chapter 5, the optical analogy and the WKB approximation provide an insightful way to understand the dynamics of quantum particles in curved spacetimes. That is, a Dirac spinor ψ can be described by:

$$\psi \approx e^{iS}\psi_0 + \mathcal{O}(\hbar) \qquad (6.31)$$

where ψ_0 is a spinor that satisfies the Dirac equation in some local inertial frame and the quantum mechanical phase S corresponds to the action of the classical particle:

$$S = m \int u_\mu dx^\mu \qquad (6.32)$$

Then, applying the optical analogy, we can understand the dynamics of quantum particles as the propagation of wave surfaces with constant S and rays given by $\partial_\mu S$. In this case, the integral curves of u^μ correspond to geodesic congruences defined by the geodesic equations. Furthermore, the spinor is described with respect to some tetrad field and we developed equations to describe the transport of spinors along geodesic lines. In what follows we will consider the dynamics of quantum test particles with the understanding that spinors can only be defined through tetrad fields and their dynamic behaviour is roughly described by the optical analogy and the WKB approximation (far from the singularity, in the region where the curvature scale R_s is relatively large compared with the Compton wavelength).

6.6 Wigner rotations

Now we have all the ingredients that are required to analyse the Wigner rotation of a qubit in motion in Schwarzschild spacetime. In particular, in this section we will consider three representative examples: (1) an equatorial radial fall, (2) an equatorial circular orbit and (3) a non-geodetic equatorial circular path.

However, before we do that, let us recall that the geodesic equations we have just derived represent the free-falling motion of spinless particles. However, because we are interested in qubits, we would like to analyse the free-falling motion of spin-$\frac{1}{2}$ particles. Of course, as discussed in sections 2.8 and 5.9, we have to consider that the spin and curvature are coupled in a non-trivial manner. Consequently, the motion of spinning particles, either classical or quantum, does not follow geodesics [10–14]. That said, we also showed that the deviation from geodetic motion is very small, of order $\hbar$, and it can be safely ignored except for the case of supermassive compact objects and/or ultra-relativistic test particles [15–18]. Therefore, in this chapter we will ignore the spin–curvature coupling, but we will discuss a couple of simplified situations in the following chapter.

6.6.1 Equatorial radial fall ($\theta = \pi/2$, $J = 0$)

As a first example, let us consider the case of a free-falling test particle dropped in Schwarzschild spacetime without angular momentum. If the test particle is dropped from rest at infinity, then the geodesics are simply given by:

$$u^t = \frac{1}{f}$$

$$u^r = -\sqrt{1 - f} \qquad (6.33)$$

$$u^\theta = 0$$

$$u^\varphi = 0$$

Clearly, we have chosen the negative sign for the radial velocity so the geodesic will represent the path of a test particle radially in-falling towards the source of the gravitational field.

The only non-zero Lorentz transformations in the local inertial frame described by the tetrad are given by:

$$\lambda^0{}_1 = \lambda^1{}_0 = -\frac{r_s}{2r^2 f} \qquad (6.34)$$

which correspond to boosts in the one-direction. Consequently, there is no Wigner rotation for this case. That is, the test particle does not change its spin state as it falls on a geodetic radial fall towards the source of the gravitational field.

6.6.2 Equatorial circular orbits ($\theta = \pi/2$, $u^r = 0$)

Let us now consider the case of a free-falling test particle moving around the source of the gravitational field in a circular orbit. The geodesics for circular orbits in the equatorial plane $\theta = \pi/2$ are given by:

$$u^t = \frac{K}{f}$$

$$u^r = 0$$

$$u^\theta = 0 \qquad (6.35)$$

$$u^\varphi = \frac{J}{r^2}$$

where the integration constants J and K are required to take the following values:

$$J^2 = \frac{1}{2}\frac{rr_s}{1 - \dfrac{3r_s}{2r}}$$

$$K = \frac{1 - \dfrac{r_s}{r}}{\sqrt{1 - \dfrac{3r_s}{2r}}} \qquad (6.36)$$

Furthermore, this value of J implies that the angular velocity is given by:

$$(u^\varphi)^2 = \frac{J^2}{r^4} = \frac{1}{2r^3} \frac{r_s}{1 - \frac{3r_s}{2r}} \tag{6.37}$$

which has a finite real value only if:

$$r > \frac{3}{2} r_s \tag{6.38}$$

That is, circular orbits in Schwarzschild metric are possible only when this condition is satisfied. Notice that this condition places the orbital radius outside the event horizon of the Schwarzschild metric given by r_s. Finally, the orbital velocity is given by:

$$v_o = r \frac{d\varphi}{dt} = r \frac{u^\varphi}{u^\tau} = \sqrt{\frac{r_s}{2r}} = \sqrt{\frac{M}{r}} \tag{6.39}$$

The non-zero Lorentz transformations in the local inertial frame defined by the tetrad field are found to be:

$$\lambda^0_{\ 1} = \lambda^1_{\ 0} = -\frac{K r_s}{2r^2 f} \tag{6.40}$$

$$\lambda^1_{\ 3} = -\lambda^3_{\ 1} = \frac{J \sqrt{f}}{r^2}$$

which correspond to boosts in the one-direction and rotations over the two-axis, respectively.

The Wigner rotation angle that corresponds to the rotation over the two-axis is given by:

$$\vartheta^1_{\ 3} = \lambda^1_{\ 3} + \frac{\lambda^1_{\ 0} u_3 - \lambda_{30} u^1}{u^0 + 1} \tag{6.41}$$

where:

$$u^a = e^a_{\ \mu} u^\mu = \left(\frac{K}{\sqrt{f}}, 0, 0, \frac{J}{r} \right) \tag{6.42}$$

and:

$$u_a = \eta_{ab} u^b$$
$$\lambda_{ab} = \eta_{ac} \lambda^c_{\ b} \tag{6.43}$$

Notice that the velocity of the particle in the local inertial frame is given by:

$$v = \frac{dx^3}{dx^0} = \frac{u^3}{u^0} = \frac{J}{K} \frac{\sqrt{f}}{r} = \sqrt{\frac{r_s}{2rf}} \tag{6.44}$$

Then, introducing all the expressions for an equatorial circular orbit on the equation for the Wigner rotation angle, we obtain:

$$\vartheta^1_3 = \frac{J\sqrt{f}}{r^2}\left(1 - \frac{Kr_s}{2rf}\frac{1}{K+\sqrt{f}}\right) \tag{6.45}$$

which only depends on the radius of the circular orbit of the test particle r and the mass of the source of the gravitational field r_s.

The total rotation angle Θ, after the test particle has moved across some proper time τ, is given by:

$$\Theta = \int \vartheta^1_3 \, d\tau$$

$$= \int \vartheta^1_3 \frac{r^2}{J} \, d\varphi$$

$$= \vartheta^1_3 \frac{r^2}{J} \int d\varphi$$

$$= \vartheta^1_3 \frac{r^2}{J} \Phi \tag{6.46}$$

where Φ is the angle traversed by the test particle during its motion. This expression means that, for a single revolution around the source of the gravitational field:

$$\Theta = 2\pi \, \vartheta^1_3 \frac{r^2}{J} \tag{6.47}$$

It is important to note that the angle Θ reflects the entire rotation undergone by the qubit's spin as it completes an orbit. There are two contributions to the value of Θ: the 'trivial rotation' and the rotation due to gravity [9]. The trivial rotation is the 2π rotation that the spin undergoes as it completes a circular orbit. Therefore, to obtain the Wigner rotation angle that is produced exclusively by pure gravitational effects, we need to subtract the trivial rotation angle of 2π:

$$\Omega = \Theta - 2\pi$$

$$= 2\pi\sqrt{f}\left(1 - \frac{Kr_s}{2rf}\frac{1}{K+\sqrt{f}}\right) - 2\pi \tag{6.48}$$

That is, once a qubit has completed an entire circular orbit in Schwarzschild spacetime, Ω is the total rotation of the spin solely due to the presence of a gravitational field. This is a purely relativistic effect due to the interaction of a spin-$\frac{1}{2}$ quantum field with a classical gravitational field.

It can be observed that Ω only depends on the ratio r_s/r. The behaviour of Ω with respect to the ratio r_s/r is shown in figure 6.1. As expected, in the weak gravitational

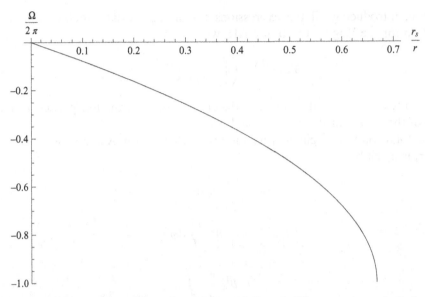

Figure 6.1. The effect of a static and isotropic gravitational field on the Wigner rotation angle Ω of a spin-$\frac{1}{2}$ qubit after completion of an entire circular orbit. In the zero-mass limit, the rotation angle is zero. On the other hand, for the smallest circular orbit allowed in Schwarzschild spacetime ($r = 3r_s/2$), the Wigner rotation angle takes the limiting value of -2π.

field limits, $M \to 0$ and $r \to \infty$, the Wigner angle due to the presence of the gravitational field goes to zero:

$$\lim_{M \to 0} \Omega = 0 \qquad \lim_{r \to \infty} \Omega = 0 \qquad (6.49)$$

On the other hand, in the limit of the circular orbit with the smallest radius ($r = 3r_s/2$), the rotation angle approaches the value of -2π:

$$\lim_{2r \to 3r_s} \Omega = -2\pi \qquad (6.50)$$

That is, for a free-falling qubit moving across the smallest circular orbit allowed by Schwarzschild spacetime, the spin gives an entire rotation after a single revolution around the source.

Finally, let us consider the case of a qubit in circular orbit right on the surface of the Earth (assuming, of course, that Earth is a spherically symmetric object with zero atmosphere). In this case, the dimensionless Earth surface potential is given by:

$$\phi_0 \equiv \frac{M_\oplus G}{R_\oplus c^2} \approx 7 \times 10^{-10} \qquad (6.51)$$

and the total Wigner rotation angle after a single orbit is given by:

$$\Omega_c \approx -5 \times 10^{-10} \qquad (6.52)$$

which is indeed a very small number.

6.6.3 General equatorial circular paths ($\theta = \pi/2$, $a^r \neq 0$)

Let us now consider the case of a qubit moving around a circular path that is not necessarily a geodesic orbit. In particular, we assume that the qubit is moving in an equatorial ($\theta = \pi/2$), circular path of radius r, with an angular velocity ω, around the source of the gravitational field. Then, the four-velocity is given by:

$$
\begin{aligned}
u^t &= E \\
u^r &= 0 \\
u^\theta &= 0 \\
u^\varphi &= \omega E
\end{aligned}
\tag{6.53}
$$

where it clearly follows that the standard angular velocity is given by:

$$
\frac{d\varphi}{dt} = \frac{u^\varphi}{u^t} = \omega
\tag{6.54}
$$

and the normalization condition for the four-velocity:

$$
g_{tt}(u^t)^2 + g_{\varphi\varphi}(u^\varphi)^2 = -fE^2 + r^2\omega^2 E^2 = -1
\tag{6.55}
$$

implies that:

$$
E = \frac{1}{\sqrt{f - r^2\omega^2}}
\tag{6.56}
$$

Furthermore, the normalization condition suggests a rather convenient parameterization of E and ω:

$$
\begin{aligned}
f E^2 &= \cosh^2 \xi \\
r^2\omega^2 E^2 &= \sinh^2 \xi
\end{aligned}
\tag{6.57}
$$

and therefore:

$$
\begin{aligned}
u^t &= \frac{\cosh \xi}{\sqrt{f}} \\
u^\varphi &= \frac{\sinh \xi}{r}
\end{aligned}
\tag{6.58}
$$

This way, the non-zero components of the four-velocity in the local inertial frame defined by the tetrad field are given by:

$$
\begin{aligned}
u^0 &= e^0{}_t(x)\, u^t = \cosh \xi \\
u^3 &= e^3{}_\varphi(x)\, u^\varphi = \sinh \xi
\end{aligned}
\tag{6.59}
$$

and the speed of the particle in this frame is given by:

$$v = \frac{dx^3}{dx^0} = \frac{u^3}{u^0} = \frac{\sinh \xi}{\cosh \xi} = \tanh \xi \tag{6.60}$$

and therefore:

$$\omega = \frac{u^\varphi}{u^t} = \frac{\sqrt{f}}{r} v \tag{6.61}$$

and:

$$\sinh \xi = \frac{v}{\sqrt{1 - v^2}}$$

$$\cosh \xi = \frac{1}{\sqrt{1 - v^2}} \tag{6.62}$$

In general, these equations do not describe a free-falling particle. That is, the test particle will require a 'rocket pack' to produce a radial force that will allow it to have a circular orbit with the specific angular velocity at a given distance from the source. Indeed, let us recall that the acceleration due to external forces other than gravity is given by:

$$a^\mu = u^\nu \, D_\nu u^\mu = u^\nu \, \partial_\nu u^\mu + u^\nu \, \Gamma^\mu_{\nu\rho} u^\rho \tag{6.63}$$

which has the only non-zero component:

$$a^r = -\frac{f}{r} \left(1 - \frac{r_s}{2r} \frac{\coth^2 \xi}{f} \right) \sinh^2 \xi \tag{6.64}$$

and indicates that, indeed, the particle will require the rocket pack aligned in the radial direction. In the specific case where the test particle has an angular velocity that corresponds exactly to the angular velocity for a circular orbit at that given distance from the source, then the acceleration will be zero.

For the general case under consideration, the non-zero local Lorentz transformations are given by:

$$\lambda^1_{\ 0} = \lambda^0_{\ 1} = -\frac{\sqrt{f}}{r} \cosh \xi \, \sinh^2 \xi \left(1 - \frac{r_s}{2rf} \right)$$

$$\lambda^1_{\ 3} = -\lambda^3_{\ 1} = +\frac{\sqrt{f}}{r} \sinh \xi \, \cos^2 \xi \left(1 - \frac{r_s}{2rf} \right) \tag{6.65}$$

which correspond to boosts in the one-direction and rotations around the two-axis, respectively.

The infinitesimal Wigner rotation that corresponds to the rotation over the two-axis is computed to be:

$$\vartheta^1_{\ 3} = \frac{\sqrt{f}}{r}\left(1 - \frac{r_s}{2rf}\right) \sinh \xi \cosh \xi$$

$$= \frac{f\omega}{\sqrt{f - r^2\omega^2}}\left(1 - \frac{r_s}{2rf}\right) \tag{6.66}$$

and the total Wigner rotation, after the particle has moved across some proper time τ is given by:

$$\Theta = \int \vartheta^1_{\ 3} \, d\tau$$

$$= \int \vartheta^1_{\ 3} \frac{1}{\omega E} \, d\varphi$$

$$= \vartheta^1_{\ 3} \frac{r}{\sinh \xi} \int d\varphi$$

$$= \vartheta^1_{\ 3} \frac{r}{\sinh \xi} \Phi \tag{6.67}$$

where Φ is the angle traversed by the test particle during its motion. Therefore, the Wigner rotation purely due to gravity after a single revolution around the source of the gravitational field is given by:

$$\Omega = 2\pi \, \vartheta^1_{\ 3} \frac{r}{\sinh \xi} - 2\pi$$

$$= 2\pi\sqrt{f}\left(1 - \frac{r_s}{2rf}\right) \cosh \xi - 2\pi \tag{6.68}$$

It can be observed that Ω depends on both, the ratio r_s/r and the speed v (through the parameter ξ).

The behaviour of Ω with respect to the ratio $r_s/r \in [0, 0.9]$ and the velocity $v \in [0, 0.95]$ is shown in figure 6.2. A few interesting features can be observed from the plot and the analytical expression of Ω. First, for a given velocity v and a circular path radius r, the rotation angle Ω always decreases as the mass of the source of the gravitational field increases. Also, right at the event horizon in the strong field limit:

$$\lim_{r \to r_s} \Omega = -\infty \tag{6.69}$$

whereas the zero mass limit is given by:

$$\lim_{M \to 0} \Omega = 2\pi(\cosh \xi - 1) = 2\pi\left(\frac{1}{\sqrt{1 - v^2}} - 1\right) \tag{6.70}$$

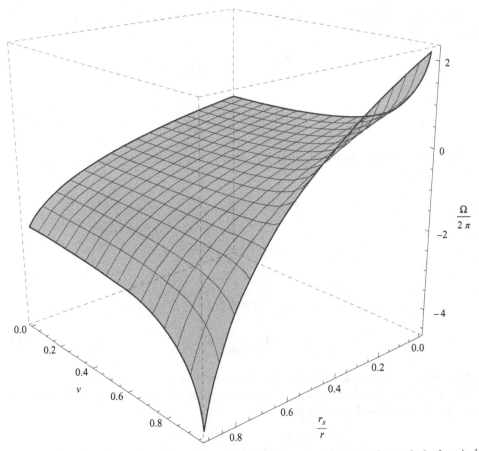

Figure 6.2. The effect of a static and isotropic gravitational field on the Wigner rotation angle Ω of a spin-$\frac{1}{2}$ qubit after moving for an entire period around an arbitrary circular path (not necessarily a free-falling circular orbit). The qubit has an angular velocity of $\omega = \sqrt{f}v/r$ in the general coordinate system and a velocity v in the local inertial frame defined by the tetrad field.

which, in general, the limit will be different from zero and its value will depend on the value of v.

Let us now consider a given value of the ratio r_s/r. The rotation angle Ω increases or decreases with the velocity v depending on the value of the ratio r_s/r as follows:

$$\frac{r_s}{r} < \frac{2}{3} \quad \Rightarrow \quad \frac{\partial}{\partial v}(\Omega) > 0$$

$$\frac{r_s}{r} = \frac{2}{3} \quad \Rightarrow \quad \frac{\partial}{\partial v}(\Omega) = 0 \qquad (6.71)$$

$$\frac{r_s}{r} > \frac{2}{3} \quad \Rightarrow \quad \frac{\partial}{\partial v}(\Omega) < 0$$

And, the following ultra-relativistic limits apply:

$$\frac{r_s}{r} < \frac{2}{3} \quad \Rightarrow \quad \lim_{v \to 1} \Omega = \infty$$

$$\frac{r_s}{r} = \frac{2}{3} \quad \Rightarrow \quad \lim_{v \to 1} \Omega = -2\pi \qquad (6.72)$$

$$\frac{r_s}{r} > \frac{2}{3} \quad \Rightarrow \quad \lim_{v \to 1} \Omega = -\infty$$

Notice that the value of the ratio $r_s/r = 2/3$ is a *fixed point* with respect to v, in the sense that Ω remains constant for all values of v. Furthermore, we recall that this value corresponds to the minimal radius for free-falling circular orbits.

Furthermore, $\Omega = 0$ if:

$$\sqrt{f}\left(1 - \frac{r_s}{2rf}\right)\cosh\xi = \frac{\sqrt{f}}{\sqrt{1 - v^2}}\left(1 - \frac{r_s}{2rf}\right) = 1 \qquad (6.73)$$

which is the condition when the net effect of the gravitational field on the Wigner rotation of the qubit is zero. Notice that $\Omega = 0$ can only be accomplished if:

$$\frac{r_s}{r} < \frac{2}{3} \qquad (6.74)$$

which corresponds to the condition for free-falling circular orbits.

To understand these results better, let us take a look at the non-relativistic limit:

$$v \ll 1 \Rightarrow \cosh\xi = \frac{1}{\sqrt{1 - v^2}} \approx 1 + \frac{v^2}{2} \qquad (6.75)$$

and, therefore, for Ω:

$$v \ll 1 \quad \Rightarrow \quad \frac{\Omega}{2\pi} \approx -1 + \sqrt{f}\left(1 - \frac{r_s}{2rf}\right)\left(1 + \frac{v^2}{2}\right)$$

$$= -1 + \sqrt{f}\left(1 - \frac{r_s}{2rf}\right) + \sqrt{f}\left(1 - \frac{r_s}{2rf}\right)\frac{v^2}{2}$$

$$= -1 + \sqrt{f} - \frac{r_s}{2r\sqrt{f}} + \frac{v^2}{2}\sqrt{f} + \frac{v^2 r_s}{4r\sqrt{f}} \qquad (6.76)$$

If we now take the case of weak gravitational fields ($r_s \ll 1$), we get:

$$v \ll 1 \quad \text{and} \quad r_s \ll 1 \Rightarrow \Omega \approx 2\pi\left(\frac{v^2}{2} - \frac{r_s}{r}\right) \qquad (6.77)$$

and although the second term is purely due to the presence of the gravitational field, the first term also contains a relativistic contribution due to the external acceleration a^μ that acts on the qubit.

If, within the context of these approximations, we also assume the case of a free-falling test particle $a^\mu = 0$, then its speed in the local inertial frame is given by equation (6.44) and can be approximated by:

$$v^2 = \frac{r_s}{2rf} \approx \frac{r_s}{2r} \tag{6.78}$$

and therefore:

$$\Omega \approx -2\pi \frac{3}{4} \frac{r_s}{r} \tag{6.79}$$

which clearly is a strictly gravitational effect.

Finally, let us take a look at the acceleration due to external (non-gravitational) forces:

$$a^r = -\frac{f}{r} \frac{v^2}{1-v^2} \left(1 - \frac{r_s}{2rfv^2}\right) \tag{6.80}$$

The acceleration has three distinctive regimes:

$$\frac{r_s}{r} < \frac{2v^2}{1+2v^2} \quad \Rightarrow \quad a^r < 0$$

$$\frac{r_s}{r} = \frac{2v^2}{1+2v^2} \quad \Rightarrow \quad a^r = 0 \tag{6.81}$$

$$\frac{r_s}{r} > \frac{2v^2}{1+2v^2} \quad \Rightarrow \quad a^r > 0$$

The middle case corresponds to free-falling qubits in circular orbits, and it is easy to check that the velocity corresponds to free-falling circular geodesics as shown in equation (6.44). In this case, it is also easy to confirm that the Wigner angle in equation (6.66) reduces to the Wigner angle for a free-falling circular geodesic in equation (6.45).

Also, if the gravitational field is weak such that the first condition in equation (6.79) holds, then the acceleration points *towards* the source of the gravitational field. That is, the rocket pack needs to 'enhance' the effects of gravity to keep the established circular path, by adding an extra centripetal force. On the other hand, if the gravitational field is strong enough so the third condition is true, then the acceleration points *away* from the source of the gravitational field. Indeed, in this case the gravitational force is so strong that the rocket pack needs to be used to reduce the centripetal force.

All the regions of interest in the phase space of the system given by the ratio r_s/r and the velocity v are shown in figure 6.3. The solid line represents the fixed point $\Omega = -2\pi$ at $r_s/r = 2/3$. Above the solid line is the region where Ω decreases with the velocity, and below is the region where Ω increases with the velocity. The dotted line corresponds to the points for which the net angle rotation is zero ($\Omega = 0$). And the dashed line represents the case of free-falling circular orbits ($a^r = 0$).

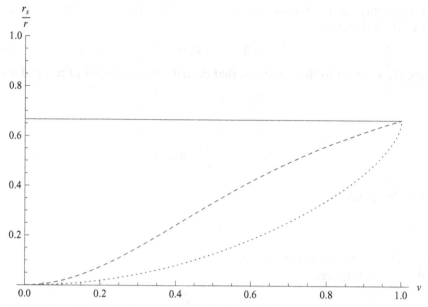

Figure 6.3. Regions of interest in the phase space of the system: fixed point (solid line), vanishing angular rotation (dotted line) and free-falling circular orbits (dashed line).

6.6.4 Geodetic precession of classical gyroscopes

As a comparison between classical and quantum phenomena, let us consider the case of a classical gyroscope in a circular orbit in Schwarzschild spacetime [1]. From section 2.9 we know that the equations of motion for this test particle are given by:

$$\frac{Du^\mu}{D\tau} = 0$$

$$\frac{DS^\mu}{D\tau} = 0 \tag{6.82}$$

and the orthogonality relation:

$$S_\mu u^\mu = 0 \tag{6.83}$$

where we have ignored the spin–curvature coupling. Furthermore, because the test particle is free-falling in a circular (geodetic) orbit, we know that:

$$u^\mu = \left(\frac{K}{f}, 0, 0, \frac{J}{r^2}\right) = \frac{K}{f}\left(1, 0, 0, \omega\right) \tag{6.84}$$

where:

$$\omega = \frac{u^\varphi}{u^t} = \frac{\mathrm{d}\varphi}{\mathrm{d}t} = \frac{Jf}{Kr^2} = \sqrt{\frac{r_s}{2r^3}} \tag{6.85}$$

is the angular velocity.

Let us assume the initial condition at $t = 0$ for the spatial part of S^μ to be a unit vector in the $\hat{\mathbf{r}}$ direction:

$$\mathbf{S}(0) = (1, 0, 0) \tag{6.86}$$

Then, the solution to the equations that describe the evolution of $\mathbf{S}$ are given by:

$$S^r(t) = \cos\tilde{\omega}t$$
$$S^\theta(t) = 0 \tag{6.87}$$
$$S^\varphi(t) = -\frac{\omega}{r\tilde{\omega}}\sin\tilde{\omega}t$$

where we have defined:

$$\tilde{\omega} \equiv \omega\frac{f}{K} \tag{6.88}$$

Clearly, the vector $\mathbf{S}$ does not return to its original orientation after a full orbit of period $T = 2\pi/\omega$. Indeed:

$$S^r(T) = \cos 2\pi\frac{\tilde{\omega}}{\omega}$$
$$S^\theta(T) = 0 \tag{6.89}$$
$$S^\varphi(T) = -\frac{\omega}{r\tilde{\omega}}\sin 2\pi\frac{\tilde{\omega}}{\omega}$$

and the vector makes an angle with the radial vector $\hat{\mathbf{r}}$ given by:

$$\Omega = 2\pi\frac{\tilde{\omega}}{\omega} - 2\pi = 2\pi\frac{\tilde{\omega} - \omega}{\omega} = -2\pi\left(1 - \frac{f}{K}\right) \tag{6.90}$$

In the weak gravity field limit, $r_s/r \ll 1$, we have:

$$\Omega = -2\pi\left(1 - \sqrt{1 - \frac{3r_s}{2r}}\right) \approx -2\pi\frac{3}{4}\frac{r_s}{r} \tag{6.91}$$

These equations are said to describe the *geodetic precession* of $\mathbf{S}$ [1, 3]. This equation should be compared with equation (6.48), which describes the Wigner rotation for a spin-$\frac{1}{2}$ particle. Furthermore, we notice that, for free-falling test particles, the classical expression for Ω in the weak field limit, equation (6.91), agrees with the quantum equation in the non-relativistic and weak field limit, equation (6.79).

6.7 Radiation damping

In this chapter we have considered the case of a spin-$\frac{1}{2}$ Dirac particle undergoing orbital motion in Schwarzschild spacetime. Furthermore, we know that Dirac spinors are commonly used to describe electrons, which are charged particles.

In addition, we know that accelerated charges radiate electromagnetic waves, and, therefore, accelerated charged particles tend to lose energy as they move [19, 20].

More specifically, the emitted radiation carries off energy, momentum and angular momentum [19]. Consequently, an electron moving in a circular orbit will radiate energy, and as it loses energy and momentum, its trajectory is affected, eventually spiralling down towards the source of the gravitational field. The effect of the emitted radiation on the motion of a charged particle is usually referred as *radiation reaction* or *radiation damping*.

In this section we will briefly discuss radiation damping for an orbiting electron. This discussion is important because we will give a rough estimate of the radiation emitted by an orbiting electron, and we will mention the difficulties in understanding the covariant expressions that describe this phenomena.

In flat spacetime, it is well known that the relativistic expression for the radiated power of an accelerated particle is given by:

$$P = -\frac{2}{3}\frac{e^2}{m^2c^3}\frac{dp^\mu}{d\tau}\frac{dp_\mu}{d\tau} \tag{6.92}$$

where p^μ is the four-momentum of an electron with electric charge e and mass m [19, 20]. This is the well-known *Larmor formula* for relativistic accelerated charges in Minkowski spacetime.

One could wrongly conclude that the generally covariant expression of the Larmor formula is given by invoking the minimal substitution rule to obtain:

$$P = -\frac{2}{3}\frac{e^2}{m^2c^3}\frac{\mathcal{D}p^\mu}{\mathcal{D}\tau}\frac{\mathcal{D}p_\mu}{\mathcal{D}\tau} \tag{6.93}$$

Indeed, this expression is obviously wrong for free-falling orbiting particles, as the covariant derivatives obey the geodesic equation and therefore are identically zero. As such, this expression leads to $P = 0$, in explicit contradiction with our expectations that accelerated charges radiate energy.

Furthermore, the radiation of an accelerated charge appears to violate the principle of equivalence. Let us recall that, according to the principle of equivalence, at every point in an arbitrary gravitational field, it is possible to choose a locally inertial coordinate system such that, in a very small region of space, the laws of nature take the same form as in an unaccelerated Cartesian coordinate system without gravitation. However, if we imagine a charged and a neutral particle in a free-falling frame, then an observer in this frame should be able to measure the energy radiated by the charged particle, and therefore he would conclude that his reference frame is being accelerated, in contradiction to the principle of equivalence.

The problem of a generally covariant expression equivalent to the Larmor formula in curved spacetime that satisfies the principle of equivalence has been extensively discussed in the literature [21–30]. In a nutshell, the accepted solution to these issues is to notice that radiation is not a generally covariant concept. As such, the concept of radiation depends on the specific frame of reference used by the observer [21, 26–27, 29, 32–33]. In other words, radiation effects are Lorentz invariant, but they are not invariant under general coordinate transformations.

Some generalizations of the Larmor formula for curved spacetimes are offered in the literature [34–37]. There are also similar generalizations for the *Lorentz–Abraham–Dirac equations*, which describe radiation damping (the effect of the radiation on the motion of the charged particle) [21–22, 28, 38–39].

In addition, the apparent paradox concerning the principle of equivalence is solved by noting that all the radiation goes into a region of space that is not accessible to the co-moving observer in the accelerated frame [40, 41]. Indeed, the co-moving observer will observe the charge as permanently at rest in an inertial frame, and therefore he will not observe any radiation.

It can be shown that, in the weak field limit, far from the event horizon, the radiated energy corresponds very closely to the classical formula for Minkowski space [22]. So, for instance, for a particle of mass m and momentum $\mathbf{p} = \gamma m \mathbf{v}$ we have:

$$E^2 = p^2 c^2 + m^2 c^4 \Rightarrow E \frac{dE}{d\tau} = p \frac{dp}{d\tau} c^2 \qquad (6.94)$$

where $p = |\mathbf{p}| = m\gamma|\mathbf{v}| = m\gamma v$ with $v = |\mathbf{v}|$, and therefore:

$$\frac{dE}{d\tau} = \frac{pc^2}{E} \frac{dp}{d\tau} = \frac{\gamma m v c^2}{\gamma m c^2} = v \frac{dp}{d\tau} \qquad (6.95)$$

where $E = \gamma m c^2$. Consequently, the radiated power can be written as:

$$P = -\frac{2}{3} \frac{e^2}{m^2 c^3} \frac{dp^\mu}{d\tau} \frac{dp_\mu}{d\tau}$$

$$= -\frac{2}{3} \frac{e^2}{m^2 c^3} \left[-\frac{1}{c^2} \left(\frac{dE}{d\tau} \right)^2 + \left(\frac{d\mathbf{p}}{d\tau} \right)^2 \right]$$

$$= -\frac{2}{3} \frac{e^2}{m^2 c^3} \left[-\beta^2 \left(\frac{d p}{d\tau} \right)^2 + \left(\frac{d\mathbf{p}}{d\tau} \right)^2 \right] \qquad (6.96)$$

In addition, for a circular orbit in a central force field, we have:

$$\frac{d\mathbf{p}}{d\tau} = -mr(u^\varphi)^2 \hat{\mathbf{r}}$$

$$= -mr \left(\frac{d\varphi}{d\tau} \right)^2 \hat{\mathbf{r}}$$

$$= -mr \left(\gamma \frac{d\varphi}{dt} \right)^2 \hat{\mathbf{r}}$$

$$= -mr \left(\gamma \frac{v}{r} \right)^2 \hat{\mathbf{r}}$$

$$= -\frac{m\gamma^2 v^2}{r} \hat{\mathbf{r}} \qquad (6.97)$$

On the other hand, for a Coulomb or Newtonian interaction, we have:

$$\frac{d\mathbf{p}}{d\tau} = \gamma \frac{d\mathbf{p}}{dt} = \gamma \mathbf{F} = -\frac{\gamma\mu}{r^2}\hat{\mathbf{r}} \tag{6.98}$$

where $\mathbf{F}$ is the force and μ is proportional to the product of charges for the Coulomb case, or proportional to the product of masses for the Newtonian case. Then, equating both expressions, we get:

$$\frac{m\gamma^2 v^2}{r} = \frac{\gamma\mu}{r^2} \Rightarrow \frac{v^2}{\gamma} = \frac{\mu}{mr} \tag{6.99}$$

Furthermore, for circular orbits, the spatial component of the four-momentum ($\mathbf{p}$) changes much more rapidly than the temporal component ($p^0 = E$) [19]. That is:

$$\left|\frac{d\mathbf{p}}{d\tau}\right| \gg \frac{1}{c}\left|\frac{dE}{d\tau}\right| \tag{6.100}$$

and consequently:

$$P = -\frac{2}{3}\frac{e^2}{m^2 c^3}\left[\left(\frac{d\mathbf{p}}{d\tau}\right)^2 - \beta^2\left(\frac{dp}{d\tau}\right)^2\right] \approx -\frac{2}{3}\frac{e^2}{m^2 c^3}\left(\frac{d\mathbf{p}}{d\tau}\right)^2 \tag{6.101}$$

Inserting the expression for circular orbits, we have:

$$P \approx -\frac{2}{3}\frac{e^2}{m^2 c^3}\left(\frac{m\gamma^2 v^2}{r}\right)^2 = -\frac{2}{3}\frac{e^2}{c^3}\frac{\gamma^4 v^4}{r^2} \tag{6.102}$$

Therefore, we can write the radiative energy loss per revolution as:

$$\delta E = T \times |P|$$

$$= \frac{2\pi}{\omega}\left(\frac{2}{3}\frac{e^2}{c^3}\frac{\gamma^4 v^4}{r^2}\right)$$

$$= \frac{2\pi r}{v}\left(\frac{2}{3}\frac{e^2}{c^3}\frac{\gamma^4 v^4}{r^2}\right)$$

$$= \frac{4\pi}{3}\frac{e^2}{r}\beta^3\gamma^4 \tag{6.103}$$

where T is the period of the circular orbit with radius r, v is the velocity, and β and γ are the standard relativistic kinematic variables.

Let us use this expression to give a rough estimate of the radiation emitted by an orbiting electron around a spherically symmetric source such as Earth. Furthermore, let us compare this value with the radiation emitted by a classical electron around a proton in a circular orbit given by *Bohr's radius* a_e under both a gravitational and an electromagnetic Coulomb interaction. As discussed before, Bohr's radius is given by:

$$a_e = \frac{\hbar}{me^2} \approx 10^{-11} \; m \tag{6.104}$$

which corresponds to the minimal circular orbit compatible with the principles of quantum mechanics.

For these circular orbits, the velocities are not ultra-relativistic, and therefore we can make the following approximation:

$$v_i \approx \sqrt{\frac{\mu_i}{mr}} \qquad (6.105)$$

where:

$$\mu_i = \begin{cases} \mu_g = GMm \\ \mu_p = Gm_p m \\ \mu_e = e^2 \end{cases} \qquad (6.106)$$

for gravitational interaction with Earth, gravitational interaction with a proton and electromagnetic interaction with a proton, respectively. In these equations, M is Earth's mass and m_p is the proton mass. Furthermore, in this approximation, let us write the total energy of the particle as:

$$E_i = \gamma_i mc^2 - \frac{\mu_i}{r} \qquad (6.107)$$

where:

$$\gamma_i = \frac{1}{\sqrt{1 - \beta_i^2}} \qquad \beta_i = \frac{v_i}{c} \qquad (6.108)$$

Then, the radiated energy per revolution is written as:

$$\delta E_i = \frac{4\pi}{3} \frac{e^2}{r} \beta_i^3 \gamma_i^4 \qquad (6.109)$$

where the numerical value of the square of the electric charge is given by $e^2 \approx 2.5 \times 10^{-28}$ kg m^3/s.

Then, the ratio of the radiated energy and the total energy for the three cases in consideration is given by:

$$\left. \frac{\delta E_e}{E_e} \right|_{a_e} \approx 1.3 \times 10^{-10}$$

$$\left. \frac{\delta E_p}{E_p} \right|_{a_e} \approx 1.4 \times 10^{-69} \qquad (6.110)$$

$$\left. \frac{\delta E_g}{E_g} \right|_{R_\oplus} \approx 4.7 \times 10^{-35}$$

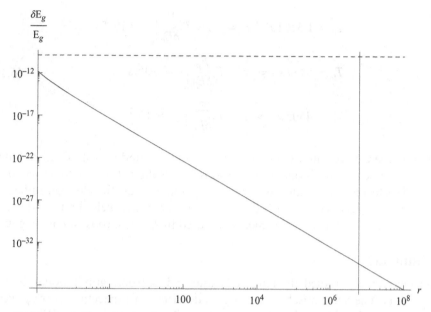

Figure 6.4. Relative energy radiated per revolution $\delta E_g/E_g$ due to the gravitational interaction between an electron and Earth with respect to the radius of the circular orbit r (solid line). Also, relative energy radiated per revolution $\delta E_e/E_e$ due to the electromagnetic interaction at Bohr's radius (dashed line).

That is, the relative energy radiated per revolution due to the gravitational interaction between a proton and an electron is about 10^{59} times smaller than the relative energy radiated per revolution due to the Coulomb interaction (in both cases calculated at the Bohr's radius $r = a_e$). Similarly, the relative energy radiated per revolution due to the gravitational interaction between an electron and Earth $(r = R_\oplus)$ is 10^{34} times greater than the relative energy radiated per revolution due to the gravitational interaction between a proton and an electron (at Bohr's radius $r = a_e$), but 10^{25} times smaller than the relative energy radiated per revolution due to the Coulomb interaction (at Bohr's radius).

The solid line in figure 6.4 shows the relative energy radiated per revolution $\delta E_g/E_g$ due to the gravitational interaction between an electron and Earth with respect to the radius of the circular orbit r. In addition, the dashed line shows the value of the relative energy radiated per revolution $\delta E_e/E_e$ due to the electromagnetic interaction at Bohr's radius. Then, in the weak field limit, the radiation in the gravitational case is much smaller than for the electromagnetic case. Recall that Earth's radius is about 6×10^6 m, and this value is marked by the vertical grid line in figure 6.4, clearly in the weak field limit. Notice that the event horizon of a black hole with the same mass as Earth is about 0.46 cm. However, let us remember that the approximations used are only valid in the regions far away from the event horizon.

The period of these circular orbits can be used to compute an upper bound for the lifetime of the orbits before the electron radiates all its energy.

$$T_e = 1.3 \times 10^{-16} s \;\Rightarrow\; \tau_e < T_e \frac{E_e}{\delta E_e}\bigg|_{a_e} \approx 10^{-6} s$$

$$T_p = 5937 s \;\Rightarrow\; \tau_e < T_p \frac{E_p}{\delta E_p}\bigg|_{a_e} \approx 10^{72} s \qquad (6.111)$$

$$T_g = 4505 s \;\Rightarrow\; \tau_e < T_g \frac{E_g}{\delta E_g}\bigg|_{R_\oplus} \approx 10^{37} s$$

That is, the electron in a Coulombian potential around a proton will radiate its energy very quickly (this is one of the reasons why Bohr's atomic model was quickly discarded). On the other hand, for the gravitational cases, the electron radiates its energy at relatively slow pace. Therefore, in the weak field limit and with a reasonable amount of orbits, the energy radiated by the electrons is quite negligible.

6.8 Summary

In this chapter we studied the effect of static and isotropic gravitational fields on spin-$\frac{1}{2}$ qubits. The Schwarzschild metric that describes this spacetime corresponds to the gravitational field produced by a spherically symmetric mass. We provided general formulas to compute the Wigner rotation of a particle in geodetic motion. In particular we discussed the case of a radial plunge and a circular orbit. We also considered the case of a general circular path, not necessarily a geodesic, which represents a qubit accelerated by some external force. The methods presented in this chapter can be easily generalized to arbitrary orbital motion [42].

Bibliography

[1] Hobson M P, Efstathiou G and Lasenby A N 2006 *General Relativity: An Introduction for Physicists* (Cambridge: Cambridge University Press)
[2] Misner C W, Thorne K S and Wheeler J A 1973 *Gravitation* (San Francisco, CA: W H Freeman)
[3] Weinberg S 1972 *Gravitation and Cosmology: Principles and Applications of The General Theory of Relativity* (New York: Wiley)
[4] Carroll S M 2004 *Spacetime and Geometry: An Introduction to General Relativity* (Reading, MA: Addison-Wesley)
[5] Earman J 1995 *Bangs, Crunches, Whimpers, and Shrieks: Singularities and Acausalities in Relativistic Spacetimes* (Oxford: Oxford University Press)
[6] Joshi P S 2008 *Gravitational Collapse and Spacetime Singularities* (Cambridge: Cambridge University Press)
[7] Chandraserkhar S 1992 *The Mathematical Theory of Black Holes* (Oxford: Oxford University Press)
[8] Wald R 1972 Gravitational spin interaction *Phys. Rev.* D **6** 406
[9] Terashima H and Ueda M 2004 Einstein–Rosen correlation in gravitational field *Phys. Rev.* A **69** 032113
[10] Papapetrou A 1951 Spinning test-particles in general relativity. I *Proc. R. Soc. London, Ser. A, Math. Phys. Sci.* **209** 248–58

[11] Corinaldesi E and Papapetrou A 1951 Spinning test-particles in general relativity. II *Proc. R. Soc. London, Ser. A, Math. Phys. Sci.* **209** 259–68

[12] Ubukhov Y N, Silenko A J and Teryaev O V 2010 Spin in stationary gravitational frames and rotating frames *The Sun, The Stars, The Universe and General Relativity* ed R Ruffini and G Vereshchagin (New York: AIP) pp 112–9

[13] Plyatsko R M, Stefanyshyn O B and Fenyk M T 2011 Mathisson Papapetrou & Dixon equations in the Schwarzschild and Kerr backgrounds *Class. Quantum Grav.* **28** 195025

[14] Plyatsko R M, Stefanyshyn O B and Fenyk M T 2010 Highly relativistic spinning particle starting near $r_{ph}^{(-)}$ in a Kerr field *Phys. Rev.* D **82** 044015

[15] Plyatsko R 2005 Ultrarelativistic circular orbits of spinning particles in a Schwarzschild field *Class. Quantum Grav.* **22** 1545–51

[16] Silenko A J 2008 Classical and quantum spins in curved spacetimes *Acta Phys. Polonica B Proc. Suppl.* **1** 87–107

[17] Plyatsko R 1998 Gravitational ultrarelativistic spin-orbit interaction and the weak equivalence principle *Phys. Rev.* D **58** 084031

[18] Singh D 2008 Perturbation method for classical spinning particle motion: I. Kerr space-time *Phys. Rev.* D **78** 104028

[19] Jackson J D 1975 *Classical Electrodynamics* 2nd edn (New York: Wiley)

[20] Landau L D and Lifshitz E M 1975 *The Classical Theory of Fields, Course of Theoretical Physics* vol **2** (Amsterdam: Elsevier)

[21] DeWitt B S and Brehme R W 1960 Radiation damping in a gravitational field *Ann. Phys.* **9** 220–59

[22] DeWitt B S and DeWitt C M 1964 Falling charges *Physics* **1** 320

[23] Eriksen E and Gron O 2007 On the energy and momentum of an accelerated charged particle and the sources of radiation *Eur. J. Phys.* **28** 401–7

[24] Fabri L 2005 Free falling electric charge in a static homogeneous gravitational field *Ann. Fond. Louis de Broglie* **30** 87–95

[25] Fil'chenkov M L 1990 Quantum radiation of a charged particle in a Schwarzschild field *Astron. Nacht.* **311** 223–6

[26] Gron O 2012 Electrodynamics of radiating charges *Adv. Math. Phys.* **2012** 528631

[27] Parrott S 1997 Radiation from a charge uniformly accelerated for all time *Gen. Relativ. Gravitation* **29** 1463–72

[28] Poisson E, Pound A and Vega I 2004 The motion of point particles in curved spacetime *Living Rev. Relativ.* **7** 1

[29] Shariati A and Khorrami M 1999 Equivalence principle and radiation by a uniformly accelerated charge *Found. Phys. Lett.* **12** 427–39

[30] Soker N and Harpaz A 2004 Radiation from a charge supported in a gravitational field *Gen. Relativ. Gravitation* **36** 315–30

[31] Ginzburg V L and Eroshenko N Yu 1995 Once again about the equivalence principle *Phys. Usp.* **33** 195–201

[32] Rohrlich F 1963 The principle of equivalence *Ann. Phys.* **22** 169–91

[33] Rosen N 1962 Field of a particle in uniform motion and uniform acceleration *Ann. Phys.* **17** 269–75

[34] Shariati A and Khorrami M 1999 Equivalence principle and radiation by a uniformly accelerated charge *Found. Phys. Lett.* **12** 427–39

[35] Fugmann W and Kretzschmar M 1991 Classical electromagnetic radiation in noninertial reference frames *Il Nuovo Cimento B* **106** 351373

[36] Hirayama T 2001 Bound and radiation fields in the Rindler frame *Prog. Theor. Phys.* **106** 71–97

[37] Hirayama T 2002 Classical radiation formula in the Rindler frame *Prog. Theor. Phys.* **108** 679–88

[38] Kretzschmar M and Fugmann W 1989 The electromagnetic field of an accelerated charge in the proper reference frame of a noninertial observer *Il Nuovo Cimento* **103** 389–412

[39] Gron O and Naess S K 2008 An electromagnetic perpetuum mobile? ArXiv:0806.0464

[40] Hobbs J M 1968 A vierbein formalism of radiation damping *Ann. Phys.* **47** 141–65

[41] De Almeida C and Saa A 2006 The radiation of a uniformly accelerated charge is beyond the horizon: a simple derivation *Am. J. Phys.* **74** 154–58

[42] Boulware D G 1980 Radiation from a uniformly accelerated charge *Ann. Phys.* **124** 169–88

[43] Bonior D 2013 Relativistic Effects on Orbiting Spin Entangled Electrons *BSc Thesis* Middle Tennessee State University

Chapter 7

Spin–curvature coupling in Schwarzschild spacetime

In this chapter we will discuss a couple of simple examples of the spin–curvature coupling in Schwarzschild spacetime. That is, we will study the case of a free-falling spin-$\frac{1}{2}$ qubit in a radial path and in a circular orbit in the Schwarzschild metric [1]. We will compute the deviation from geodetic motion at order $\hbar$, and the resulting change in the Wigner angle. This problem has been discussed before in the literature [2]. However, we offer a different, simpler and more detailed approach. Despite its simplicity, this example illustrates the complex dynamics of particles with spin in curved spacetime.

7.1 Spinor components

Let us consider a two-spinor of the form:

$$\chi = \begin{pmatrix} \alpha \\ \beta \end{pmatrix} \tag{7.1}$$

where $\alpha, \beta \in \mathbb{C}$. In general, the orientation of the spin in a local inertial coordinate system is given by the vector:

$$\mathbf{s} = (\sin \zeta \cos \xi, \, \sin \zeta \sin \xi, \, \cos \zeta) \tag{7.2}$$

where the two spherical angular variables ξ and ζ are related to the spinor components by:

$$\alpha = \cos(\zeta/2)$$
$$\beta = e^{i\xi} \sin(\zeta/2) \tag{7.3}$$

doi:10.1088/978-1-627-05330-3ch7

Then, we can write the spin vector as:

$$\begin{aligned}
\mathbf{s} &= (\sin\zeta\cos\xi,\ \sin\zeta\sin\xi,\ \cos\zeta) \\
&= (2\Re(\alpha^*\beta),\ 2\Im(\alpha^*\beta),\ \alpha\alpha^* - \beta\beta^*) \\
&= (s_1,\ s_2,\ s_3)
\end{aligned} \tag{7.4}$$

where $\Re$ and $\Im$ denote the real and imaginary components of their arguments, respectively.

Thus, for instance, the computational basis $\{|0\rangle, |1\rangle\}$ is given by:

$$|0\rangle = \begin{pmatrix} 1 \\ 0 \end{pmatrix} \Rightarrow \begin{pmatrix} \zeta \\ \xi \end{pmatrix} = \begin{pmatrix} 0 \\ 0 \end{pmatrix} \Rightarrow \mathbf{s}_0 = (0,\ 0,\ 1)$$

$$|1\rangle = \begin{pmatrix} 0 \\ 1 \end{pmatrix} \Rightarrow \begin{pmatrix} \zeta \\ \xi \end{pmatrix} = \begin{pmatrix} \pi \\ 0 \end{pmatrix} \Rightarrow \mathbf{s}_1 = (0,\ 0,\ -1) \tag{7.5}$$

which shows that the three-axis is the quantization axis.

On the other hand, the spinors on the 'diagonal basis':

$$|+\rangle = \frac{1}{\sqrt{2}} \begin{pmatrix} 1 \\ 1 \end{pmatrix} \Rightarrow \begin{pmatrix} \zeta \\ \xi \end{pmatrix} = \begin{pmatrix} \pi/2 \\ 0 \end{pmatrix} \Rightarrow \mathbf{s}_+ = (1,\ 0,\ 0)$$

$$|-\rangle = \frac{1}{\sqrt{2}} \begin{pmatrix} 1 \\ -1 \end{pmatrix} \Rightarrow \begin{pmatrix} \zeta \\ \xi \end{pmatrix} = \begin{pmatrix} \pi \\ -\pi \end{pmatrix} \Rightarrow \mathbf{s}_- = (-1,\ 0,\ 0) \tag{7.6}$$

correspond to a quantization axis parallel to the one-direction.

Finally, for spinors in the '$+_i$' diagonal basis we get:

$$|+_i\rangle = \frac{1}{\sqrt{2}} \begin{pmatrix} 1 \\ i \end{pmatrix} \Rightarrow \begin{pmatrix} \zeta \\ \xi \end{pmatrix} = \begin{pmatrix} \pi/2 \\ \pi/2 \end{pmatrix} \Rightarrow \mathbf{s}_{+i} = (0,\ 1,\ 0)$$

$$|-_i\rangle = \frac{1}{\sqrt{2}} \begin{pmatrix} 1 \\ -i \end{pmatrix} \Rightarrow \begin{pmatrix} \zeta \\ \xi \end{pmatrix} = \begin{pmatrix} \pi \\ -\pi/2 \end{pmatrix} \Rightarrow \mathbf{s}_{-i} = (0,\ -1,\ 0) \tag{7.7}$$

which corresponds to a quantization axis parallel to the two-direction.

7.2 Constant spinors in spherical coordinates

Let us first consider the expressions that describe spinors in flat space–time in spherical coordinates. The covariant derivative will be different from the standard derivative because we are not using Cartesian coordinates. Indeed, we recall that the spinor connection is given by:

$$\Gamma_\mu = -\frac{1}{8}\omega_\mu{}^{ab}[\gamma_a, \gamma_b] \tag{7.8}$$

and after some algebra its components are found to be:

$$\Gamma_t = 0$$
$$\Gamma_r = 0$$
$$\Gamma_\theta = \frac{i}{2}\begin{pmatrix} \sigma_3 & 0 \\ 0 & \sigma_3 \end{pmatrix} \tag{7.9}$$
$$\Gamma_\varphi = -\frac{i\sin\theta}{2}\begin{pmatrix} \sigma_2 & 0 \\ 0 & \sigma_2 \end{pmatrix} + \frac{i\cos\theta}{2}\begin{pmatrix} \sigma_1 & 0 \\ 0 & \sigma_1 \end{pmatrix}$$

where σ_i are the Pauli matrices in their standard representation. The φ component can be written is a simplified way as:

$$\Gamma_\varphi = \frac{i}{2}\begin{pmatrix} 0 & e^{i\theta} & 0 & 0 \\ e^{-i\theta} & 0 & 0 & 0 \\ 0 & 0 & 0 & e^{i\theta} \\ 0 & 0 & e^{-i\theta} & 0 \end{pmatrix} \tag{7.10}$$

Let us consider now a constant four-spinor ψ_c whose expression may change at each point in spacetime due to the structure of the spherical coordinates. We know that its covariant derivative has to be zero, as there are no gravitational fields involved. That is:

$$\mathcal{D}_\mu \psi_c = (\partial_\mu - \Gamma_\mu)\psi_c = 0 \tag{7.11}$$

which implies two non-trivial equations:

$$(\partial_\theta - \Gamma_\theta)\psi_c = 0$$
$$(\partial_\varphi - \Gamma_\varphi)\psi_c = 0 \tag{7.12}$$

For the θ component:

$$\partial_\theta\psi_c = \frac{i}{2}\begin{pmatrix} \sigma_3 & 0 \\ 0 & \sigma_3 \end{pmatrix}\psi_c \tag{7.13}$$

As this expression implies two linearly independent equations on the components of ψ_c (the diagonal matrix implies the exact same equation for the upper two and the lower two components of ψ_c), we can write without loss of generality:

$$\psi_c = \begin{pmatrix} \chi_c \\ \chi_c \end{pmatrix} \tag{7.14}$$

where χ_c is a two-component spinor:

$$\chi_c = \begin{pmatrix} \chi_1 \\ \chi_2 \end{pmatrix} \tag{7.15}$$

that satisfies:

$$\partial_\theta \chi_c = \frac{i}{2}\sigma_3 \chi_c \qquad (7.16)$$

and therefore:

$$\partial_\theta \chi_1 = \frac{i}{2}\chi_1$$
$$\partial_\theta \chi_2 = -\frac{i}{2}\chi_2 \qquad (7.17)$$

that has the solutions:

$$\chi_1 = A_1(\varphi)\, e^{i\theta/2}$$
$$\chi_2 = A_2(\varphi)\, e^{-i\theta/2} \qquad (7.18)$$

where A_1 and A_2 are functions of φ that remain to be determined.

Proceeding to the equations for the φ component, we get:

$$\partial_\varphi \chi_1 = \frac{i}{2}e^{i\theta}\chi_2$$
$$\partial_\varphi \chi_2 = \frac{i}{2}e^{-i\theta}\chi_1 \qquad (7.19)$$

and inserting the expressions obtained from the θ component equation, we obtain:

$$e^{i\theta/2}\partial_\varphi A_1 = \frac{i}{2}e^{i\theta/2}A_2$$
$$e^{-i\theta/2}\partial_\varphi A_2 = \frac{i}{2}e^{-i\theta/2}A_1 \qquad (7.20)$$

which imply:

$$\partial_\varphi A_1 = \frac{i}{2}A_2$$
$$\partial_\varphi A_2 = \frac{i}{2}A_1 \qquad (7.21)$$

with two linearly independent solutions given by:

$$A_1 = A_2 = e^{i\varphi/2}$$
$$A_1 = -A_2 = e^{-i\varphi/2} \qquad (7.22)$$

Therefore, there are two two-spinors that satisfy the covariant differential equation:

$$\chi_c^+ = \frac{1}{\sqrt{2}} \begin{pmatrix} e^{i/2(\varphi+\theta+\alpha)} \\ e^{i/2(\varphi-\theta+\alpha)} \end{pmatrix} \qquad \chi_c^- = \frac{1}{\sqrt{2}} \begin{pmatrix} e^{i/2(-\varphi+\theta+\alpha)} \\ -e^{i/2(-\varphi-\theta+\alpha)} \end{pmatrix} \qquad (7.23)$$

where α is a constant global phase and the spinors are orthogonal and properly normalized to unity:

$$\chi_c^{+\dagger}\chi_c^- = \chi_c^{-\dagger}\chi_c^+ = 0$$
$$\chi_c^{+\dagger}\chi_c^+ = \chi_c^{-\dagger}\chi_c^- = 1 \qquad (7.24)$$

Then, any linear combination of the form:

$$\psi = \frac{1}{\sqrt{2}} \begin{pmatrix} \alpha_1\chi_c^+ + \beta_1\chi_c^- \\ \alpha_2\chi_c^+ + \beta_2\chi_c^- \end{pmatrix} \qquad (7.25)$$

with:

$$\alpha_i, \beta_i \in \mathbb{C} \qquad |\alpha_i|^2 + |\beta_i|^2 = 1 \qquad i = 1, 2 \qquad (7.26)$$

will satisfy:

$$\mathcal{D}_\mu \psi = 0 \qquad (7.27)$$

as originally required.

For the specific example of a spinor in the orbital plane, $\theta = \pi/2$, and setting the global phase to zero, $\alpha = 0$, we have:

$$\chi_c^+ = \frac{e^{i\varphi/2}}{2} \begin{pmatrix} 1+i \\ 1-i \end{pmatrix} \qquad \chi_c^- = \frac{e^{-i\varphi/2}}{2} \begin{pmatrix} 1+i \\ -1+i \end{pmatrix} \qquad (7.28)$$

Notice that these spinors form an orthonormal basis, and then, for instance, the computational basis spinor can be obtained by making the right selection of the parameters α, β. Indeed, consider as an example the spinor in the computational basis at $\varphi = 0$:

$$\begin{aligned} \alpha_1 = \beta_1 = \frac{1-i}{2} \\ \alpha_2 = \beta_2 = 0 \end{aligned} \qquad \Rightarrow \qquad \psi = \begin{pmatrix} 1 \\ 0 \\ 0 \\ 0 \end{pmatrix} \qquad (7.29)$$

Then, for an arbitrary angle φ, this constant spinor takes the form:

$$\psi = \frac{1}{2} \begin{pmatrix} e^{i\varphi/2} + e^{-i\varphi/2} \\ i\left(-e^{i\varphi/2} + e^{-i\varphi/2}\right) \\ 0 \\ 0 \end{pmatrix}$$

$$= \begin{pmatrix} \cos\varphi/2 \\ \sin\varphi/2 \\ 0 \\ 0 \end{pmatrix} \tag{7.30}$$

just as we could have expected.

7.3 The Lemaitre tetrad

As mentioned in chapter 5, the computation of the deviation from geodetic motion of a free-falling particle is simplified if we use a tetrad field in which the spinor is at rest. So, let us consider the *Lemaitre tetrad field*, which describes inertial observers in a free radial fall from rest at infinity in Schwarzschild spacetime.

7.3.1 Tetrad fields

The non-zero components of the Lemaitre tetrad field are given by:

$$\tilde{e}_0^{\ t} = \frac{1}{f}$$

$$\tilde{e}_0^{\ r} = -\sqrt{1-f}$$

$$\tilde{e}_1^{\ r} = 1$$

$$\tilde{e}_1^{\ t} = -\frac{\sqrt{1-f}}{f} \tag{7.31}$$

$$\tilde{e}_2^{\ \theta} = \frac{1}{r}$$

$$\tilde{e}_3^{\ \varphi} = \frac{1}{r\sin\theta}$$

and the inverse components:

$$\tilde{e}^0{}_t = 1$$

$$\tilde{e}^0{}_r = \frac{\sqrt{1-f}}{f}$$

$$\tilde{e}^1{}_r = \frac{1}{f} \qquad (7.32)$$

$$\tilde{e}^1{}_t = \sqrt{1-f}$$

$$\tilde{e}^2{}_\theta = r$$

$$\tilde{e}^3{}_\varphi = r\sin\theta$$

In the above expressions, we have used tildes to help us distinguish between the Lemaitre and the standard tetrad field components used in the previous chapter.

It can be observed that, for a stationary observer in the local inertial frame of the tetrad field:

$$u^a = (1, 0, 0, 0) \qquad (7.33)$$

its four-velocity in the general coordinate system is given by:

$$u^\mu = \tilde{e}_a{}^\mu u^a = \left(\frac{1}{f}, -\sqrt{1-f}, 0, 0 \right) \qquad (7.34)$$

which corresponds to a free-falling particle moving in a radial path toward the source of the gravitational field.

Furthermore, it can be easily verified that the Lemaitre tetrad field leads to the Schwarzschild metric:

$$g^{\mu\nu}(x) = \tilde{e}_a{}^\mu(x)\, \tilde{e}_b{}^\nu(x)\, \eta^{ab} \qquad (7.35)$$

where $g^{\mu\nu}$ is the Schwarzschild metric tensor. In addition, after some algebra it is possible to show that the local Lorentz transformation that transforms the standard tetrad field $e_a{}^\mu(x)$ into the Lemaitre tetrad field $\tilde{e}_a{}^\mu(x)$ is given by:

$$e_a{}^\mu(x) = \tilde{e}_b{}^\mu(x)\, \Lambda^b{}_a(x) \qquad (7.36)$$

where the numerical elements of $\Lambda^b{}_a$ are given in matrix form by:

$$\Lambda^b{}_a = \begin{pmatrix} \cosh\xi & \sinh\xi & 0 & 0 \\ \sinh\xi & \cosh\xi & 0 & 0 \\ 0 & 0 & 1 & 0 \\ 0 & 0 & 0 & 1 \end{pmatrix} \qquad (7.37)$$

which corresponds to a boost in the one-direction (parallel to the $\hat{r}$ axis) where:

$$\cosh \xi = \frac{1}{\sqrt{f}}$$

$$\sinh \xi = \sqrt{\frac{1-f}{f}}$$

(7.38)

and the speed of the boost is given by:

$$v = \tanh \xi = \sqrt{1-f}$$

(7.39)

This makes sense: the 'hovering' tetrad field $e_a{}^\mu(x)$ has to be boosted along the $\hat{r}$ axis to account for the free-falling observer that moves with $\tilde{e}_a{}^\mu(x)$.

7.3.2 Connection one-forms

The connection one-forms are given by:

$$\omega_\mu{}^{ab}(x) = \tilde{e}^{a\alpha}(x) \, \nabla_\mu \tilde{e}^b{}_\alpha(x)$$

(7.40)

and the non-zero connection one-forms are found to be:

$$\omega_t{}^1{}_0 = \omega_t{}^0{}_1 = \frac{\dot{f}}{2}$$

$$\omega_r{}^1{}_0 = \omega_r{}^0{}_1 = \frac{\dot{f}}{2f\sqrt{1-f}}$$

$$\omega_\theta{}^2{}_0 = \omega_\theta{}^0{}_2 = -\sqrt{1-f}$$

$$\omega_\theta{}^1{}_2 = -\omega_\theta{}^2{}_1 = -1$$

$$\omega_\varphi{}^0{}_3 = \omega_\varphi{}^3{}_0 = -\sqrt{1-f}\sin\theta$$

$$\omega_\varphi{}^1{}_3 = -\omega_\varphi{}^3{}_1 = -\sin\theta$$

$$\omega_\varphi{}^2{}_3 = -\omega_\varphi{}^3{}_2 = -\cos\theta$$

(7.41)

7.3.3 The spinor connection

The spinor connection is given by:

$$\Gamma_\mu = -\frac{1}{8}\omega_\mu{}^{ab}[\gamma_a, \gamma_b]$$

(7.42)

and after some algebra its components are found to be:

$$\Gamma_t = -\frac{\dot{f}}{4}\begin{pmatrix} 0 & \sigma_1 \\ \sigma_1 & 0 \end{pmatrix}$$

$$\Gamma_r = -\frac{\dot{f}}{4f\sqrt{1-f}}\begin{pmatrix} 0 & \sigma_1 \\ \sigma_1 & 0 \end{pmatrix}$$

$$\Gamma_\theta = \frac{\sqrt{1-f}}{2}\begin{pmatrix} 0 & \sigma_2 \\ \sigma_2 & 0 \end{pmatrix} + \frac{i}{2}\begin{pmatrix} \sigma_3 & 0 \\ 0 & \sigma_3 \end{pmatrix}$$

$$\Gamma_\varphi = \frac{\sin\theta\sqrt{1-f}}{2}\begin{pmatrix} 0 & \sigma_3 \\ \sigma_3 & 0 \end{pmatrix} - \frac{i\sin\theta}{2}\begin{pmatrix} \sigma_2 & 0 \\ 0 & \sigma_2 \end{pmatrix} + \frac{i\cos\theta}{2}\begin{pmatrix} \sigma_1 & 0 \\ 0 & \sigma_1 \end{pmatrix}$$

(7.43)

where σ_i are the Pauli matrices in their standard representation.

7.4 Radial fall

A test particle in geodetic radial motion dropped from infinity is characterized by the four-velocity:

$$u^\mu = \left(\frac{1}{f}, -\sqrt{1-f}, 0, 0\right) \tag{7.44}$$

which corresponds to a static test particle in the Lemaitre tetrad:

$$u^a = \tilde{e}^a_{\ \mu}u^\mu = (1, 0, 0, 0) \tag{7.45}$$

Also, from chapter 5 we know that the general motion of a spin-$\frac{1}{2}$ particle is described by the four-velocity:

$$v^\mu = u^\mu + \delta u^\mu \tag{7.46}$$

where u^μ is the geodetic four-velocity:

$$u^\mu D_\mu u^\nu = 0 \tag{7.47}$$

and δu^μ is the deviation from the geodesic path given by equation (5.301):

$$\delta u^\mu = \frac{\hbar}{mi} g^{\mu\nu} \bar{\chi}_0 D_\nu \chi_0 + \mathcal{O}(\hbar^2) \tag{7.48}$$

where χ_0 is the four-spinor in the local inertial frame in which the particle is at rest. In this frame we can write the general form of χ_0 as:

$$\chi_0 = \begin{pmatrix} \alpha \\ \beta \\ 0 \\ 0 \end{pmatrix} \tag{7.49}$$

where α and β are complex numbers normalized as:

$$\alpha\alpha^* + \beta\beta^* = 1 \tag{7.50}$$

and therefore χ_0 represents a superposition of spin-up and spin-down positive energy eigenstates. Notice that, because the spinor is being described in the frame in which it is at rest, the *lower components* (i.e. the third and fourth components) are identically zero, whereas the *upper components* (i.e. the first and second components) are different from zero.

It is easy to check that, in this case, the covariant derivative of the spinor is identically zero. Indeed, all the terms in the spinor connection that involve the gravitational field involve matrices that combine lower with upper spinor components. As χ_0 has no lower components, then the covariant derivative is zero[1].

Therefore, for a free-falling spinor following a geodetic radial path, the non-geodetic component of the four-velocity is given by:

$$\delta u^\mu = 0 \tag{7.51}$$

and the resulting motion is completely geodetic.

7.5 The co-moving tetrad field

It is convenient to define a set of tetrads that move along the inertial observer [6]. These *co-moving tetrads* satisfy the *co-moving conditions*:

$$\mathrm{d}x^a = e^a_{\ \mu}\mathrm{d}x^\mu = 0 \qquad a = 1, 2, 3$$

$$\mathrm{d}x^0 = e^0_{\ \mu}\mathrm{d}x^\mu = \mathrm{d}S \tag{7.52}$$

$$e^{\ \mu}_0 = u^\mu$$

as well as the standard tetrad conditions:

$$e^{\ \mu}_a e^b_{\ \mu} = \delta^b_a$$

$$e^{\ \mu}_a e^{\ \nu}_b \eta^{ab} = g^{\mu\nu} \tag{7.53}$$

In brief, we can describe the co-moving frame as the inertial frame in which accelerated observers are instantaneously at rest [3–5]. We immediately notice, for instance, that the Lemeitre tetrad is a special case of the co-moving tetrad field.

---

[1] Notice that the only terms in the spinor connection that do not combine upper and lower components are those that merely describe the spinor in spherical coordinates. As shown in the second section of the present chapter, these terms do not contribute to the covariant derivative with the right expression of the spinor χ_0 in spherical coordinates.

In Schwarzschild spacetime, in the equatorial plane ($\theta = \pi/2$), the non-trivial components of the co-moving tetrad field are given by:

$$e_0^{\ r} = u^r$$

$$e_0^{\ \varphi} = u^\varphi$$

$$e_0^{\ t} = u^t$$

$$e_2^{\ \theta} = 1/r$$

$$e_3^{\ r} = u^r u^t \sqrt{\frac{f}{f u^t u^t - 1}}$$

$$e_3^{\ \varphi} = u^\varphi u^t \sqrt{\frac{f}{f u^t u^t - 1}} \tag{7.54}$$

$$e_3^{\ t} = \sqrt{\frac{f u^t u^t - 1}{f}}$$

$$e_1^{\ r} = u^\varphi r \sqrt{\frac{f}{f u^t u^t - 1}}$$

$$e_1^{\ \varphi} = -\frac{u^r}{r} \frac{1}{\sqrt{(f u^t u^t - 1)f}}$$

It is easy to check that these tetrad components satisfy the standard and co-moving tetrads conditions. Notice that we are using a different orientation of the local inertial frame axes to those used in [6]. The transformation rule to go from the published convention to ours is simply given by the permutation:

$$1 \rightarrow 2 \qquad 2 \rightarrow 3 \qquad 3 \rightarrow 1 \tag{7.55}$$

which denotes the indices in the local inertial frame. We have chosen the present convention to identify the axes of the local inertial frames to have a consistent notation throughout the entire book.

7.6 General motion in the equatorial plane

Let us consider a test particle in arbitrary geodesic motion in the equatorial plane ($\theta = \pi/2$) with four-velocity given by:

$$u^\mu = (u^t, u^r, 0, u^\varphi) \tag{7.56}$$

As shown in chapter 5, for a spin-$\frac{1}{2}$ Dirac spinor, the force due to the spin–curvature coupling is given by:

$$f^\alpha = \frac{\hbar}{4} g^{\alpha\mu} u^\nu R_{\mu\nu\gamma\delta} \overline{\psi}_0 \sigma^{\gamma\delta} \psi_0$$

$$= \frac{\hbar}{4} g^{\alpha\mu} u^\nu R_{\mu\nu\gamma\delta} e_a^{\ \gamma} e_b^{\ \delta} \overline{\psi}_0 \sigma^{ab} \psi_0 \tag{7.57}$$

where ψ_0 is the spinor as described in a co-moving frame, $R_{\mu\nu\gamma\delta}$ is the curvature tensor and:

$$\sigma^{ab} = \frac{i}{2}\left[\gamma^a, \gamma^b\right] \tag{7.58}$$

is the commutator of Dirac matrices in the standard representation. As discussed before, this force makes the test particle deviate from its geodetic motion.

The components of the force can be calculated using the Riemman tensor components for the Schwarzschild metric presented in the previous chapter. After some tensor algebra, the final result is found to be:

$$f^t = \frac{\hbar}{4f}\frac{3r_s}{r^2}\, u^r u^\varphi\, \overline{\psi}_0 \sigma^{31} \psi_0$$

$$f^r = -\frac{\hbar}{4}\frac{3r_s f}{r^2}\, u^\varphi u^t\, \overline{\psi}_0 \sigma^{31} \psi_0 \tag{7.59}$$

$$f^\theta = -\frac{\hbar}{4}\frac{3r_s}{r^2}\sqrt{\frac{f}{fu^t u^t - 1}}\, u^t u^\varphi u^\varphi\, \overline{\psi}_0 \sigma^{23} \psi_0 - \frac{\hbar}{4}\frac{r_s}{r^3}\frac{1}{\sqrt{(fu^t u^t - 1)f}}\, u^r u^\varphi\, \overline{\psi}_0 \sigma^{12} \psi_0$$

$$f^\varphi = 0$$

Alternatively, using the spin components defined in the opening section of the present chapter, we get:

$$f^t = \frac{\hbar}{4f}\frac{3r_s}{r^2}\, u^r u^\varphi\, s_2$$

$$f^r = -\frac{\hbar}{4}\frac{3r_s f}{r^2}\, u^\varphi u^t\, s_2 \tag{7.60}$$

$$f^\theta = -\frac{\hbar}{4}\frac{3r_s}{r^2}\sqrt{\frac{f}{fu^t u^t - 1}}\, u^t u^\varphi u^\varphi\, s_1 - \frac{\hbar}{4}\frac{r_s}{r^3}\frac{1}{\sqrt{(fu^t u^t - 1)f}}\, u^r u^\varphi\, s_3$$

$$f^\varphi = 0$$

These results agree with what has been reported in the literature for classical particles [6]. Indeed, the spatial components of the spin–curvature force in the local inertial frame of the tetrads for a particle of spin $\hbar/2$ in the s_2 direction are given by:

$$f^1 = e^1_{\ r} f^r = \frac{3r_s}{2r}\sqrt{\frac{f}{fu^t u^t - 1}} u^\varphi u^\varphi u^t \left(\frac{\hbar}{2}\right)$$

$$f^2 = e^2_{\ \theta} f^\theta = 0 \tag{7.61}$$

$$f^3 = e^3_{\ t} f^t + e^3_{\ r} f^r = -\frac{3r_s}{2r^2}\frac{u^r u^\varphi}{\sqrt{(fu^t u^t - 1)f}} \left(\frac{\hbar}{2}\right)$$

Table 7.1. Spin–curvature force for three different qubit states in the equatorial plane ($\theta = \pi/2$).

Qubit State	Spin Component	f^r	f^θ	f^φ
$\lvert+\rangle, \lvert-\rangle$	s_1	0	$\neq 0$	0
$\lvert+_i\rangle, \lvert-_i\rangle$	s_2	$\neq 0$	0	0
$\lvert 0\rangle, \lvert 1\rangle$	s_3	0	$\neq 0$	0

where $f^{1,2,3}$ are the forces as measured by a co-moving observer (recall that our indices differ from those in [6] by a permutation).

A couple of general observations can be made from these expressions. First, if the motion is radial, then $u^\varphi = 0$, and consequently $f^\alpha = 0$. That is, the spin–curvature interaction in radial motion does not change the geodetic motion of the test particle. This is in complete agreement with the specific result obtained in a previous section. Second, if the motion has an angular φ component, $u^\varphi \neq 0$, then the geodetic motion of the test particle may be deviated along the r and θ directions, depending on the particle's spin. These results are summarized in table 7.1.

Furthermore, it can be observed that in the absence of a gravitational field source, $r_s = 0$, and, as expected, the force is zero $f^\alpha = 0$. Similarly, in the weak field limit at large distances, the force tends to zero. Thus we have the following limiting behaviour:

$$\lim_{r_s \to 0} f^\alpha = 0 \qquad \lim_{r \to \infty} f^\alpha = 0 \qquad (7.62)$$

7.7 Circular orbits

As a specific example, let us consider the case of geodetic circular orbits in the equatorial plane $\theta = \pi/2$ in Schwarzschild spacetime. The non-zero components of the four-velocity are given by:

$$u^t = \frac{K}{f}$$

$$u^\varphi = \frac{J}{r^2} \qquad (7.63)$$

where the integration constants J and K are required to take the following values:

$$J^2 = \frac{1}{2} \frac{r r_s}{1 - \dfrac{3 r_s}{2r}}$$

$$K = \frac{1 - \dfrac{r_s}{r}}{\sqrt{1 - \dfrac{3 r_s}{2r}}} \qquad (7.64)$$

Then, the non-trivial components of the spin–curvature interaction are:

$$f^r = -\frac{\hbar}{4} \frac{3 r_s f}{r^2} \, u^\varphi u^t \, s_2$$

$$f^\theta = -\frac{\hbar}{4} \frac{3 r_s}{r^2} \sqrt{\frac{f}{f u^t u^t - 1}} \, u^t u^\varphi u^\varphi \, s_1$$

(7.65)

which can be rewritten as:

$$f^r = -\frac{\hbar}{r_s^2} \frac{3}{8\sqrt{2}} \frac{1 - r_s/r}{1 - 3r_s/2r} \left(\frac{r_s}{r}\right)^{7/2} s_2$$

$$f^\theta = -\frac{\hbar}{r_s^3} \frac{3}{8\sqrt{2}} \frac{\sqrt{1 - r_s/r}}{1 - 3r_s/2r} \left(\frac{r_s}{r}\right)^{9/2} s_1$$

(7.66)

Now let us look at the square of the magnitude of the spin–curvature four-force:

$$|f_s|^2 = g_{\mu\nu} f^\mu f^\nu$$

$$= g_{rr} f^r f^r + g_{\theta\theta} f^\theta f^\theta$$

$$= \left(\frac{\hbar}{r_s^2} \frac{3}{8\sqrt{2}}\right)^2 \frac{1 - r_s/r}{(1 - 3r_s/2r)^2} \left(\frac{r_s}{r}\right)^7 s_2^2$$

$$+ \left(\frac{\hbar}{r_s^2} \frac{3}{8\sqrt{2}}\right)^2 \frac{1 - r_s/r}{(1 - 3r_s/2r)^2} \left(\frac{r_s}{r}\right)^7 s_1^2$$

$$= \left(\frac{\hbar}{r_s^2} \frac{3}{8\sqrt{2}}\right)^2 \frac{1 - r_s/r}{(1 - 3r_s/2r)^2} \left(\frac{r_s}{r}\right)^7 (s_1^2 + s_2^2)$$

(7.67)

and, therefore, both components contribute equally to the total magnitude of the force (depending, of course, on the orientation of the spin s_1 or s_3). In addition, it can be observed that the spin–curvature force is not only suppressed by a $\hbar$ factor, but it also decays more rapidly than the $1/r^2$ Newtonian gravitational force. Indeed, for large r:

$$f_s \propto \frac{\sqrt{1 - r_s/r}}{1 - 3r_s/2r} \left(\frac{r_s}{r}\right)^{7/2} \approx \mathcal{O}\left(r^{-7/2}\right) = r^{-3/2} \mathcal{O}\left(r^{-2}\right) < \mathcal{O}\left(r^{-2}\right)$$

(7.68)

Consequently, the spin–curvature coupling is much weaker than the Newtonian gravitational attraction for spinless particles.

7.8 Non-geodetic motion

Let us consider the deviation from the circular geodetic motion of a test particle in the equatorial plane with the spin in the 'up' direction in the computational basis:

$$\psi = \begin{pmatrix} 1 \\ 0 \\ 0 \\ 0 \end{pmatrix} \tag{7.69}$$

and, therefore, from the discussion in the first two sections of this chapter, the spin components are given by:

$$s_1 = 2\cos\varphi/2 \sin\varphi/2 = \sin\varphi$$

$$s_2 = 0 \tag{7.70}$$

$$s_3 = \cos^2\varphi/2 - \sin^2\varphi/2 = \cos\varphi$$

For a circular orbit, the angle φ in terms of the $t = x^0$ time component in the local inertial frame is given by:

$$\varphi = \frac{\Theta\omega}{2\pi}t + \alpha \tag{7.71}$$

where α is a constant phase factor that can be set to zero, Θ is the total rotation after a single orbit computed in the previous chapter and given by equation (6.47):

$$\Theta = 2\pi\sqrt{f}\left(1 - \frac{Kr_s}{2rf}\frac{1}{K + \sqrt{f}}\right) \tag{7.72}$$

and ω is the rotation frequency given by:

$$\omega = \frac{u^\varphi}{u^t} = \sqrt{\frac{r_s}{2r^3}} \tag{7.73}$$

Indeed, a single rotation has a period of:

$$T = \frac{2\pi}{\omega} \tag{7.74}$$

and the total rotation angle of the spinor after a single orbit is given by Θ, in agreement with the results discussed in the previous chapter. For clarity, the behaviour of Θ with respect to r_s/r is shown in figure 7.1. The vertical grid line represents the minimum radius of a circular orbit in Schwarzschild spacetime.

Therefore, let us define:

$$\tilde{\omega} = \frac{\Theta\omega}{2\pi} \tag{7.75}$$

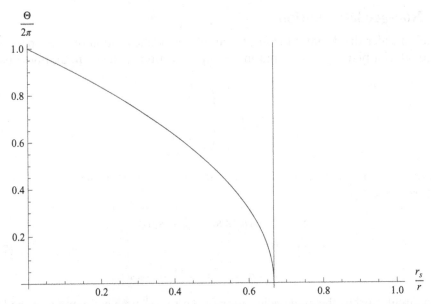

Figure 7.1. Behaviour of the angle Θ with respect to r_s/r. The vertical grid line represents the minimum radius of a circular orbit in Schwarzschild spacetime.

so we can write the spin vector components as:

$$s_1 = \sin(\tilde{\omega}t)$$

$$s_2 = 0 \qquad (7.76)$$

$$s_3 = \cos(\tilde{\omega}t)$$

and then we arrive at the following expression for the spin–curvature force:

$$f^\theta = -\frac{\hbar}{4}\frac{3r_s}{r^3}\sqrt{f}\, u^t u^\varphi \cos(\tilde{\omega}t) \qquad (7.77)$$

To understand the dynamics better, figure 7.2 shows the behaviour of $\sin(\tilde{\omega}T)$ and $\cos(\tilde{\omega}T)$ after completion of an entire orbit, $T = 2\pi/\omega$. That is, the figure shows the plots of $\sin\Theta$ (solid line) and $\cos\Theta$ (dashed line), with respect to r_s/r.

These expressions suggest that the trigonometric function 'modulates' the spin–curvature interaction. Let us now consider the value of the spin–curvature interaction after completion of an entire orbit. Figure 7.3 shows the behaviour of the relative difference of the absolute values of $\sin\Theta$ and $\cos\Theta$ with respect to r_s/r. Furthermore, for weak gravitational fields, $s_1 \approx 0$ and $s_3 \approx 1$, which implies that $f^\theta \neq 0$. That is, the spin–curvature interaction for weak gravitational fields will consist of a force along the θ axis. This makes sense, for weak gravitational fields, $\Theta \approx 2\pi$ after one orbit, which means that the spin is mostly rotating with frequency ω. Therefore, the spin is mostly aligned with the $\hat{\varphi}$ direction (i.e. parallel to the three-direction in the local inertial frame).

On the other hand, as we increase the intensity of the gravitational field, the angle Θ will become relatively smaller than 2π owing to the associated Wigner rotation. In this situation, the spin is not always aligned with the $\hat{\varphi}$ direction, but slowly

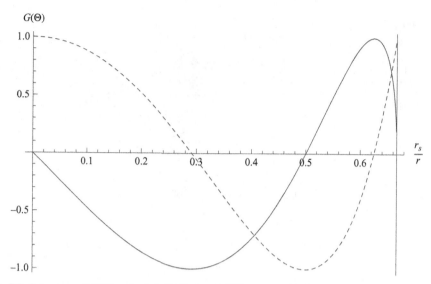

Figure 7.2. Behaviour of $G(\Theta) = \sin\Theta$ (solid line) and $G(\Theta) = \cos\Theta$ (dashed line) with respect to r_s/r. The vertical grid line represents the minimum radius of a circular orbit in Schwarzschild spacetime.

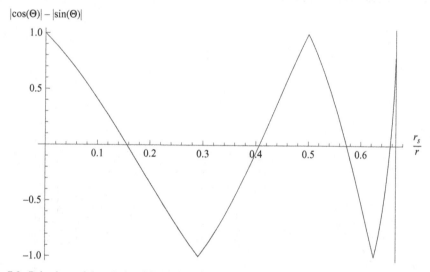

Figure 7.3. Behaviour of the relative difference of the absolute values of $\sin\Theta$ and $\cos\Theta$ with respect to r_s/r. The vertical grid line represents the minimum radius of a circular orbit in Schwarzschild spacetime.

incorporates a component along the $\hat{r}$ direction (parallel to the one-direction in the local inertial frame). As a result, the spin component will not be completely aligned with the $\hat{\varphi}$ direction, and therefore the spin–curvature interaction will decrease.

Figure 7.4 shows the behaviour of $2r_s^2 f^\theta/\hbar$ with respect to r_s/r. The vertical grid line represents the minimum radius of a circular orbit in Schwarzschild spacetime. As a comparison, the dashed line shows a commensurate force with a $(r_s/r)^2$ scaling. A more telling graph is shown in figure 7.5, which plots the absolute value

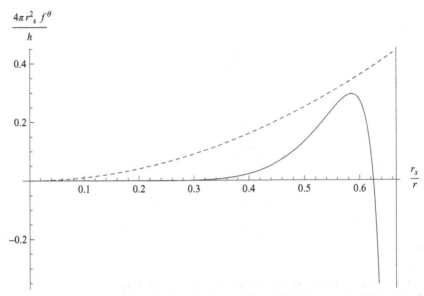

Figure 7.4. Behaviour of $2r_s^2 f^\theta/\hbar$ with respect to r_s/r. The vertical grid line represents the minimum radius of a circular orbit in Schwarzschild spacetime.

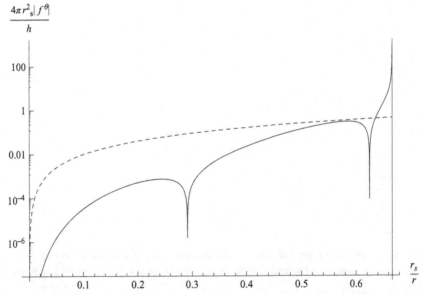

Figure 7.5. Behaviour of $2r_s^2 |f^\theta|/\hbar$ with respect to r_s/r. The vertical grid line represents the minimum radius of a circular orbit in Schwarzschild spacetime.

of $2r_s^2 f^\theta/\hbar$ with respect to r_s/r in a logarithmic scale. This figure shows the periodic behaviour due to the trigonometric function in the f^θ equation. Furthermore, the figure clearly shows that in the weak field limit, the scaling of the strength of the spin–curvature interaction is weaker than the classical Newtonian central force going as r^{-2}. Finally, it can be observed that the force diverges as we approach the circular orbit with the minimal radius.

In addition, we also know that:

$$\frac{f^\alpha}{m} = \frac{\mathcal{D}v^\alpha}{\mathcal{D}\tau}$$

$$= v^\beta \mathcal{D}_\beta v^\alpha$$

$$= \left(u^\beta + \delta u^\beta\right)\mathcal{D}_\beta\left(u^\alpha + \delta u^\alpha\right)$$

$$= \delta u^\beta \mathcal{D}_\beta u^\alpha + u^\beta \mathcal{D}_\beta \delta u^\alpha \qquad (7.78)$$

where v^α is the corrected four-velocity, m is the mass of the test particle, u^α is the geodetic four-velocity, and we have ignored terms in $\hbar^2$. Then, for the θ component in the case of circular orbits:

$$\frac{f^\theta}{m} = u^\beta \mathcal{D}_\beta \delta u^\theta$$

$$= u^t \mathcal{D}_t \delta u^\theta + u^\varphi \mathcal{D}_\varphi \delta u^\theta$$

$$= u^t \partial_t \delta u^\theta + u^\varphi \partial_\varphi \delta u^\theta + u^\varphi \Gamma^\theta_{\varphi\varphi} \delta u^\varphi$$

$$= u^t \partial_t \delta u^\theta \qquad (7.79)$$

Equating both expressions of f^θ gives:

$$u^t \partial_t \delta u^\theta = -\frac{\hbar}{4m}\frac{3r_s}{r^3}\sqrt{f}\,u^t u^\varphi \cos(\tilde{\omega}t) \qquad (7.80)$$

and consequently:

$$\partial_t \delta u^\theta = -\frac{\hbar}{4m}\frac{3r_s}{r^3}\sqrt{f}\,u^\varphi \cos(\tilde{\omega}t) \qquad (7.81)$$

Therefore:

$$\delta u^\theta = -\frac{\hbar}{4m}\frac{3r_s}{r^3}\frac{\sqrt{f}}{\tilde{\omega}}\,u^\varphi \sin(\tilde{\omega}t)$$

$$= -\frac{\hbar}{4m}\frac{3r_s}{r^3}\frac{2\pi}{\Theta}\sqrt{f}\,u^t \sin(\tilde{\omega}t) \qquad (7.82)$$

where we have ignored the integration constant.

Thus, the expression for the non-geodetic correction to the four-velocity is given by:

$$\delta u^\theta = -\frac{\hbar}{4m}\frac{3r_s}{r^3}\sqrt{\frac{1-r_s/r}{1-3r_s/2r}}\frac{2\pi}{\Theta}\sin(\tilde{\omega}t) \qquad (7.83)$$

In addition, the associated velocity is given by:

$$\omega^{\theta} = \frac{\delta u^{\theta}}{u^t} = -\frac{\hbar}{4m} \frac{3r_s}{r^3} \sqrt{f} \, \frac{2\pi}{\Theta} \sin(\tilde{\omega} t) \tag{7.84}$$

We observe that in the weak gravitational field limits:

$$\lim_{r \to \infty} \delta u^{\theta} = 0 \qquad \lim_{r_s \to 0} \delta u^{\theta} = 0 \tag{7.85}$$

as expected.

The total magnitude of the four-velocity correction is given by:

$$|\delta v|^2 = g_{\theta\theta} \delta u^{\theta} \delta u^{\theta}$$

$$= r^2 \left(\frac{\hbar}{4m} \frac{3r_s}{r^3} \sqrt{\frac{1 - r_s/r}{1 - 3r_s/2r}} \right)^2 \left(\frac{2\pi}{\Theta} \right)^2 \sin^2(\tilde{\omega} t)$$

$$= \left(\frac{\hbar}{4m} \frac{3r_s}{r^2} \right)^2 \frac{1 - r_s/r}{1 - 3r_s/2r} \left(\frac{2\pi}{\Theta} \right)^2 \sin^2(\tilde{\omega} t) \tag{7.86}$$

and therefore:

$$\delta v = \frac{\hbar}{4m} \frac{3r_s}{r^2} \frac{2\pi}{\Theta} \sqrt{\frac{1 - r_s/r}{1 - 3r_s/2r}} \sin(\tilde{\omega} t) \tag{7.87}$$

and the associated velocity is given by:

$$\delta v_t = \frac{\delta v}{u^t} = \frac{\hbar}{4m} \frac{3r_s}{r^2} \frac{2\pi}{\Theta} \sqrt{1 - \frac{r_s}{r}} \sin(\tilde{\omega} t) \tag{7.88}$$

The behaviour of the non-geodetic correction $\delta v \times 2mr_s/\hbar$ is shown in figure 7.6. Also, the behaviour of the non-geodetic angular velocity $\delta v_t \times 2mr_s/\hbar$ is shown in figure 7.7. In both figures, the vertical grid line shows the position of the minimal circular orbit. It can be observed that these quantities diverge as we approach the minimum radius for a circular orbit: $r \to 3r_s/2r$. This only means that for strong gravitational fields and small orbits, the Wentzel–Kramers–Brillouin (WKB) expansion that we used to approximate the correction to the geodesics breaks down, and the expressions are no longer accurate.

As a reference, the multiplicative factor for an electron on Earth's gravitational field is approximately given by:

$$\frac{\hbar}{2mr_s} \approx 0.04 \tag{7.89}$$

in meter/second units.

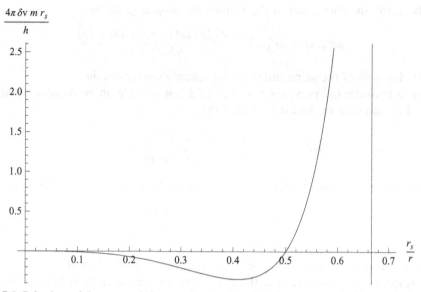

Figure 7.6. Behaviour of the non-geodetic correction $\delta v \times 2mr_s/\hbar$ with respect to r_s/r. The vertical grid line represents the minimum radius of a circular orbit in Schwarzschild spacetime.

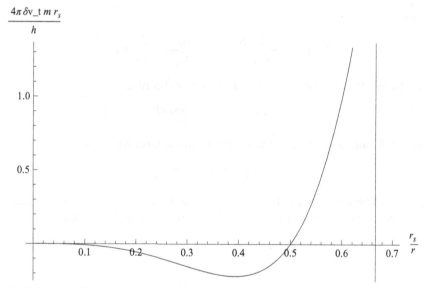

Figure 7.7. Behaviour of the non-geodetic angular velocity, $\delta v_t \times 2mr_s/\hbar$ with respect to r_s/r. The vertical grid line represents the minimum radius of a circular orbit in Schwarzschild spacetime.

7.9 Wigner rotations

The correction to the Lorentz transformation due to the non-geodetic deviation in the local inertial frame is given by:

$$\delta\lambda^a_{\;b}(x) = \delta a^a(x)u_b(x) - u^a(x)\delta a_b(x) - e^a_{\;\mu}(x)\delta u^\beta(x)\nabla_\beta e^{\;\mu}_b(x) \qquad (7.90)$$

Therefore, the correction to the Wigner rotation is given by:

$$\delta\vartheta^a{}_b(x) = \delta\lambda^a{}_b(x) + \frac{\delta\lambda^a{}_0(x)\, u_b(x) - \delta\lambda_{b0}(x)\, u^a(x)}{u^0(x) + 1} \tag{7.91}$$

which clearly is of the same order as the velocity correction δu^μ.

Let us consider the previous example, of a test particle on an equatorial circular orbit. For this case we had determined that:

$$\delta u^\theta = -\frac{\hbar}{4m} \frac{3r_s}{r^3} \sqrt{\frac{1 - r_s/r}{1 - 3r_s/2r}} \frac{2\pi}{\Theta} \sin(\tilde\omega t) \tag{7.92}$$

and therefore the only non-trivial corrections to the Lorentz transformations are:

$$\delta\lambda^0{}_2(x) = \delta\lambda^2{}_0(x) = -r\,\delta u^\theta$$
$$\delta\lambda^1{}_2(x) = -\delta\lambda^2{}_1(x) = \sqrt{f}\,\delta u^\theta \tag{7.93}$$

which correspond to boosts along the two-direction and rotations over the three-axis, respectively. Furthermore, the correction to the differential Wigner rotation angle is given by:

$$\delta\vartheta^1{}_2(x) = -\delta\vartheta^2{}_1(x) = \delta\lambda^1{}_2(x) \tag{7.94}$$

and therefore:

$$\delta\vartheta^1{}_2(x) = -\frac{\hbar}{4m} \frac{3r_s}{r^3} \frac{1 - r_s/r}{\sqrt{1 - 3r_s/2r}} \frac{2\pi}{\Theta} \sin(\tilde\omega t) \tag{7.95}$$

We observe that in the weak gravitational field limits:

$$\lim_{r\to\infty} \delta\vartheta^1{}_2(x) = 0 \qquad \lim_{r_s\to 0} \delta\vartheta^1{}_2(x) = 0 \tag{7.96}$$

and in the strong field limit at the minimal radius circular orbit:

$$\lim_{2r\to 3r_s} \delta\vartheta^1{}_2(x) = -\infty \tag{7.97}$$

The total Wigner rotation angle Ω solely due to the spin–curvature interaction is obtained by integration over the path of the test particle. In this case:

$$\begin{aligned}
\Omega &\approx \int_{\tau_i}^{\tau_f} \delta\vartheta^1{}_2 \, d\tau \\
&= \int_{\varphi_i}^{\varphi_f} \delta\vartheta^1{}_2 \, \frac{d\tau}{d\varphi} \, d\varphi \\
&= \frac{\delta\vartheta^1{}_2}{u^\varphi} \int_{\varphi_i}^{\varphi_f} d\varphi \\
&= \frac{\delta\vartheta^1{}_2}{u^\varphi} \, \Phi
\end{aligned} \tag{7.98}$$

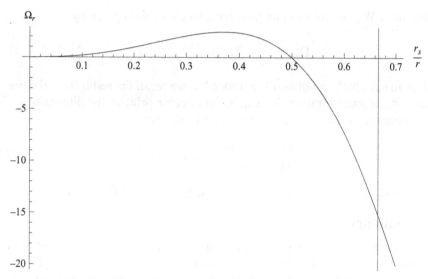

Figure 7.8. Behaviour of the Wigner rotation angle due to the spin–curvature interaction, Ω_r, with respect to r_s/r.

where Φ is the total angle traversed by the test particle.

After completion of a single orbit over a rotation period $T = 2\pi/\omega$, we have:

$$\Omega = -\frac{\hbar}{4m}\frac{3r_s}{r^3}\frac{1-r_s/r}{\sqrt{1-3r_s/2r}}\frac{2\pi}{\Theta}\sin(\Theta)\frac{r^2}{J}2\pi$$

$$= -\frac{\hbar}{2mr_s}\frac{12\pi^2}{\sqrt{2}}\left(\frac{r_s}{r}\right)^{3/2}\left(1-\frac{r_s}{r}\right)\frac{\sin\Theta}{\Theta} \qquad (7.99)$$

To help visualize the behaviour of the Wigner rotation Ω, let us define the following variable:

$$\Omega_r = \Omega\frac{2mr_s}{\hbar} = \frac{12\pi^2}{\sqrt{2}}\left(\frac{r_s}{r}\right)^{3/2}\left(1-\frac{r_s}{r}\right)\frac{\sin\Theta}{\Theta} \qquad (7.100)$$

which corresponds to the Wigner rotation after one period $T = 2\pi/\omega$. Figure 7.8 shows the behaviour of Ω_r with respect to r_s/r. It can be observed that Ω_r remains finite for all circular orbits ($r \geqslant 3r_s/2$).

As an example, let us consider the case of a qubit in circular orbit right on the surface of the Earth (assuming, of course, that Earth is a spherically symmetric object with zero atmosphere). In this case, the dimensionless Earth surface potential is given by:

$$\phi_0 \equiv \frac{M_\oplus G}{R_\oplus c^2} \approx 7\times 10^{-10} \qquad (7.101)$$

and the total Wigner rotation angle after a single orbit is given by:

$$\Omega_{sc} \approx \frac{\hbar}{2mr_s} \, 8 \times 10^{-22} \approx 3 \times 10^{-23} \qquad (7.102)$$

which is an extremely small number. Indeed, if we recall the result from the previous chapter, the Wigner rotation for a qubit in circular orbit at the dimensional Earth surface potential value is given by Ω_c, and therefore:

$$\frac{\Omega_{sc}}{\Omega_c} \approx \frac{3 \times 10^{-23}}{5 \times 10^{-10}} \approx 6 \times 10^{-14} \qquad (7.103)$$

which produces an extremely small effect on the phase of the qubit.

7.10 Summary

In this chapter we discussed the spin–curvature coupling for a Dirac spinor in Schwarzschild spacetime. We derived expressions for the deviation to geodetic motion δu^μ and the force associated with the spin–curvature coupling. As expected, these results not only depend on the curvature of the spacetime, but also on the orientation of the spin. Finally, we analysed numerically the behaviour for a simple example and concluded the effect is very small.

These effects may be negligible within the context of quantum information applications such as communications, sensing and computation. Nevertheless, it is crucial to understand all the factors that may influence the dynamics of qubits in curved spacetime.

Bibliography

[1] Lanzagorta M 2013 Wigner rotation due to spin-curvature interaction for a qubit in a Schwarzschild field forthcoming, available upon request
[2] Alsing P M, Stephenson G J and Kilian P 2009 Spin-induced non-geodesic motion, gyroscopic precession, Wigner rotation and EPR correlations of massive spin-$\frac{1}{2}$ particles in a gravitational field arXiv:0902.1396v1 [quant-ph]
[3] Plyatsko R 1998 Gravitational ultrarelativistic spin-orbit interaction and the weak equivalence principle *Phys. Rev.* D **58** 084031
[4] Bolos V J 2006 Lightlike simultaneity, comoving observers and distances in general relativity *J. Geom. Phys.* **56** 813–29
[5] Bolos V J 2007 Intrinsic definitions of relative velocity in general relativity *Commun. Math. Phys.* **273** 217–236
[6] Schutz B 2009 *A First Course in General Relativity* 2nd edn (Cambridge: Cambridge University Press)

Chapter 8

Qubits in Kerr spacetime

The Kerr metric is an exact solution to Einstein's field equations that represents the stationary and axisymmetric spacetime produced by a spherically symmetric rotating object of mass M and angular momentum Ma [1, 2]. This solution can be shown to be unique if one demands that (1) the spacetime tends to the Minkowski form as $r \to \infty$ and (2) the geometry is non-singular outside of a smooth closed convex event horizon.

Although the Kerr metric is more sophisticated than the Schwarzschild metric, it remains tractable and many computations can be done analytically. The importance of the Kerr metric is that it represents the gravitational field produced by a rotating spherical object such as a planet or a star. Thus, the Kerr metric is important for understanding the effects of a gravitational field on Earth-based quantum computation and communication technologies. That said, more sophisticated rotating frames can be found in the literature [3].

Furthermore, the Kerr solution exhibits a purely relativistic effect known as *frame dragging*, which describes how a rotating mass tends to 'drag' the surrounding spacetime as it rotates through spacetime. As we will discuss in this chapter, frame dragging will affect the spin of qubits moving in a Kerr spacetime [4].

8.1 The metric tensor

A stationary spacetime means that the metric components do not depend on the time-like coordinate, but the line element is not necessarily invariant under reflection transformations of the time-like component. Furthermore, the axisymmetric requirement implies that the metric components do not depend on a specific space-like component.

The Kerr metric for a stationary and axisymmetric spacetime is better expressed in terms of the *Boyer–Lindquist coordinates* (t, r, θ, φ). The rotation axis is chosen

doi:10.1088/978-1-627-05330-3ch8

along the $\hat{z}$ axis. In natural units $(G = 1, c = 1)$, the line element that corresponds to the Kerr metric is given by:

$$dS^2 = -\frac{\rho^2 \Delta}{\Sigma^2} dt^2 + \frac{\Sigma^2 \sin^2\theta}{\rho^2}(d\varphi - \omega dt)^2 + \frac{\rho^2}{\Delta} dr^2 + \rho^2 d\theta^2 \qquad (8.1)$$

where:

$$\rho^2 = r^2 + a^2\cos^2\theta$$

$$\Delta = r^2 - rr_s + a^2$$

$$\Sigma^2 = (r^2 + a^2)^2 - a^2\Delta\sin^2\theta \qquad (8.2)$$

$$r_s = 2M$$

$$\omega = \frac{r_s r a}{\Sigma^2}$$

Clearly, the Kerr metric reduces to the Schwarzschild metric in the no-rotation limit:

$$\lim_{a \to 0} dS^2 = -f\, dt^2 + \frac{1}{f}\, dr^2 + r^2(d\theta^2 + \sin^2\theta\, d\varphi^2) \qquad (8.3)$$

whereas before:

$$f = 1 - \frac{r_s}{r}$$

As a consequence, the analyses presented in the following sections also apply to the Schwarzschild metric by setting the limit $a \to 0$.

We also notice that in the zero gravitational field limit $(M \to 0)$, the space-like Boyer–Lindquist coordinates do not reduce to the standard set of spherical coordinates. Indeed, in this limit we have:

$$\lim_{M \to 0} dS^2 = -dt^2 + (r^2 + a^2)\sin^2\theta\, d\varphi^2 + \frac{\rho^2}{r^2 + a^2} dr^2 + \rho^2 d\theta^2 \qquad (8.4)$$

which reduces to the standard expression for the line element in Minkowski space if we make the change of coordinates:

$$x = \sqrt{r^2 + a^2}\, \sin\theta\, \cos\varphi$$

$$y = \sqrt{r^2 + a^2}\, \sin\theta\, \sin\varphi \qquad (8.5)$$

$$z = r\cos\theta$$

Then, as expected, the space-like Boyer–Lindquist coordinates reduce to the standard spherical coordinates in the no-rotation limit $(a \to 0)$.

8.2 Structure of Kerr spacetime

From the expression of the line element and the geometric interpretation of the Boyer–Lindquist coordinates, it can be observed that Kerr spacetime has an intrinsic singularity found at:

$$\rho = 0 \quad \Rightarrow \quad \begin{cases} r = 0 \\ \theta = \pi/2 \end{cases} \tag{8.6}$$

Notice that, although the intrinsic singularity for the Schwarzschild spacetime was a point, the intrinsic singularity in Kerr spacetime corresponds to a ring of radius a lying in the equatorial plane $\theta = \pi/2$ [2]. Indeed, from the Boyer–Lindquist coordinates, the intrinsic singularity is located at:

$$x = a\cos\varphi$$
$$y = a\sin\varphi \tag{8.7}$$
$$z = 0$$

In addition, as expected, in the no-rotation limit, the radius of this ring reduces to zero, in agreement with the Schwarzschild solution. We denote the ring singularity by σ.

In addition, the Kerr spacetime has two event horizons that are related to coordinate singularities [2, 3]. These occur on the surfaces defined by:

$$\Delta = 0 \tag{8.8}$$

which implies the two event horizon radii:

$$r_\pm = \frac{r_s}{2} \pm \sqrt{\frac{r_s^2}{4} - a^2} \tag{8.9}$$

Because of Penrose's cosmic censorship hypothesis, one has to demand the existence of event horizons that cover the true singularity, thereby avoiding troublesome naked singularities [5]. From the expression above, event horizons in Kerr spacetime exist if the term inside the square root is positive. That is:

$$\frac{r_s^2}{4} > a^2 \quad \Rightarrow \quad -0.5 < \frac{a}{r_s} < 0.5 \tag{8.10}$$

In addition, the Kerr geometry has two *stationary limit surfaces* $S^\pm$ defined by $g_{tt}(S^\pm) = 0$. These are *infinite redshift surfaces* where the rotation of the compact object is so strong that any test particle is forced to rotate with the source, even if it has an arbitrarily large angular momentum [2]. For the case of the Kerr metric, these surfaces are found at:

$$S^\pm = \frac{r_s}{2} \pm \sqrt{\frac{r_s^2}{4} - a^2\cos^2\theta} \tag{8.11}$$

Consequently, the Kerr spacetime geometry has the following structure:

$$\sigma \subseteq S^- \subseteq r_- \subseteq r_+ \subseteq S^+ \subseteq r_s \tag{8.12}$$

In the following discussions we will concentrate on qubits travelling in regions exterior to the Schwarzschild radius sphere r_s and the stationary limit surface, where the Boyer–Lindquist coordinates are well defined. It is possible, however, to analyse the interior of r_s and S^+ by using *Eddington–Finkelstein coordinates*, which provide an analytical continuation that extends the range of validity of the equations [3].

8.3 Tetrad fields and connection one-forms

To analyse the dynamics of a qubit in Kerr spacetime, we need to select a convenient tetrad field and to compute the associated connection one-forms. Later on, we will derive the equations for geodesics in Kerr spacetimes; however, in contrast to the method for the Schwarzschild metric case, here we will use a variational approach that does not require the explicit computation of the affine connection. For future reference, however, we include the expressions for the non-zero affine connections [6].

8.3.1 Affine connection

Let us recall that, because the Kerr metric has no torsion, the affine connection is related to the metric tensor by:

$$\Gamma^\alpha_{\mu\nu} = \frac{1}{2} g^{\sigma\alpha} \left(\frac{\partial g_{\nu\sigma}}{\partial x^\mu} + \frac{\partial g_{\mu\sigma}}{\partial x^\nu} - \frac{\partial g_{\nu\mu}}{\partial x^\sigma} \right) \tag{8.13}$$

Therefore, the non-zero components of the affine connection are computed to be:

$$\Gamma^t_{rt} = M(r^2 + a^2)(r^2 - a^2\cos^2\theta)/\Sigma^2\Delta$$
$$\Gamma^t_{\theta t} = -2Mra^2\cos\theta\sin\theta/\Sigma^2$$
$$\Gamma^t_{r\varphi} = aM\sin^2\theta(a^4\cos^2\theta - r^2a^2\cos^2\theta - r^2a^2 - 3r^4)/\Sigma^2\Delta$$
$$\Gamma^t_{\theta\varphi} = 2Mra^3\sin^3\theta\cos\theta/\Sigma^2$$
$$\Gamma^r_{tt} = M(r^2 - a^2\cos^2\theta)\Delta/\Sigma^3$$
$$\Gamma^r_{\varphi t} = -aM\sin^2\theta(r^2 - a^2\cos^2\theta)\Delta/\Sigma^3$$
$$\Gamma^r_{rr} = (ra^2\sin^2\theta - M(r^2 - a^2\cos^2\theta))/\Sigma\Delta$$
$$\Gamma^r_{\theta r} = -a^2\cos\theta\sin\theta/\Sigma$$
$$\Gamma^r_{\theta\theta} = -r\Delta/\Sigma$$
$$\Gamma^r_{\varphi\varphi} = \Delta\sin^2\theta(Ma^2\sin^2\theta(r^2 - a^2\cos^2\theta) - r\Sigma^2)/\Sigma^3 \qquad (8.14)$$
$$\Gamma^\theta_{tt} = -2Mra^2\sin\theta\cos\theta/\Sigma^3$$
$$\Gamma^\theta_{\varphi t} = 2Mra\sin\theta\cos\theta(r^2 + a^2)/\Sigma^3$$
$$\Gamma^\theta_{rr} = a^2\sin\theta\cos\theta/\Sigma\Delta$$
$$\Gamma^\theta_{r\theta} = r/\Sigma$$
$$\Gamma^\theta_{\theta\theta} = -a^2\sin\theta\cos\theta/\Sigma$$
$$\Gamma^\theta_{\varphi\varphi} = -\cos\theta\sin\theta(\Sigma^2\Delta + 2Mr(r^4 + a^2 + 2r^2))/\Sigma^3$$
$$\Gamma^\varphi_{rt} = Ma(r^2 - a^2\cos^2\theta)/\Sigma^2\Delta$$
$$\Gamma^\varphi_{\theta t} = -2Mra\cot\theta/\Sigma^2$$
$$\Gamma^\varphi_{r\varphi} = ((r - M)\Sigma^2 - M(r^2 + a^2)(r^2 - a^2\cos^2\theta))/\Sigma^2\Delta$$
$$\Gamma^\varphi_{\theta\varphi} = \cot\theta + 2Mra^2\cos\theta\sin\theta/\Sigma^2$$

8.3.2 Tetrad fields

It is convenient to choose the tetrad field that defines a local inertial frame in each point of Kerr spacetime with the following non-zero components [7]:

$$e_0^{\,t}(x) = \frac{1}{W}$$

$$e_1^{\,r}(x) = \frac{\sqrt{\Delta}}{\rho}$$

$$e_2^{\,\theta}(x) = \frac{1}{\rho} \tag{8.15}$$

$$e_3^{\,\varphi}(x) = \frac{W}{\sqrt{\Delta}\sin\theta}$$

$$e_3^{\,t}(x) = \frac{a\sin\theta}{\sqrt{\Delta}}\left(W - \frac{1}{W}\right)$$

where:

$$W = \sqrt{1 - \frac{rr_s}{\rho^2}} \tag{8.16}$$

Let us remark once more that the choice of tetrad field is somewhat arbitrary, and it could be defined in a different way. For instance, the literature offers at least one different alternative [8].

In the present case, the tetrad represents the local inertial frame of a 'hovering' observer. Indeed, let us assume an observer at rest in the local inertial frame:

$$u^a = (1, 0, 0, 0) \tag{8.17}$$

Then, in the general coordinate system:

$$u^\mu = e_a^{\,\mu}(x)\, u^a$$

$$= \left(\frac{1}{W}, 0, 0, 0\right) \tag{8.18}$$

which represents a hovering observer. Such an observer would require a rocket pack to be able to withstand the gravitational attraction and the rotation induced by the frame dragging.

8.3.3 Connection one-forms

For our choice of tetrad field, the non-trivial associated connection one-forms are given by:

$$\omega_{\mu 01}(x) = \frac{\sqrt{\Delta}}{\rho^3 W}(M + r(W^2 - 1))(\delta^0_\mu - a\sin^2\theta\,\delta^3_\mu)$$

$$\omega_{\mu 02}(x) = \frac{a\sin 2\theta}{2\rho^3}\left(W - \frac{1}{W}\right)((r^2 + a^2)\delta^3_\mu - a\delta^0_\mu)$$

$$\omega_{\mu 03}(x) = \frac{a}{\rho^2 W^2\sqrt{\Delta}}(M\sin\theta\,\delta^1_\mu + (W^2 - 1)(r\sin\theta\,\delta^1_\mu - \Delta\cos\theta\,\delta^2_\mu))$$

$$\omega_{\mu 12}(x) = -\frac{\sqrt{\Delta}}{\rho}\left(\frac{a^2\sin 2\theta}{2\Delta}\delta^1_\mu + r\delta^3_\mu\right)$$

$$\omega_{\mu 13}(x) = \frac{\sin\theta}{\rho W}\left(\frac{a}{\rho^2}(M - r + rW^2)(\delta^0_\mu - a\sin^2\theta\,\delta^3_\mu) - rW^2\delta^3_\mu\right)$$

$$\omega_{\mu 23}(x) = \frac{\sqrt{\Delta}\cos\theta}{\rho W}\left(\frac{a}{\rho^2}(W^2 - 1)(a\sin^2\theta\,\delta^3_\mu - \delta^0_\mu) - \delta^3_\mu\right)$$

(8.19)

In this chapter we will only consider orbital motion in the equatorial plane ($\theta = \pi/2$). For this specific case, the non-zero connection one-forms associated with our choice of tetrad field simplify to:

$$\omega_t{}^0{}_1(x) = \omega_t{}^1{}_0(x) = \frac{r_s}{2r^3}\sqrt{\frac{\Delta}{f}}$$

$$\omega_\varphi{}^0{}_1(x) = \omega_\varphi{}^1{}_0(x) = -\frac{ar_s}{2r^3}\sqrt{\frac{\Delta}{f}}$$

$$\omega_r{}^0{}_3(x) = \omega_r{}^3{}_0(x) = \frac{ar_s}{2r^2 f\sqrt{\Delta}}$$

(8.20)

$$\omega_\theta{}^1{}_2(x) = -\omega_\theta{}^2{}_1(x) = -\frac{\sqrt{\Delta}}{r}$$

$$\omega_t{}^1{}_3(x) = -\omega_t{}^3{}_1(x) = -\frac{ar_s}{2r^3\sqrt{f}}$$

$$\omega_\varphi{}^1{}_3(x) = -\omega_\varphi{}^3{}_1(x) = \frac{a^2 r_s}{2r^3\sqrt{f}} - \sqrt{f}$$

8.4 Geodesics

The geodesics in a gravitational field can be obtained using the Lagrangian:

$$\mathcal{L} = m\dot{x}^\mu \dot{x}^\nu g_{\mu\nu}(x) \tag{8.21}$$

and the associated Euler–Lagrange equations:

$$\frac{d}{d\tau}\left(\frac{\partial \mathcal{L}}{\partial \dot{x}^\mu}\right) = 0 \tag{8.22}$$

where:

$$\dot{x}^\mu = \frac{dx^\mu}{d\tau} = u^\mu(x) = (u^t(x), u^r(x), u^\theta(x), u^\varphi(x)) \tag{8.23}$$

normalized as:

$$u^\mu(x)u_\mu(x) = -1 \tag{8.24}$$

for massive test particles. This method of deriving the geodesics avoids the explicit computation of the affine connection for the Kerr metric, a rather cumbersome undertaking [2].

It is important to note that, in contrast to orbits in Schwarzschild spacetime, non-equatorial orbits in Kerr spacetime are not constrained to a plane [2]. Then, for simplicity, we will restrict our discussion to equatorial trajectories ($\theta = \pi/2$).

In the equatorial plane the geodesic equations for free-falling massive particles are found to be:

$$u^t(x) = \frac{ar_s}{r\Delta}\left(\frac{K}{\omega} - J\right)$$

$$u^r(x) = \pm\sqrt{(K^2 - 1) + \frac{r_s}{r} + \frac{a^2(K^2 - 1) - J^2}{r^2} + \frac{r_s(J - aK)^2}{r^3}} \tag{8.25}$$

$$u^\theta(x) = 0$$

$$u^\varphi(x) = \frac{ar_s}{r\Delta}\left(K + \frac{rfJ}{ar_s}\right)$$

Because the Kerr metric tensor is independent of t and φ, K and J are integration constants associated with the covariant components of the particle's four-momentum that are conserved along the geodesic:

$$p_t = K \Rightarrow \frac{dp_t}{d\tau} = 0$$

$$p_\varphi = J \Rightarrow \frac{dp_\varphi}{d\tau} = 0 \tag{8.26}$$

These two conserved quantities correspond to energy and angular momentum conservation, respectively [2, 3].

8.5 Quantum dynamics

As discussed in chapters 5 and 6, the optical analogy and the Wentzel–Kramers–Brillouin (WKB) approximation provide an insightful way to understand the dynamics of quantum particles in curved spacetimes. To refresh the core of those ideas, here we will merely repeat the principal outcome of the argument. That is, a Dirac spinor ψ can be described by:

$$\psi \approx e^{iS}\psi_0 + \mathcal{O}(\hbar) \tag{8.27}$$

where ψ_0 is a spinor that satisfies the Dirac equation in some local inertial frame and the quantum mechanical phase S corresponds to the action of the classical particle:

$$S = m \int u_\mu \mathrm{d}x^\mu \tag{8.28}$$

Then, applying the optical analogy, we can understand the dynamics of quantum particles as the propagation of wave surfaces with constant S and rays given by $\partial_\mu S$ (as long as we are not too close to the singularity). In this case, the integral curves of u^μ correspond to geodesic congruences defined by the geodesic equations. Furthermore, in chapter 5 we developed a theory to describe the transport of spinors along geodesic lines.

8.6 Wigner rotations

In this section we will explore the effect of frame dragging on qubit states. For simplicity, we will limit our discussion of the effects of frame dragging in Kerr spacetime to equatorial radial falls and equatorial circular orbits. However, the equations presented above are general and can be applied to situations that are more sophisticated.

8.6.1 Equatorial radial fall ($\theta = \pi/2$, $J = 0$)

A radial fall is a free-falling test particle with zero angular momentum ($J = 0$) and dropped from rest at $r \to \infty$ ($K = 1$). The geodesics are given by:

$$u^t(x) = \frac{ar_s}{\Delta r\omega}$$

$$u^r(x) = -\sqrt{\frac{r_s}{r} + \frac{a^2 r_s}{r^3}} \tag{8.29}$$

$$u^\theta(x) = 0$$

$$u^\varphi(x) = \frac{ar_s}{\Delta r}$$

which clearly reduce to the Schwarzschild geodesics in the no-rotation limit $(a \rightarrow 0)$.

In the local inertial frame defined by the tetrad field, we obtain the following components for the four-velocity:

$$u^0 = \frac{1}{\sqrt{f}}$$

$$u^1 = -\frac{r}{\sqrt{\Delta}} \sqrt{\frac{r_s}{r}\left(1 + \frac{a^2}{r^2}\right)}$$

$$u^2 = 0$$

$$u^3 = -\frac{a}{\sqrt{\Delta}}\left(\sqrt{f} - \frac{1}{\sqrt{f}}\right)$$

(8.30)

These geodesic equations show that a free-falling test particle with zero angular momentum $(J = 0)$ acquires an angular velocity. This is because the spacetime is being dragged as the source of the gravitational field rotates [1, 2]. Indeed, for the case under consideration:

$$\frac{d\varphi}{dt} = \frac{u^\varphi(x)}{u^t(x)} = \omega$$

(8.31)

And in the local inertial frame, the tangential velocity is given by:

$$v_t = \frac{u^3}{u^0} = \frac{a}{\sqrt{\Delta}}\left(\frac{r_s}{r}\right)$$

(8.32)

In addition, the radial velocity v_r in the general coordinate frame is also affected by the angular momentum of the gravitational field source:

$$v_r = \frac{dr}{dt} = \frac{u^r(x)}{u^t(x)} = -\frac{\Delta r\omega}{ar_s}\sqrt{\frac{r_s}{r} + \frac{a^2 r_s}{r^3}}$$

(8.33)

where we recall from equation (8.2) that ω depends on the mass M and the angular momentum a of the source of the gravitational field, as well as the radial distance r.

The behaviour of v_t, the tangential velocity observed in the local inertial frame, with respect to r/r_s is shown in figure 8.1 The two lines correspond to the maximum value of $a/r_s = \pm 0.5$. Similarly, the behaviour of the radial velocity observed in the general coordinate system, v_r, with respect to r_s/r is shown in figure 8.2. The dashed line corresponds to the radial velocity in Schwarzschild spacetime $(a = 0)$, whereas the solid line represents the maximum value of the rotation parameter $a/r_s = \pm 0.5$.

As we discussed before, these geodesic equations represent the free-falling motion of spinless particles. However, we need to recall that spin and curvature are coupled in a non-trivial manner. As a consequence, the motion of spinning particles, either

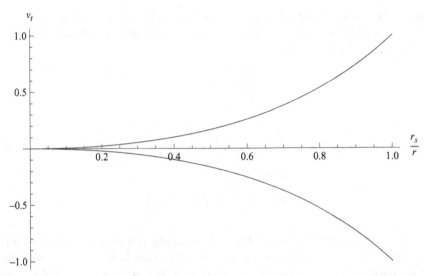

Figure 8.1. Tangential velocity v_t in the local inertial frame induced by frame dragging on a free-falling particle with zero angular momentum in the equatorial plane. The two lines correspond to the maximum value of $a/r_s = \pm 0.5$.

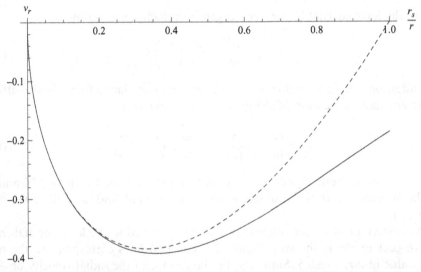

Figure 8.2. Radial velocity v_r in the general coordinate frame with respect to r_s/r. The dashed line corresponds to the radial velocity in Schwarzschild spacetime ($a = 0$), whereas the solid line represents the maximum value of the rotation parameter $a/r_s = \pm 0.5$.

classical or quantum, does not follow geodesics [9–13]. That said, it is also known that the deviation from geodetic motion is very small, except for the case of supermassive compact objects and/or ultra-relativistic test particles [14–17]. In any event, here we will ignore the spin–curvature coupling effects.

Using the geodesic equations we can calculate the six non-zero Lorentz transformations that describe the motion of the test particle in the local inertial frame:

$$\lambda^0_{\ 1}(x) = \lambda^1_{\ 0}(x) = \frac{ar_s^2}{2\Delta r^4}\left(a - \frac{1}{\omega}\right)\sqrt{\frac{\Delta}{f}}$$

$$\lambda^0_{\ 3}(x) = \lambda^3_{\ 0}(x) = \frac{ar_s}{2r^2 f\sqrt{\Delta}}\sqrt{\frac{r_s}{r}\left(1 + \frac{a^2}{r^2}\right)} \tag{8.34}$$

$$\lambda^1_{\ 3}(x) = -\lambda^3_{\ 1}(x) = \frac{a^2 r_s^2}{2\Delta r^4\sqrt{f}}\left(\frac{1}{\omega} - a\right) + \frac{ar_s\sqrt{f}}{\Delta r}$$

These Lorentz transformations represent boosts on the one- and three-directions, and rotations over the two-axis, respectively. Notice that the boost in the three-direction and the accompanying rotation over the two-direction are exclusively due to frame dragging effects.

The associated infinitesimal Wigner rotation is found to be:

$$\vartheta^1_{\ 3}(x) = \frac{a^2 r_s^2}{2\Delta r^4\sqrt{f}}\left(\frac{1}{\omega} - a\right) + \frac{ar_s\sqrt{f}}{\Delta r}$$

$$+ \frac{ar_s^2}{2\Delta r^4(f + \sqrt{f})}\left(a^2 f + r^2 + \frac{ar_s}{\omega r}\right) \tag{8.35}$$

And the total angular rotation of the particle's spin, which is solely due to gravitational effects, is given by:

$$\Omega = \int_{\tau_i}^{\tau_f} \vartheta^1_{\ 3}(x)\mathrm{d}\tau = \int_{r_i}^{r_f} \frac{\vartheta^1_{\ 3}(x)}{u^r(x)}\,\mathrm{d}r \tag{8.36}$$

where the integration takes place over the path of the free-falling particle.

Figure 8.3 shows the rotation angle Ω induced by frame dragging on a free-falling particle with zero angular momentum launched from rest at infinity. Thus, the integration over the radial coordinate is taken from infinite to r. As previously discussed, the minimum value of r is taken to be the stationary limit surface S^+, which takes the value of r_s in the equatorial plane.

Furthermore, to guarantee the existence of event horizons and avoid naked singularities, the ratio a/r_s is only allowed to take values between -0.5 and 0.5. When $a = 0$ the spacetime geometry reduces to Schwarzschild and a radially falling particle with zero angular momentum does not experience any rotation ($\Omega = 0$). As the rotation of the compact body increases, so does the Wigner rotation angle. That is, in this case the spin rotation angle Ω is completely due to frame dragging. The two solid lines in figure 8.3 correspond to the upper and lower bounds of the angular momentum of the compact object. The maximal rotation is reached at the stationary limit surface, which in this case coincides with the Schwartzchild radius, and has a maximal value of $|\Omega(r = S^+)| \approx 3.1828$.

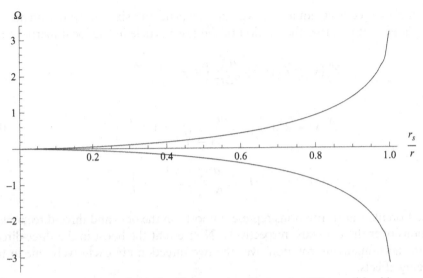

Figure 8.3. Total spin rotation angle Ω induced by frame dragging on a free-falling particle with zero angular momentum in the equatorial plane. The two solid lines correspond to the upper and lower bounds of the angular momentum of the compact object ($a = 0.5$, -0.5, respectively).

8.6.2 Equatorial circular orbits ($\theta = \pi/2$, $u^r = 0$)

The geodesics for a circular orbit ($u^r = 0$) of constant radius r in the equatorial plane ($\theta = \pi/2$) are given by:

$$u^t(x) = \frac{ar_s}{\Delta r}\left(\frac{K}{\omega} - J\right)$$

$$u^r(x) = 0$$

$$u^\theta(x) = 0 \tag{8.37}$$

$$u^\varphi(x) = \frac{1}{\Delta}\left(\frac{aKr_s}{r} + fJ\right)$$

Notice how the angular four-velocity in the $\hat{\varphi}$ direction contains a term that explicitly depends on the angular momentum a of the source of the gravitational field. This is the contribution of frame dragging towards the four-momentum of the test particle.

In the local inertial frame, the two non-trivial components of the four-velocity are given by:

$$u^0 = \frac{K}{f}$$

$$u^3 = \frac{aK}{\sqrt{\Delta}}\left(\frac{1}{\sqrt{f}} - \sqrt{f}\right) + \sqrt{\frac{f}{\Delta}}J \tag{8.38}$$

For circular orbits, the integration constants K and J are required to take the following values:

$$K = \frac{1 - \dfrac{r_s}{r} \mp a\sqrt{\dfrac{r_s}{2r^3}}}{\sqrt{1 - \dfrac{3r_s}{2r} \mp 2a\sqrt{\dfrac{r_s}{2r^3}}}}$$

$$J = \mp \frac{1 + \dfrac{a^2}{r^2} \pm 2a\sqrt{\dfrac{r_s}{2r^3}}}{\sqrt{1 - \dfrac{3r_s}{2r} \mp 2a\sqrt{\dfrac{r_s}{2r^3}}}} \sqrt{\frac{rr_s}{2}}$$

(8.39)

where the upper and lower signs correspond to counter-rotating and co-rotating circular orbits, respectively. It can be observed that the values of a and r_s have a further restriction. Indeed, from the equations for K and J we have the condition for circular orbits in the Kerr metric:

$$1 - \frac{3r_s}{2r} \mp 2a\sqrt{\frac{r_s}{2r^3}} > 0 \ \Rightarrow \ \pm a < \left(1 - \frac{3r_s}{2r}\right)\sqrt{\frac{r^3}{2r_s}}$$

(8.40)

for counter- and co-rotating orbits, respectively. In addition, this bound has to be combined with the bound from Penrose's cosmic censorship hypothesis. Furthermore, as in the previous example, we will ignore the $\mathcal{O}(\hbar)$ non-geodesic motion induced by the coupling between the spin and the curvature.

Thus, figure 8.4 shows the tangential velocity observed in the local inertial frame v_t as a function of r_s/r, where:

$$v_t = \frac{u^3}{u^0}$$

(8.41)

The solid lines represent the velocity at the maximal values of $a/r_s = \pm 0.5$. The dotted line corresponds to the Schwarzschild case ($a = 0$). The two dashed lines correspond to the limits on a imposed by the circular orbit condition.

The Lorentz transformations that describe the motion of the particle in the local inertial frame are:

$$\lambda^0{}_1(x) = \lambda^1{}_0(x) = \frac{ar_s^2}{2r^4\sqrt{\Delta f}}\left(J - \frac{K}{\omega} + aK + \frac{fJr}{r_s}\right)$$

$$\lambda^1{}_3(x) = -\lambda^3{}_1(x) = \frac{a^2 r_s^2}{2\Delta r^4\sqrt{f}}\left(\frac{K}{\omega} - J\right)$$

$$- \frac{ar_s}{\Delta r}\left(K + \frac{fJr}{ar_s}\right)\left(\frac{a^2 r_s}{2r^3\sqrt{f}} - \sqrt{f}\right)$$

(8.42)

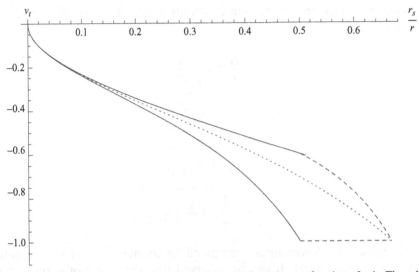

Figure 8.4. Tangential velocity observed in the local inertial frame v_t as a function of r_s/r. The solid lines represent the velocity at the maximal values of $a/r_s = \pm 0.5$. The dotted line corresponds to the Schwarzschild case ($a = 0$). The two dashed lines correspond to the limits on a imposed by the circular orbit condition.

These transformations correspond to a boost in the one-direction and a rotation over the two-direction, respectively.

The associated infinitesimal Wigner rotation is:

$$\vartheta^1{}_3(x) = \frac{ar_s^2 \sqrt{f}}{2r^4\Delta(K+\sqrt{f})}\left(J - \frac{K}{\omega} + aK + \frac{fJr}{r_s}\right)\times\left(-\frac{aK}{f} + aK + J\right)$$
$$+ \frac{a^2 r_s^2}{2\Delta r^4 \sqrt{f}}\left(\frac{K}{\omega} - J\right) - \frac{ar_s}{\Delta r}\left(K + \frac{fJr}{ar_s}\right)\left(\frac{a^2 r_s}{2r^3 \sqrt{f}} - \sqrt{f}\right) \qquad (8.43)$$

which depends on the angular momentum a and the gravitational ratio r_s/r. Therefore, the total Wigner rotation per circular orbit of radius r is:

$$\Theta = \int_{\tau_1}^{\tau_2} \vartheta^1{}_3(x)d\tau = \int_0^{2\pi} \frac{\vartheta^1{}_3(x)}{u^\varphi(x)}d\varphi = 2\pi\frac{\vartheta^1{}_3}{u^\varphi} \qquad (8.44)$$

and we have used the fact that, for circular orbits, ϑ^1_3 and u^ϕ have fixed values, independent of the coordinates.

To analyse the gravitational effects on the spin of the orbiting particle, we need to subtract the 2π due to the trivial rotation of the particle around the compact object:

$$\Omega = \Theta - 2\pi = 2\pi\left(\frac{\vartheta^1{}_3}{u^\varphi} - 1\right) \qquad (8.45)$$

The possible values for the Wigner rotation angle (per orbit) due to gravity Ω for a particle moving on an equatorial circular orbit in Kerr spacetime are shown in figure 8.5. The two solid lines correspond to the upper and lower bounds imposed by the existence of event horizons $r_\pm$. That is, the upper and lower bounds corresponds to the maximal angular momentum of the compact object:

$$a^\pm = \pm \frac{r_s}{2} \tag{8.46}$$

Clearly, the two solid lines correspond to the counter-rotating and co-rotating cases, respectively.

The bounds on the angular momentum of the compact object imposed by the dynamics of the circular orbit further restrict the space of values for Ω. The two maximal values correspond to:

$$a_\circ^\pm = \pm \left(1 - \frac{3r_s}{2r}\right)\sqrt{\frac{r^3}{2r_s}} \tag{8.47}$$

and are shown as dashed lines in figure 8.5. These bounds generalize the well-known gravitational effect that forbids circular orbits of arbitrarily small radius in Schwarzschild spacetimes. Notice the dotted line in figure 8.5, which represents the case where the Kerr metric reduces to the Schwarzschild metric ($a = 0$) and takes the value Ω_0.

The spin dynamic described by figure 8.5 is expected. Even in the case of a non-rotating compact object ($a = 0$), the gravitational field affects the particle's spin by Ω_0.

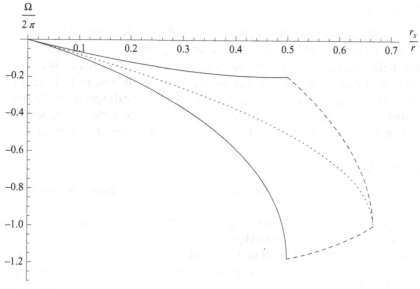

Figure 8.5. Possible values for the gravitationally induced Wigner rotation angle Ω (per orbit) for a particle moving in an equatorial circular orbit in Kerr spacetime. The two solid lines correspond to the upper and lower bounds imposed by the existence of both event horizons, Ω_+ and Ω_-, respectively. The two dashed lines correspond to the upper and lower bounds of the angular momentum a imposed by the dynamics of circular orbits. The dotted line corresponds to the Wigner rotation for circular orbits in Schwarzschild spacetime (Ω_0).

If the particle co-rotates with the compact object, $a > 0$, then the spin rotation angle increases: $\Omega_+ > \Omega_0$. If the particle counter-rotates with the compact object, $a < 0$, and the spin rotation angle decreases: $\Omega_- < \Omega_0$. However, notice that except for $r \to \infty$, there is no circular orbit that completely mitigates the gravitational effects on the spin: $\Omega \neq 0$. In addition, the figure makes clear that, in the weak field limits ($r_s \to 0$, $r \to \infty$), the Wigner rotation entirely due to frame dragging is much smaller than the Wigner rotation produced by a non-rotating source of the same mass. However, for the co-rotating case, the Wigner rotation due to frame dragging is much larger than the Wigner rotation due to gravity alone as we approach the event horizon.

Also, the spin rotation angle is bounded for all circular orbits: $|\Omega| \lesssim 1.2$. The spin rotation is additive on completion of a single orbit. Thus, the spin rotation angle after N orbits around the compact object is given by: $\Omega_N = N\Omega$.

Furthermore, let us consider the angular velocity ω_s produced by the non-geodetic motion that results from the spin–curvature coupling. If we require the spin–curvature correction to be of about the same value as the maximum angular velocity due to frame dragging (for an electron):

$$\delta u^\varphi = \frac{\hbar}{mr^2} \approx u^\varphi = \frac{ar_s}{\Delta r} \quad \Rightarrow \quad \frac{r_s}{r} \approx 21 \tag{8.48}$$

which implies the electron is well inside the event horizon and extremely close to the singularity.

8.7 Summary

In this chapter we discussed the dynamics of qubits in Kerr spacetime. That is, we studied the Wigner rotations that affect a spin-$\frac{1}{2}$ particle in the presence of a gravitational field produced by a spherically symmetric rotating mass. We limited the discussion to a radial fall and circular orbits, as these allowed us to have analytical formulae. In both cases, we showcased the effects of frame dragging on the spin of the test particle. We presented all the formulae necessary to tackle more sophisticated trajectories, but they will most probably require some form of numerical modelling.

Bibliography

[1] Chandraserkhar S 1992 *The Mathematical Theory of Black Holes* (Oxford: Oxford University Press)

[2] Hobson M P, Efstathiou G and Lasenby A N 2006 *General Relativity: An Introduction for Physicists* (Cambridge: Cambridge University Press)

[3] Carter B 1968 Global structure of the Kerr family of gravitational fields *Phys. Rev.* **174** 1559

[4] Lanzagorta M 2012 Effect of gravitational frame dragging on orbiting qubits arXiv:1212.2200 [quant-ph]

[5] Penrose R 1969 Gravitational collapse: the role of general relativity *Rev. Nuovo Cimento, Ser. 1, Num. Spec.* **1** 252

[6] Farooqui A, Kamran N and Panangaden P 2013 An exact expression for photon polarization in Kerr geometry arXiv:1306.6292v1 [math-ph]

[7] Iyer B R and Kumar A 1977 Dirac equation in Kerr space-time *Pramana* **8** 500–11

[8] Bardeen J M, Press W H and Teukolsky S A 1972 Rotating black holes: locally nonrotating frames, energy extraction, and scalar synchrotron radiation *Astrophys. J.* **178** 347–69

[9] Papapetrou A 1951 Spinning test-particles in general relativity. I *Proc. R. Soc. London, Ser. A, Math. Phys. Sci.* **209** 248–58

[10] Corinaldesi E and Papapetrou A 1951 Spinning test-particles in general relativity. II *Proc. R. Soc. London, Ser. A, Math. Phys. Sci.* **209** 259–68

[11] Ubukhov Y N, Silenko A J and Teryaev O V 2010 Spin in stationary gravitational frames and rotating frames *The Sun, The Stars, The Universe and General Relativity* ed R Ruffini and G Vereshchagin (New York: AIP) pp 112–9

[12] Plyatsko R M, Stefanyshyn O B and Fenyk M T 2011 Mathisson Papapetrou & Dixon equations in the Schwarzschild and Kerr backgrounds *Class. Quantum Grav.* **28** 195025

[13] Plyatsko R M, Stefanyshyn O B and Fenyk M T 2010 Highly relativistic spinning particle starting near $r_{ph}^{(-)}$ in a Kerr field *Phys. Rev.* D **82** 044015

[14] Plyatsko R 2005 Ultrarelativistic circular orbits of spinning particles in a Schwarzschild field *Class. Quantum Grav.* **22** 1545–51

[15] Silenko A J 2008 Classical and quantum spins in curved spacetimes *Acta Phys. Polonica B Proc. Suppl.* **1** 87–107

[16] Plyatsko R 1998 Gravitational ultrarelativistic spin-orbit interaction and the weak equivalence principle *Phys. Rev.* D **58** 084031

[17] Singh D 2008 Perturbation method for classical spinning particle motion: I. Kerr space-time *Phys. Rev.* D **78** 104028

Chapter 9

Quantum information processing in curved spacetimes

In this chapter we will discuss some simple instances of quantum information processing in the presence of classical gravitational fields. In particular, we will explore how Wigner rotations due to spacetime curvature affect the performance of a simple steganographic quantum communication protocol, quantum teleportation, Einstein–Podolsky–Rosen (EPR) experiments, entanglement and quantum computations. Finally, we present a brief discussion of possible implementations of quantum gravitometers.

Before we begin, however, let us remark that our emphasis is the analysis of spin-based quantum information. In other words, we are mostly interested in the behaviour of spin in the presence of a classical gravitational field, and how spin-based quantum information processing technologies are affected by spacetime curvature. That said, it is important to note that some other approaches have been proposed by the scientific community. For instance, there has been some interest in the study of entanglement between the traverse excitations of two Bose–Einstein condensates in the presence of gravity and non-inertial frames [1, 2].

9.1 Unitary qubit transformations

As we discussed in previous chapters, the effect of a gravitational field on a Dirac spinor is given through the Wigner rotation[1]. That is, the net effect of the gravitational field can be described by the unitary transformation $\delta\hat{U}_g$ that represents the differential local Lorentz transformation $\delta\Lambda(x)$ in the local inertial frame associated with the chosen tetrad field that represents the spacetime metric $g_{\mu\nu}$. Thus, the differential effect of a gravitational field on a spinor $\psi_{p,\sigma}$ looks like:

$$\delta\hat{U}_g\,\psi_{p,\sigma} \propto \sum_\alpha \delta D_{\sigma\alpha}\,\psi_{\delta\Lambda p,\alpha} \qquad (9.1)$$

---

[1] We will use the vector notation described on the first section of chapter 4.

where $\delta D_{\alpha\beta}$ is the unitary operator that represents the infinitesimal Wigner rotation and it is given by:

$$\delta\hat{D} = I + \frac{i}{2}\left(\vartheta_{23}\sigma_x + \vartheta_{31}\sigma_y + \vartheta_{12}\sigma_z\right)d\tau \qquad (9.2)$$

where $\sigma_{x,y,z}$ are the Pauli matrices and ϑ_{ij} are the infinitesimal Wigner rotation angles. Then, for a finite Wigner rotation:

$$\hat{D} = \mathcal{T} \exp\left(\frac{i}{2}\int_{\tau_i}^{\tau_f} \boldsymbol{\sigma}\cdot(\vartheta_{23}, \vartheta_{31}, \vartheta_{12})\,d\tau\right) \qquad (9.3)$$

where $\mathcal{T}$ is the time ordering operator. In the case where the Wigner rotation takes place along a single direction (e.g. the y direction), there is no need for $\mathcal{T}$ and the unitary operator simplifies to:

$$\hat{D} = \exp\left(\frac{i}{2}\int_{\tau_i}^{\tau_f} \sigma_y\,\vartheta_{31}\,d\tau\right)$$

$$= \exp\left(\frac{i\,\sigma_y}{2}\int_{\tau_i}^{\tau_f} \vartheta_{31}\,d\tau\right)$$

$$= \exp\left(\frac{i\,\sigma_y}{2}\Omega\right) \qquad (9.4)$$

where Ω is the total Wigner rotation integrated over the entire path of the particle. As discussed in section 5.7, it is important to remark that the state $\psi_{p,\sigma}$ in equation (9.1) does not represent a quantum particle of momentum p^μ localized at some point x^μ in spacetime. Instead, these expressions are used to represent extended states of helicity σ and definite momentum p^a if observed from the position x^a of the local inertial frame described by the tetrad field that produces the appropriate metric tensor.

In the examples discussed in the previous three chapters, the tetrad field and axes were chosen in such a way that the only non-zero differential Wigner rotation angle was ϑ_{13} (movement takes place on the equatorial plane $\theta = \pi/2$). This is a rotation over the two-axis, which is parallel to the $\hat{\theta}$-axis. Therefore, the corresponding unitary qubit rotation is given by:

$$\hat{D} = e^{i\sigma_y\Omega/2}$$

$$= \cos\left(\frac{\Omega}{2}\right) + i\sigma_y\sin\left(\frac{\Omega}{2}\right)$$

$$= \begin{pmatrix} \cos\left(\Omega/2\right) & \sin\left(\Omega/2\right) \\ -\sin\left(\Omega/2\right) & \cos\left(\Omega/2\right) \end{pmatrix} \qquad (9.5)$$

Let us choose the quantization axis parallel to the three-direction, i.e. parallel to the $\hat{\varphi}$ axis[2]. Then, the effect of the unitary qubit transformation on the two positive energy Dirac states (spin up and spin down) is given by:

$$\hat{U}_g|p, +\rangle = \cos\left(\frac{\Omega}{2}\right)|\Lambda p, +\rangle + \sin\left(\frac{\Omega}{2}\right)|\Lambda p, -\rangle$$

$$\hat{U}_g|p, -\rangle = -\sin\left(\frac{\Omega}{2}\right)|\Lambda p, +\rangle + \cos\left(\frac{\Omega}{2}\right)|\Lambda p, -\rangle$$

(9.6)

where in general Ω and Λp will depend on the path of the test particle and on the final time τ_f in which the integral is performed.

Thus, we can define the computational basis as:

$$|0\rangle \leftrightarrow |p, +\rangle = |p\rangle \otimes |+\rangle \qquad |1\rangle \leftrightarrow |p, -\rangle = |p\rangle \otimes |-\rangle \qquad (9.7)$$

which is affected by a gravitational field as:

$$\hat{U}_g|0\rangle = \cos\left(\frac{\Omega}{2}\right)|0\rangle + \sin\left(\frac{\Omega}{2}\right)|1\rangle$$

$$\hat{U}_g|1\rangle = -\sin\left(\frac{\Omega}{2}\right)|0\rangle + \cos\left(\frac{\Omega}{2}\right)|1\rangle$$

(9.8)

and we have ignored the momentum-dependent part of the state. This is a perfectly fine assumption, as long as we limit our discussion to momentum eigenvectors.

9.2 Qubit states

Let us first consider a general one-qubit state given by:

$$|\psi\rangle = \alpha|0\rangle + \beta|1\rangle \implies \rho = |\psi\rangle\langle\psi| \qquad (9.9)$$

Under the influence of a gravitational field, this state is transformed into:

$$|\psi\rangle \to |\tilde{\psi}\rangle = e^{i\sigma_y \Omega/2}|\psi\rangle \qquad (9.10)$$

with an associated density matrix:

$$\begin{aligned} \tilde{\rho} &= |\tilde{\psi}\rangle\langle\tilde{\psi}| \\ &= e^{i\sigma_y \Omega/2}|\psi\rangle\langle\psi|e^{-i\sigma_y \Omega/2} \\ &= e^{i\sigma_y \Omega/2}\rho\, e^{-i\sigma_y \Omega/2} \end{aligned} \qquad (9.11)$$

---

[2] For qubits moving in the equatorial plane in circular paths around the source of the gravitational field, this choice of the quantization axis will lead to helicity states.

Then, the *trace distance* between ρ and $\tilde{\rho}$ is given by:

$$D(\rho, \tilde{\rho}) = \frac{1}{2}\text{Tr}|\rho - \tilde{\rho}|$$

$$= \frac{1}{2}\text{Tr}|\rho - e^{i\sigma_y\Omega/2}\rho\,e^{-i\sigma_y\Omega/2}|$$

$$= 0 \tag{9.12}$$

That is, the trace distance is always zero, regardless of the state ρ and the value of Ω. On the other hand, the *fidelity* between ψ and $\tilde{\psi}$ is given by:

$$F(\psi, \tilde{\psi}) = |\langle\psi|\tilde{\psi}\rangle|^2$$
$$= |\langle\psi|e^{i\sigma_y\Omega/2}|\psi\rangle|^2$$
$$= \left|(\alpha^*\,\beta^*)\begin{pmatrix}\cos(\Omega/2) & \sin(\Omega/2) \\ -\sin(\Omega/2) & \cos(\Omega/2)\end{pmatrix}\begin{pmatrix}\alpha \\ \beta\end{pmatrix}\right|^2$$
$$= |\cos(\Omega/2) + 2i\Im(\alpha^*\beta)\sin(\Omega/2)|^2$$
$$= \cos^2(\Omega/2) + 4\Im^2(\alpha^*\beta)\sin^2(\Omega/2) \tag{9.13}$$

and for an infinitesimal rotation angle $\delta\Omega$:

$$F(\psi, \tilde{\psi}) \approx 1 + \delta\Omega^2\left(\frac{4\Im^2(\alpha^*\beta) - 1}{4}\right) \tag{9.14}$$

Notice that, if the qubit is such that:

$$\Im^2(\alpha^*\beta) = \frac{1}{4} \tag{9.15}$$

then:

$$F(\psi, \tilde{\psi}) \approx 1 \tag{9.16}$$

which means that ψ and $\tilde{\psi}$ are nearly the same state. This is to be expected, as the transformation is a rotation over the 'y' axis.

If we consider $\psi_\perp$ as a state orthogonal to ψ:

$$|\psi_\perp\rangle = -\beta^*|0\rangle + \alpha^*|1\rangle \Rightarrow \langle\psi|\psi_\perp\rangle = 0 \tag{9.17}$$

then the fidelity between $\tilde{\psi}$ and $\psi_\perp$ gives a measure of the probability of confusing the measurement of the two states. That is, it gives an estimate of the error probability induced by gravity. In this simple example, the fidelity between $\tilde{\psi}$ and $\psi_\perp$ is given by:

$$F(\tilde{\psi}, \psi_\perp) = \sin^2(\Omega/2)(1 - 4\Im^2(\alpha^*\beta)) \tag{9.18}$$

which for an infinitesimal rotation angle $\delta\Omega$ takes the form:

$$F(\tilde{\psi}, \psi_\perp) \approx \delta\Omega^2\left(\frac{1 - 4\Im^2(\alpha^*\beta)}{4}\right) \tag{9.19}$$

As before, we notice that, if the qubit is such that:

$$\Im^2(\alpha^*\beta) = \frac{1}{4} \tag{9.20}$$

then:

$$F(\tilde{\psi}, \psi_\perp) \approx 0 \tag{9.21}$$

which means that $\tilde{\psi}$ and $\psi_\perp$ are nearly orthogonal.

Let us now consider the 1-qubit uniform superposition, which gives equal probability of measuring '0' or '1':

$$|\psi\rangle = \frac{|0\rangle + |1\rangle}{\sqrt{2}} \tag{9.22}$$

In the presence of gravity, this state will 'drift' into:

$$|\psi\rangle \rightarrow \frac{\cos\Omega/2 + \sin\Omega/2}{\sqrt{2}}|0\rangle + \frac{\cos\Omega/2 - \sin\Omega/2}{\sqrt{2}}|1\rangle \tag{9.23}$$

where, in general, the Wigner rotation angle is a function of time $\Omega = \Omega(t)$. The behaviour of the probability of measuring '1' and '0' with respect to Ω is shown in figure 9.1. It can be observed that, depending on the sign of Ω, either the probability of measuring '1' or '0' will be amplified, while the other is reduced.

In figure 9.2 we show the behaviour of the probability of measuring '1' and '0' with respect to r_s/r for a qubit after completion of a circular orbit of radius r in the Schwarzschild metric produced by an object of mass $r_s/2$ using the expression for Ω given in equation (6.48).

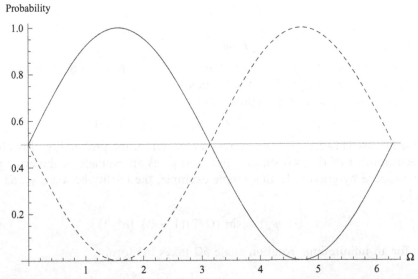

Figure 9.1. Drifting of the probability of measuring '1' (solid line) and '0' (dashed line) for a 1-qubit state in the uniform superposition in the presence of a gravitational field described by the Wigner rotation angle $\Omega(t)$.

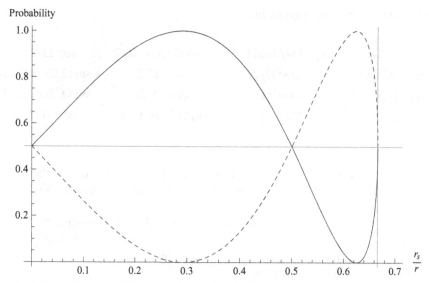

Figure 9.2. Drifting of the probability of measuring '1' (solid line) and '0' (dashed line) for a 1-qubit state in the uniform superposition after completion of a circular orbit of radius r in the Schwarzschild spacetime produced by an object of mass $r_s/2$.

Notice that, if the Wigner rotation angle is very small, $\Omega \ll 1$, then:

$$P_0 = |\langle 0|\psi\rangle|^2 \approx \frac{1}{2} + \frac{\Omega}{2}$$

$$P_1 = |\langle 1|\psi\rangle|^2 \approx \frac{1}{2} - \frac{\Omega}{2}$$

(9.24)

where P_0 and P_1 are the probabilities of measuring '0' and '1', respectively. Therefore, the measurement error is approximately given by $\Omega/2$.

Finally, let us consider the 2-qubit state uniform superposition:

$$|\psi\rangle = \frac{|00\rangle + |01\rangle + |10\rangle + |11\rangle}{2}$$

(9.25)

which under the presence of gravity gets transformed into:

$$|\psi\rangle \rightarrow \frac{(\cos \Omega/2 - \sin \Omega/2)^2}{2}|00\rangle + \frac{\cos^2 \Omega/2 - \sin^2 \Omega/2}{2}|01\rangle$$

$$+ \frac{\cos^2 \Omega/2 - \sin^2 \Omega/2}{2}|10\rangle + \frac{(\cos \Omega/2 + \sin \Omega/2)^2}{2}|11\rangle$$

(9.26)

and the rotation matrix is given by:

$$
\begin{pmatrix}
\cos^2 \Omega/2 & -\sin \Omega/2 \cos \Omega/2 & -\sin \Omega/2 \cos \Omega/2 & \sin^2 \Omega/2 \\
\cos \Omega/2 \sin \Omega/2 & \cos^2 \Omega/2 & -\sin^2 \Omega/2 & -\sin \Omega/2 \cos \Omega/2 \\
\cos \Omega/2 \sin \Omega/2 & -\sin^2 \Omega/2 & \cos^2 \Omega/2 & -\sin \Omega/2 \cos \Omega/2 \\
\sin^2 \Omega/2 & \sin \Omega/2 \cos \Omega/2 & \sin \Omega/2 \cos \Omega/2 & \cos^2 \Omega/2
\end{pmatrix}
$$

$$(9.27)$$

where we have assumed that the two qubits are close enough so that their Wigner angles are exactly the same. The drifting of the state with respect to the Wigner angle is shown in figure 9.3.

In figure 9.4 we show the behaviour of the probability of measuring '00', '11' and '01' or '10' with respect to r_s/r for a qubit after completion of a single circular orbit of radius r in the Schwarzschild metric produced by an object of mass $r_s/2$ using the expression for Ω given in equation (6.48).

We notice that the probability of measuring '00' and '11' is amplified and diminished in a periodic fashion. On the other hand, the probability of measuring '01' or '10' is always smaller than the original 1/4 probability of the uniform superposition. This suggests the intriguing possibility that maybe drifting qubits could be used to measure gravitational fields. This is certainly the case if the qubits

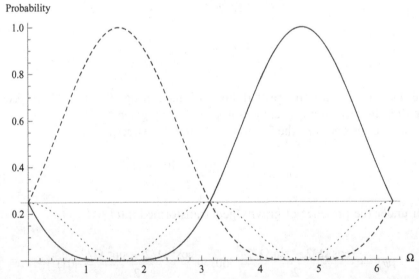

Figure 9.3. Drifting of the probability of measuring '00' (solid line), '01' and '10' (dotted line) and '11' (dashed line) for a 2-qubit state in the uniform superposition in the presence of a gravitational field described by the Wigner rotation angle $\Omega(t)$.

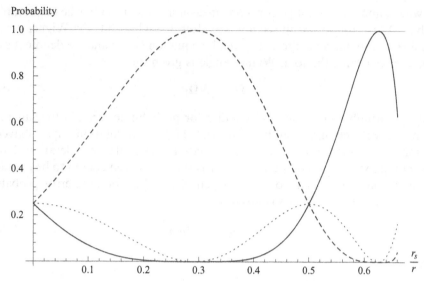

Figure 9.4. Drifting of the probability of measuring '00' (solid line), '01' and '10' (dotted line) and '11' (dashed line) for a 2-qubit state after completion of a circular orbit of radius r in the Schwarzschild spacetime produced by an object of mass $r_s/2$.

are near a black hole and the Wigner angle is large. Indeed, if $\Omega \approx \pi/2$, then the measurement of the '11' qubit state will be very close to 1. Later on, we will explore in more detail how to use Wigner rotations in the design of quantum gravitometers.

9.3 Quantum communications

Let us consider the steganographic quantum communications protocol described in section 4.2. That is, we encode classical logical information in the quantum states:

$$|\tilde{0}\rangle = \alpha|0\rangle + \beta|1\rangle \qquad |\tilde{1}\rangle = -\beta^*|0\rangle + \alpha^*|1\rangle \qquad (9.28)$$

that are being exchanged between Alice and Bob. These states are properly normalized to the unity:

$$|\alpha|^2 + |\beta|^2 = 1 \qquad (9.29)$$

To extract the logical information out of the quantum states, Alice and Bob are required to perform measurements in the right basis as described by the pair of complex numbers $\{\alpha, \beta\}$.

Let us limit ourselves to the case of a single qubit in an equatorial circular orbit around a spherically symmetric mass in a spacetime defined by the Schwarzschild metric (which is a rather good approximation to Earth's gravitational field). Let us also assume the very simple situation in which Alice prepares and sends to Bob several qubits at the time. Alice is in a ground station and Bob is aboard a satellite travelling in a circular orbit around Earth. However, Bob will keep the qubits stored somewhere, and he will not perform any measurements until after he has completed N orbits. As we found out in section 6.6, after a single orbit, the Wigner rotation angle due to gravitational effects is given by equation (6.48) and we denote it as Ω_1. Then, after N orbits, the total Wigner angle is given by:

$$\Omega_N = N\Omega_1 \qquad (9.30)$$

All the formulae for the capacity and error probabilities remain the same as in section 4.3, with the understanding that here the Wigner rotation angle is given by equation (9.30). Also, notice that these equations take into consideration that the quantization axis for the computational basis states has been chosen to be parallel to the $\hat{\varphi}$ direction, which leads to helicity eigenstates. Therefore, the error probability for this channel after N orbits is given by:

$$\epsilon = b_N^2 \left(1 - 4\delta^2\right) \qquad (9.31)$$

where:

$$\delta = \Im\left(\alpha^*\beta\right) \qquad (9.32)$$

and:

$$b_N = \sin\left(\frac{\Omega_N}{2}\right) \qquad (9.33)$$

Then, unless the unitary operation that inverts the gravitational Wigner rotation is applied to the orbiting state, $\epsilon \neq 0$ implies errors in the quantum communication protocol.

Let us also notice that, in a similar manner as relativistic quantum information in inertial frames, the error depends not only on the structure of the gravitational field embedded in the Wigner angle Ω but also on the structure of the basis used for the communication protocol. For example, for the specific case of a qubit moving on a circular orbit on Schwarzschild spacetime, there is a privileged basis that makes the communication protocol immune to the effects of gravitation. Indeed, if $\delta = 1/4$, then the error vanishes $\epsilon = 0$. For instance, this is the case if the classical information is encoded in the $+_i$ diagonal basis $\{|+_i\rangle, |-_i\rangle\}$, which are the eigenvectors of σ_y.

The behaviour of ϵ with respect to the gravitational ratio r_s/r for the computational basis after a single orbit in Schwarzschild spacetime is shown in figure 9.5. We can observe that if the gravitational field is strong enough, say at about $r_s/r \approx 0.157$, then the error is $\epsilon \approx 0.5$ and the channel capacity for this quantum communications protocol will decrease to zero.

Let us consider now a 'practical' example. The mass of Earth is about $r_\oplus \approx 0.443$ cm and the radius about $R_\oplus \approx 6.4 \times 10^3$ km. A typical low-orbit satellite has an orbital altitude of about $R \approx 2\,000$ km. Therefore, the gravitational ratio is of about:

$$\frac{r_\oplus}{R_\oplus + R} \approx 5 \times 10^{-10} \qquad (9.34)$$

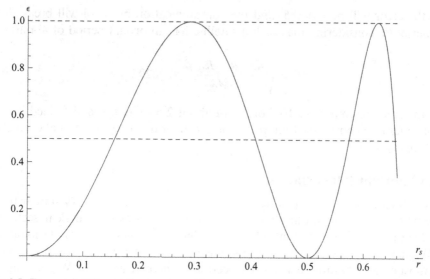

Figure 9.5. Error probability ϵ in the steganographic quantum communications protocol described in the text in function of the gravitational ratio r_s/r for the computational basis after completion of a single circular orbit in Schwarzschild spacetime.

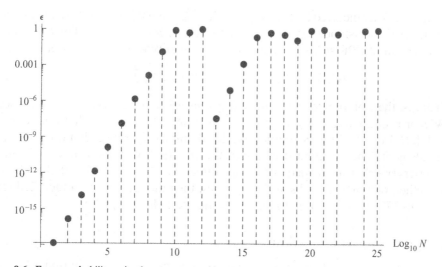

Figure 9.6. Error probability ϵ in the steganographic quantum communications protocol for a low Earth satellite in function of the number of orbits N.

and after a single orbit, the associated error is of about:

$$\epsilon \approx 1.4 \times 10^{-18} \qquad (9.35)$$

The behaviour of the error probability ϵ with the number of orbits N is shown in figure 9.6. It can be observed that, for instance, if Bob's satellite performs 10^{10} orbits, then the error will be $\epsilon \approx 0.5$, and the communication protocol will break down. Furthermore, considering that such a satellite has an orbital period of about:

$$T \approx \sqrt{\frac{4\pi^2 (R + R_{\oplus})^3}{GM_{\oplus}}} \approx 7629\,s \qquad (9.36)$$

then the satellite completes 10^{10} orbits in about 2.5 million years! Indeed, for this simple quantum communication protocol, the error due to gravitational corrections is extremely small.

9.4 Quantum teleportation

The analysis of quantum teleportation in a gravitational field is very similar to that presented in section 4.4 for inertial frames. Of course, the Wigner angle in section 4.4 was due to Lorentz transformations between inertial frames. In a gravitational field, the Wigner rotation takes the form discussed in the previous three chapters. Notice that most of the equations discussed in section 4.4 involve a specific Wigner angle Ω_p or the associated error ϵ. Therefore, all these equations remain valid for the case of quantum teleportation in the presence of a gravitational field.

9.5 EPR experiments

Similar errors are incurred in the study of EPR experiments in Schwarzschild and Kerr spacetimes. For instance, let us consider $(\mathcal{T}, \mathcal{S}, \mathcal{Q}, \mathcal{R})$ as a set of directions for the measurement of EPR states [3]. Then, Bell's inequality will take the modified form:

$$\langle \mathcal{QS} \rangle + \langle \mathcal{RS} \rangle + \langle \mathcal{RT} \rangle - \langle \mathcal{QT} \rangle = 2\sqrt{2}\cos^2 \Omega \qquad (9.37)$$

That is, the violation to Bell's inequality *appears* to be reduced by the presence of a Wigner rotation angle $\Omega \neq 0$. However, it is important to note that this is not the case [3]. Indeed, the Wigner rotation is a local unitary operation that does not affect entanglement. The apparent decrease in the violation of Bell's inequality is due to the observer using an inadequate set of directions $(\mathcal{T}, \mathcal{S}, \mathcal{Q}, \mathcal{R})$. If the observer knows the exact orbital paths of the particles, then they can derive a modified set of directions $(\mathcal{T}', \mathcal{S}', \mathcal{Q}', \mathcal{R}')$ that maximally violate Bell's inequality.

9.6 Quantum computation

Let us now consider the case of a quantum computer in a gravitational field [4]. For a quantum computer on the surface of the Earth, the Wigner rotation is of about 2.2×10^{-9} per day, or 2.5×10^{-14} per second. Therefore, after one day, the error fidelity between qubits will lead to an error probability of about $\epsilon_d \approx 5 \times 10^{-18}$. Similarly, the error probability after one second is about $\epsilon_s \approx 6 \times 10^{-28}$.

This is an interesting result. Qubits in a quantum computer interact with the Earth's gravitational field. Consequently, their state will slowly change with time, even if the computer is in the idle state. This is a consequence of the direct coupling between spin-based quantum information and classical gravitational fields. That is, as qubits gravitate, the quantum information that they encode is constantly changing as time goes by. Clearly, this is in strong contrast to the much more abstract concept of classical information encoded in a bit, where we expect the state of the classical computer to be constant in time (even in the presence of a gravitational field).

Furthermore, even though the error probability is a small number, it is known that even an arbitrary small uncorrected error probability will undermine the complexity of most (if not all) quantum algorithms [5]. To this end, we can estimate the computational complexity of a quantum algorithm in a gravitational field using a simple, but powerful, error analysis method previously described in the literature [5].

9.6.1 General iterative algorithm

Let us imagine a large quantum computer running an algorithm that starts on some initial quantum state, and then performs a large number of iterations of some quantum operation.

Suppose the quantum algorithm in consideration requires a single multi-qubit gate $\hat{U}$ that is applied m times to a given quantum state. The system has n qubits and can be used to represent $N = 2^n$ possible states. In the absence of gravity, and ignoring any possible sources of noise and error, the first computational step looks like:

$$\rho^{(1)} = \hat{U}\rho^{(0)}\hat{U}^\dagger \tag{9.38}$$

and the last computational step after m iterations is given by:

$$\rho^{(m)} = \hat{U}^m\rho^{(0)}\hat{U}^{\dagger m} \tag{9.39}$$

We will model the presence of a weak gravitational field in the following fashion. After one iteration the state will look like:

$$\rho^{(1)} = \hat{D}\hat{U}\rho^{(0)}\hat{U}^\dagger\hat{D}^\dagger \tag{9.40}$$

where $\hat{D}$ is the unitary operation that represents the Wigner rotation between computational steps. Then, after two iterations the state will be given by:

$$\rho^{(2)} = \hat{D}\hat{U}\hat{D}\hat{U}\rho^{(0)}\hat{U}^\dagger\hat{D}^\dagger\hat{U}^\dagger\hat{D}^\dagger \tag{9.41}$$

and so on. Furthermore, for the case under consideration, the Wigner rotation can be written as:

$$\hat{D} = \bigotimes_{j=1}^{n} e^{i\hat{\boldsymbol{\Sigma}}_j \cdot \boldsymbol{\Omega}_j/2} \tag{9.42}$$

where $\boldsymbol{\Omega}_j$ is the vector of Wigner rotation angles for the jth qubit and $\hat{\boldsymbol{\Sigma}}_j$ is the vector of spin rotation generators that acts on the jth qubit.

In general, each qubit will be in a different position in spacetime, and their spins may be aligned in different directions. To ease the analysis, let us assume not only that the gravitational field is weak, but also that all qubits undergo the exact same Wigner rotation. We also assume that the quantum computer can operate at 1 MHz.

That is, it can perform one operation in 10^{-6} seconds. Then, we assume that there is an Ω such that:

$$\Omega \approx \Omega_j \quad \forall j \implies \Omega \approx 2.5 \times 10^{-20} \tag{9.43}$$

Then, we can write $\hat{D}$ as:

$$\hat{D} \approx 1 + \frac{i\Omega}{2} \sum_{j=1}^{n} \hat{\Sigma}_j + \mathcal{O}(\Omega^2) \tag{9.44}$$

where Ω is the Wigner rotation angle after 10^{-6} seconds and $\hat{\Sigma}_j$ is a Hermitian operator that acts in a non-trivial manner only on the jth qubit:

$$\hat{\Sigma}_j = \underbrace{\mathbb{I} \otimes \mathbb{I} \otimes \ldots \otimes \mathbb{I}}_{j-1} \otimes \sigma \otimes \underbrace{\mathbb{I} \otimes \mathbb{I} \otimes \ldots \otimes \mathbb{I}}_{n-j} \tag{9.45}$$

where the underbrace denotes the number of single-qubit identity operators in the tensor product and σ is the spin rotation generator.

Consequently, at order $\mathcal{O}(\Omega)$ the state after one iteration can approximately be written as:

$$\begin{aligned}
\rho^{(1)} &\approx \left(1 + \frac{i\Omega}{2} \sum_{j=1}^{n} \hat{\Sigma}_j\right) \hat{U}\rho^{(0)}\hat{U}^\dagger \left(1 - \frac{i\Omega}{2} \sum_{k=1}^{n} \hat{\Sigma}_k^\dagger\right) + \mathcal{O}(\Omega^2) \\
&\approx \hat{U}\rho^{(0)}\hat{U}^\dagger + \frac{i\Omega}{2} \sum_{j=1}^{n} \left(\hat{\Sigma}_j \hat{U}\rho^{(0)}\hat{U}^\dagger - \hat{U}\rho^{(0)}\hat{U}^\dagger \hat{\Sigma}_j^\dagger\right) + \mathcal{O}(\Omega^2)
\end{aligned} \tag{9.46}$$

If we define Ξ as:

$$\begin{aligned}
\Xi(\rho) &\equiv \frac{i}{2} \sum_{j=1}^{n} \left(\hat{\Sigma}_j \hat{U}\rho\hat{U}^\dagger - \hat{U}\rho\hat{U}^\dagger \hat{\Sigma}_j^\dagger\right) \\
&= \frac{i}{2} \sum_{j=1}^{n} \left[\hat{\Sigma}_j, U\rho\hat{U}^\dagger\right]
\end{aligned} \tag{9.47}$$

then we can write $\rho^{(1)}$ at first order in Ω as:

$$\rho^{(1)} \approx \hat{U}\rho^{(0)}\hat{U}^\dagger + \Omega \, \Xi(\rho^{(0)}) \tag{9.48}$$

Then, the second iteration will look like:

$$\begin{aligned}
\rho^{(2)} &\approx \hat{U}\rho^{(1)}\hat{U}^\dagger + \Omega \, \Xi(\rho^{(1)}) \\
&\approx \hat{U}\hat{U}\rho^{(0)}\hat{U}^\dagger\hat{U}^\dagger + \Omega\hat{U}\,\Xi(\rho^{(0)})\hat{U}^\dagger + \Omega\Xi(\hat{U}\rho^{(0)}\hat{U}^\dagger) + \mathcal{O}(\Omega^2) \\
&\approx \hat{U}\hat{U}\rho^{(0)}\hat{U}^\dagger\hat{U}^\dagger + \Omega\left(\hat{U}\,\Xi(\rho^{(0)})\hat{U}^\dagger + \Xi(\hat{U}\rho^{(0)}\hat{U}^\dagger)\right) + \mathcal{O}(\Omega^2)
\end{aligned} \tag{9.49}$$

and after m iterations the state will be of the form:

$$\rho^{(m)} \approx \left(\hat{U}\right)^m \rho^{(0)} \left(\hat{U}^\dagger\right)^m + \Omega \underbrace{(\ldots)}_{m} + \mathcal{O}(\Omega^2) \tag{9.50}$$

where the underbrace denotes how many n-sum terms Ξ are contained inside the parenthesis.

At this point it should be clear that the presence of the gravitational field will affect the probability of measuring the final state of the system. The exact deviation from the expected performance of the quantum algorithm will depend in a non-trivial manner on the type and number of quantum gates, data qubits, ancillary qubits and error correction encoding used to implement the computation. Notice that even though the gravitational effects are encoded in a unitary transformation, to find and implement the exact inverse transformation at each computational step may not be a feasible task (remember that these are rotations over extremely small angles: $\Omega \approx 10^{-20}$).

Therefore, we make the conjecture that the probability of measuring the right result of the computation after m iterations is approximately bounded by:

$$p \gtrsim (1 - n\epsilon)^m \qquad n\epsilon \lesssim 1 \tag{9.51}$$

where:

$$\epsilon \approx \frac{\Omega}{2} \sqrt{1 - 4\Im^2(\alpha^*\beta)} \tag{9.52}$$

and α and β are related to the choice of the quantization basis being used. In general, it will be difficult to determine the *exact* value of α and β as a small gradient in the field will shift their values. However, we know that in the worst case the error is bounded by:

$$\epsilon \lesssim \Omega/2 \tag{9.53}$$

In such a case, the probability of having an error in the computation of the quantum algorithm is bounded by:

$$e \lesssim 1 - (1 - n\epsilon)^m \tag{9.54}$$

The algorithm can be iterated to increase the probability of success to satisfy any desired threshold. For example, after k runs of the entire algorithm, the worst case bound for the probability of error is reduced to:

$$e^k \lesssim \left(1 - (1 - n\epsilon)^m\right)^k \tag{9.55}$$

and k is chosen such that:

$$e^k \approx \delta \tag{9.56}$$

where δ is the maximum error probability desired for the computational process. That is, k is given by:

$$k \lesssim \frac{\log \delta}{\log(1 - (1 - n\epsilon)^m)} \tag{9.57}$$

and in the asymptotic limit for large m we have:

$$k \lesssim -\log\delta \times \left(\frac{1}{1-n\epsilon}\right)^m = \mathcal{O}\left(\left(\frac{1}{1-n\epsilon}\right)^m\right) \tag{9.58}$$

and the *true complexity* of the algorithm (i.e. the total number of iterations necessary to complete the computational task with error probability δ) is:

$$\mathcal{O}(m \times k) \approx \mathcal{O}\left(m \times \left(\frac{1}{1-n\epsilon}\right)^m\right) \tag{9.59}$$

where we recall that in complexity theory, the big $\mathcal{O}$ notation strictly refers to the worst-case complexity in the asymptotic limit. In other words, as the number of qubits n grows, approaching the value of $1/\epsilon$, the asymptotic algorithmic complexity becomes exponential in the number of iterations m.

Therefore, because of the presence of gravity, the complexity of the algorithm changes from m to an exponential term in m. Even though ϵ is a very small number, if m is large enough, then the complexity will be exponentially large in m. That is, under the assumptions and conjectures established, *gravity exponentially degrades the time complexity of iterative quantum algorithms*.

Let us emphasize that this is a formal result that emerges from the conceptual underpinnings of complexity theory. Indeed, complexity theory is the analysis of algorithms in the asymptotic limit, where the number of iterations tends to infinity. As such, the true complexity of iterative quantum algorithms under the influence of a gravitational field is exponential.

At this point, one could be tempted to try to estimate how large m should be so that the complexity degradation is observable. However, this question falls outside the scope of complexity theory, which is concerned with the large asymptotic limits. In addition, complexity results completely ignore multiplicative constants, which may be important for predicting the runtime performance of an algorithm. In any event, these runtime coefficients for quantum computation remain unknown.

In what follows we will consider the effect of gravitation on the two most important quantum algorithms known so far: Grover's algorithm and Shor's algorithm.

9.6.2 Grover's algorithm

Grover's algorithm is an *amplitude amplification algorithm* [6, 7]. These types of algorithm basically consist of a series of qubit operations that transform a quantum state in such a way that, after a certain number of iterations, the measurement of some states is much more likely than others. That is, the amplitude is amplified for those states that convey an answer to some computational problem.

In the case of Grover's algorithm, the problem is to find an item in a completely unsorted and unstructured database. Suppose there are n qubits spanning a super-position of $N = 2^n$ states, and we are looking for some specific state in the super-position. Starting from the uniform superposition, the probability of measuring the right item is $1/N$. However, after applying $\sqrt{N}$ iterations of the *Grover operator*, the measurement of the superposition will result in the desired state with high probability ($\approx 2/3$). Therefore, the complexity of Grover's algorithm is $\mathcal{O}(\sqrt{N})$. This result has to be compared with the best-known classical method, the exhaustive search, with complexity $\mathcal{O}(N)$.

Two-qubit case

It is instructive to show the behaviour of Grover's algorithm in the presence of gravity for a superposition of 2-qubits. This example is particularly enlightening because in the absence of gravity, the solution to the search problem is achieved with probability one after a single iteration of Grover's operator [6]. Consequently, the simulation of the entire algorithm can be performed analytically.

The algorithm starts with the 2-qubit uniform superposition:

$$|\psi\rangle = \frac{|00\rangle + |01\rangle + |10\rangle + |11\rangle}{2} \qquad (9.60)$$

and the *Grover operator* is given by:

$$\hat{G} = \hat{M} \times \hat{O} \qquad (9.61)$$

where $\hat{M}$ is the *inversion around the mean* and $\hat{O}$ is the *oracle*. For the 2-qubit case, the inversion around the mean operator is given by:

$$\hat{M} = -\frac{1}{2}\begin{pmatrix} 1 & -1 & -1 & -1 \\ -1 & 1 & -1 & -1 \\ -1 & -1 & 1 & -1 \\ -1 & -1 & -1 & 1 \end{pmatrix} \qquad (9.62)$$

If the item we are looking for is the '00' element, then the oracle takes the form:

$$\hat{O} = \begin{pmatrix} -1 & 0 & 0 & 0 \\ 0 & 1 & 0 & 0 \\ 0 & 0 & 1 & 0 \\ 0 & 0 & 0 & 1 \end{pmatrix} \qquad (9.63)$$

In the absence of gravity, the state after a single Grover iteration is given by:

$$\hat{G}|\psi\rangle = |00\rangle \qquad (9.64)$$

and therefore the probability of measuring '00' is exactly 1. On the other hand, classical brute-force search for an element in a database of four elements may take up to four operations (four comparisons, one for each element of the dataset).

However, in the presence of gravity, the operator that is applied to the uniform state is approximated by:

$$\hat{G} \times \hat{D}|\psi\rangle \qquad (9.65)$$

where $\hat{D}$ is the 2-qubit Wigner rotation matrix shown in equation (9.27). That is, the quantum register is initialized, and then, some time goes by before the application of the Grover operator $\hat{G}$. Therefore, in the presence of a gravitational field, the probability of measuring the state 'ab' is given by P_{ab}, and takes the following values:

$$P_{00} = \cos^4\left(\frac{\Omega}{2}\right)$$

$$P_{01} = \frac{1}{4}\sin^2\Omega$$

$$P_{01} = \frac{1}{4}\sin^2\Omega \tag{9.66}$$

$$P_{01} = \sin^4\left(\frac{\Omega}{2}\right)$$

which are clearly normalized to 1.

The behaviour of the probabilities of measuring '00', '01' and '10', and '11' with respect to the Wigner rotation angle Ω is shown in figure 9.7. In this case, Ω represents the Wigner angle obtained in the time between two consecutive operations. That is, if the quantum computer operates at 1 Hz, then the Wigner angle is the rotation due to the drifting of the qubit in a gravitational field during the time span of 1 second. Similarly, in figure 9.8, we show the probabilities of measuring '00', '01' and '10', and '11' for a quantum computer that has completed a circular orbit in Schwarzschild spacetime.

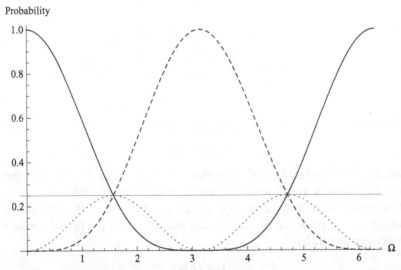

Figure 9.7. Probability of measuring '00' (solid line), '01' and '10' (dotted line) and '11' (dashed line) for a 2-qubit state after a single iteration of the Grover operator in the presence of a gravitational field described by the Wigner rotation angle $\Omega(t)$.

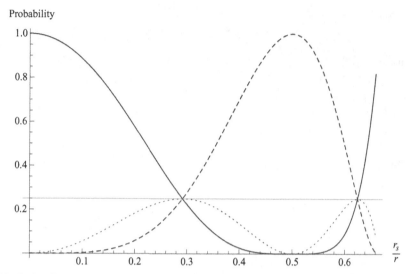

Figure 9.8. Probability of measuring '00' (solid line), '01' and '10' (dotted line) and '11' (dashed line) for a 2-qubit state after a single iteration of the Grover operator for a quantum computer in a circular orbit of radius r in the Schwarzschild spacetime produced by a body of mass $r_s/2$.

It can be observed that, in the presence of a gravitational field, the probability of success of Grover's algorithm will decrease. If the quantum computer is near a black hole with a strong gravitational field such that $\Omega \approx \pi/2$ during a clock tick of the quantum computer, then the probability of success is no better than the classical search. Furthermore, if $\Omega \approx \pi$, then the probability of success is nearly zero.

If we apply a second Grover iteration to the state, we obtain the probabilities shown in figure 9.9. This makes evident that further applications of the Grover iteration will not improve the end result. Similarly, in figure 9.10, we show the probabilities of measuring '00', '01' and '10', and '11' for such a quantum computer in a circular orbit in Schwarzschild spacetime. Therefore, we need to run the entire algorithm several more times to get the correct answer with high probability.

N-qubit case
As we have seen, gravity affects the performance of Grover's algorithm. Furthermore, following the analysis presented in the previous section we can see that, because of gravity, the true complexity of Grover's quantum algorithm searching a database of N elements is not given by $\mathcal{O}(\sqrt{N})$, but by:

$$\mathcal{O}\left(\sqrt{N} \times \left(\frac{1}{1-n\epsilon} \right)^{\sqrt{N}} \right) \tag{9.67}$$

which is exponential in N. As usual, the number of qubits is given by $n = \log N$.

As we mentioned before, complexity theory is the asymptotic analysis of computational processes and ignores constant multiplicative factors. Consequently,

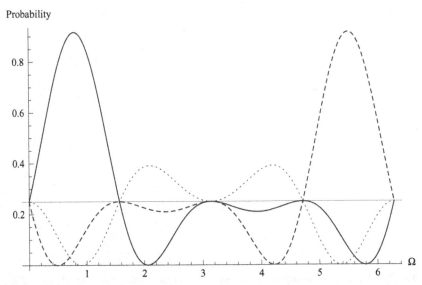

Figure 9.9. Probability of measuring '00' (solid line), '01' and '10' (dotted line) and '11' (dashed line) for a 2-qubit state after two iterations of the Grover operator in the presence of a gravitational field described by the Wigner rotation angle $\Omega(t)$.

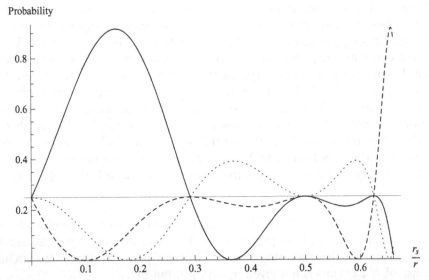

Figure 9.10. Probability of measuring '00' (solid line), '01' and '10' (dotted line) and '11' (dashed line) for a 2-qubit state after two iterations of the Grover operator for a quantum computer in a circular orbit of radius r in the Schwarzschild spacetime produced by a body of mass $r_s/2$.

from a formal point of view, we cannot make any reliable estimates about how large N should be so that the complexity degradation is observable. Nevertheless, we can try to give a conceptual meaning to this complexity result by assuming that *all* runtime constants are equal to 1 and considering the case of finite N.

Taking such a deviation from complexity theory, let us also assume that $n\epsilon$ is small enough, so that we can use a Taylor series to estimate the value of k:

$$k \approx \left(\frac{1}{1-n\epsilon}\right)^{\sqrt{N}} \approx 1 + n\epsilon\sqrt{N} \tag{9.68}$$

where we have assumed that the gravitational error is very small:

$$n\epsilon \ll 1 \tag{9.69}$$

Then, the number of iterations to complete Grover's algorithm is given by:

$$\mathcal{O}\left(\sqrt{N} + n\epsilon N\right) = \mathcal{O}(nN) = \mathcal{O}(N\log N) \tag{9.70}$$

We can see that the term $n\epsilon N$ becomes important when:

$$n\epsilon N \approx \sqrt{N} \tag{9.71}$$

which implies:

$$\sqrt{N}\log N \approx \frac{1}{\epsilon} \approx 10^{20} \tag{9.72}$$

The behaviour of the algorithmic complexity for Grover's algorithm is illustrated in figure 9.19. The solid line shows the scaling of Grover's algorithm in the presence of a gravitational field (scaling as $\mathcal{O}(\sqrt{N} + \epsilon N \log N)$), whereas the dashed line corresponds to the scaling of Grover's algorithm without gravity (scaling as $\mathcal{O}(\sqrt{N})$). It can be observed that for large values of the number of elements N, the number of iterations does not scale as $\sqrt{N}$, but increases with $N \log N$.

The vertical grid line in figure 9.11 corresponds to $N \approx 10^{36}$, whereas the horizontal grid line corresponds to 10^{18} computational steps. It is important to remark,

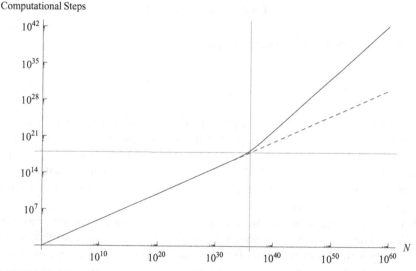

Figure 9.11. Number of computational steps in function of the number of elements N necessary to complete Grover's algorithm with (solid line) and without (dashed line) gravity.

once more, that this plot is conceptual, as the running time constants have been assumed to be equal to one. Consequently, in the worst case and under all the assumptions and conjectures previously established (a 1 GHz quantum computer in Earth's gravitational field), the maximum size of the database that can be searched is about 10^{36} elements (about 120 computational qubits). For a higher number of elements, the complexity of Grover's algorithm is polynomial in the number of elements and exponential in the number of qubits.

9.6.3 Shor's algorithm

Let us remember that Shor's algorithm can factorize with probability close to one an n-bits co-prime number in $\mathcal{O}(\log^3 N)$ computational steps. For the case of Shor's algorithm, the true complexity due to the presence of gravity is given by:

$$\mathcal{O}\left(\log^3 N \times \left(\frac{1}{1 - \epsilon \log N} \right)^{\log^3 N} \right) \tag{9.73}$$

or, equivalently, in terms of number of qubits:

$$\mathcal{O}\left(n^3 \times \left(\frac{1}{1 - n\epsilon} \right)^{n^3} \right) \tag{9.74}$$

which is exponential in n. Similarly, if $n\epsilon$ is very small, the number of iterations necessary to complete Shor's algorithm, when all runtime constants are of order 1, is given by:

$$\mathcal{O}(n^3 + \epsilon n^7) = \mathcal{O}(n^7) \tag{9.75}$$

The breakout of complexity occurs when:

$$n \to \frac{1}{\epsilon} \ \Rightarrow \ n \to 10^{20} \tag{9.76}$$

That is, in the worst case, the complexity of Shor's algorithm will become exponential for more than 10^{20} qubits. The behaviour of the number of iterations with respect to the number of qubits n is shown in figure 9.12. Therefore, for standard cryptoanalytical applications of Shor's algorithm, gravity should not pose a problem. Once again, this plot is conceptual, as the running time constants have been assumed to be equal to one.

Clearly, Shor's algorithm is more resilient to gravitational effects than Grover's algorithm. The reason is that Shor's algorithm has a linear complexity whereas Grover's algorithm is exponential (in terms of the number of qubits n). Therefore, Grover's algorithm requires more time to compute the solution, and the gravitational effects will accumulate during this period.

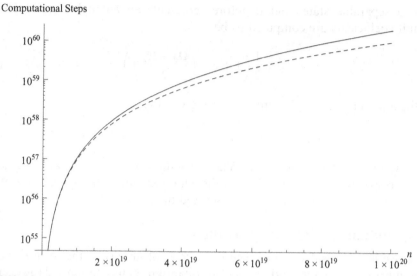

Computational Steps

Figure 9.12. Number of computational steps in function of the number of qubits n necessary to complete Shor's algorithm with (solid line) and without (dashed line) gravity.

9.7 Gravity-induced entanglement

If we are not limited to qubits represented as momentum eigenstates, then the effect of gravity is a non-local unitary transformation that may induce entanglement. This is similar to the case discussed in section 4.4 for entanglement induced by Lorentz transformations between inertial frames.

As before, let us assume that we have a qubit in a superposition of states such as:

$$|\psi\rangle = \frac{1}{\sqrt{2}}(|p\rangle + |q\rangle) \otimes |+\rangle \tag{9.77}$$

which is a separable state and therefore not entangled. However, if the components of the state are affected in a different manner by the gravitational field, then the state transforms into:

$$
\begin{aligned}
|\psi\rangle \to |\tilde{\psi}\rangle = \hat{U}_g|\psi\rangle \\
= \frac{1}{\sqrt{2}}\left(\cos\left(\frac{\Omega_p}{2}\right)|\Lambda(x)p\rangle + \cos\left(\frac{\Omega_q}{2}\right)|\Lambda(x)q\rangle\right) \otimes |+\rangle \\
+ \frac{1}{\sqrt{2}}\left(\sin\left(\frac{\Omega_p}{2}\right)|\Lambda(x)p\rangle + \sin\left(\frac{\Omega_q}{2}\right)|\Lambda(x)q\rangle\right) \otimes |-\rangle
\end{aligned}
\tag{9.78}
$$

where $\Lambda(x)$ is a local Lorentz transformation and the momentum variables are defined in the local inertial frames determined by a given tetrad field. In general, this is not a separable state, and therefore represents an entangled state. The two Schmidt coefficients are computed to be:

$$\lambda_{\pm} = \frac{1}{2}\left(1 \pm \cos\left(\frac{\Omega_q - \Omega_p}{2}\right)\right) \tag{9.79}$$

and the amount of entanglement is simply given by:

$$E(|\tilde{\psi}\rangle) = -|\lambda_+|^2\log(|\lambda_+|^2) - |\lambda_-|^2\log(|\lambda_-|^2) \tag{9.80}$$

where Ω_p and Ω_q correspond to the Wigner rotations for the states with momentum p and q, respectively. The behaviour of the entanglement with respect to the Wigner rotations is exactly the same as that discussed in section 4.4.

9.8 Quantum sensing and gravimetry

So far, we have considered the 'negative' effects of gravity in the form of noise in communication systems and errors in quantum computations. However, the coupling between quantum information and classical gravitational fields can be exploited to design novel quantum gravitometers: quantum sensors that measure minute deviations of the Earth's gravitational field.

Two different implementations are possible: interferometric quantum gravimetry and algorithmic quantum gravimetry. As both of these techniques rely on the Wigner angle, one could refer to them as instances of *Wigner gravimetry*. In this section we will briefly discuss these novel approaches to gravimetry.

Before that, it is important to notice that these approaches are radically different from current techniques. Indeed, the most precise measurements of a gravitational field currently use atom interferometers that do not rely on Wigner rotations. In reality, the use of atom interferometry to measure the gravitational field is not very different from what Galileo did back in 1589. According to the history records, Galileo threw heavy objects from the Tower of Pisa and established the gravitational acceleration g through the classical formula:

$$d = \frac{gT^2}{2} \tag{9.81}$$

where d is the distance travelled by the falling objects in time T [8].

In atom interferometry-based gravimetry, however, the objects being dropped are heavy atoms [9–14]. The gravitational acceleration g is therefore calculated through the expression:

$$\Delta\phi\,\frac{\lambda}{2} \approx \frac{gT^2}{2} \tag{9.82}$$

where λ is the de Broglie wavelength:

$$\lambda = \frac{2\pi}{mv} \tag{9.83}$$

in natural units $\hbar=1$, where m and v are the mass and velocity of the atom, respectively. In the above equation, $\Delta\phi$ is the phase shift that is measured using an interferometric experimental set-up. It is important to mention that this phase is completely unrelated to the Wigner rotation. Indeed, $\Delta\phi$ is related to the laser-induced oscillations on the state of the atom with respect to two energy levels[3]. The precision of this technique appears to be limited to about:

$$\frac{\delta g}{g} \approx 10^{-9} \tag{9.84}$$

because a higher precision requires smaller λ, and therefore one needs heavier atoms, which are difficult to manage in this type of interferometric set-up.

9.8.1 Interferometric quantum gravimetry

Interferometric quantum gravitometers use quantum interferometers to determine the relative phase variation due to the Wigner rotation produced by the gravitational field. The detection performance can be amplified by exploiting the entanglement found in NOON states [15–17]. Indeed, let us consider a Mach–Zender interferometer like that shown in figure 9.13. If the following NOON state is sent through the input ports:

$$|\Psi\rangle_{\text{in}} = \frac{1}{\sqrt{2}}(|N\rangle_1|0\rangle_2 + |0\rangle_1|N\rangle_2) \tag{9.85}$$

then we expect that the state received in the output ports will be:

$$|\Psi\rangle_{\text{out}} = \frac{1}{\sqrt{2}}(|N\rangle_1|0\rangle_2 + e^{iN\phi}|0\rangle_1|N\rangle_2) \tag{9.86}$$

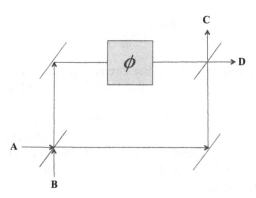

Figure 9.13. Mach–Zender interferometer with input ports A and B, output ports C and D, and a phase delay φ in one of the arms.

---

[3] A good discussion of the gravitational effects on the spectrum of the hydrogen atom can be found in the literature [18].

where the indices 1 and 2 refer to the lower and upper arms of the interferometer, respectively[4].

We can write the NOON state in terms of the creation and annihilation operators as:

$$|\Psi\rangle_{out} = \frac{1}{\sqrt{2}} \left(\frac{(a_1^\dagger)^N}{\sqrt{N!}} |0\rangle_1 |0\rangle_2 + e^{iN\phi} \frac{(a_2^\dagger)^N}{\sqrt{N!}} |0\rangle_1 |0\rangle_2 \right)$$

$$= \frac{1}{\sqrt{2N!}} \left((a_1^\dagger)^N + e^{iN\phi}(a_2^\dagger)^N \right) |0\rangle_1 |0\rangle_2 \tag{9.87}$$

To measure the phase shift ϕ, the detector has to implement a measurement of the following observable:

$$\hat{A}_D = |N0\rangle\langle 0N| + |0N\rangle\langle N0|$$

$$= \frac{1}{N!} \left((a_1^\dagger)^N |0\rangle\langle 0|(a_2^\dagger)^N + (a_2^\dagger)^N |0\rangle\langle 0|(a_1^\dagger)^N \right) \tag{9.88}$$

With this set-up, the amount of 'noise' is given by:

$$\Delta^2 A_D = \left(\langle \hat{A}_D^2 \rangle - \langle \hat{A}_D \rangle^2 \right)$$

$$= \sin^2 N\phi \tag{9.89}$$

whereas the phase responsivity is:

$$\frac{d\langle \hat{A}_D \rangle}{d\phi} = -N \sin N\phi \tag{9.90}$$

Then, the phase estimation error is approximately given by:

$$\delta\phi \approx \frac{\Delta A_D}{\left| \frac{d\langle \hat{A}_D \rangle}{d\phi} \right|} = \frac{1}{N} \tag{9.91}$$

That is, this interferometric phase measurement procedure using highly entangled states can reach the Heisenberg limit and beat the standard quantum limit (with a $1/\sqrt{N}$ scaling) [15, 17].

Therefore, if we need a sensitivity of $\delta\phi \approx 10^{-27}$, we would need a state made of $N \approx 10^{27}$ qubits entangled with the quantum vacuum. On the other hand, if we use standard quantum interferometry techniques without entanglement, we would require $N \approx 10^{54}$ qubits injected in the interferometer. Needless to say, the feasibility of producing these highly entangled states is a huge scientific and technical challenge that remains to be solved.

---

[4] It is important to note that, as discussed in chapter 5, the quantum vacuum states at both ends of the interferometers, measured at different points of time, may actually be *different*. In such a case, we need to account for the Bogolubov transformations that relate both quantum vacuum states.

9.8.2 Algorithmic quantum gravimetry

A completely different approach to quantum gravimetry is the use of a quantum computer running a modified version of Grover's algorithm. This technique exploits the fact that the power of Grover's algorithm resides in the way it performs amplitude amplification. That is, the oracle marks the correct answer and the algorithm amplifies the state until it can be measured with an adequate probability.

As we have seen several times, the coupling between gravity and spin depends on the orientation of the spin. Therefore, the oracle can be designed in such a way that marks the state that will receive maximum rotation due to the presence of gravity. Then, the application of a certain number of Grover iterations will amplify the state we are looking for. In a sense, we are exploiting the gravitational error introduced in a quantum computation (described in section 9.5) to measure the strength of the gravitational field. Alternatively, we could use the gravitational field as the oracle in Grover's algorithm.

Two-qubit gravitometer

Let us consider a 2-qubit system that is initialized to be in the uniform superposition:

$$|\psi\rangle = \frac{|00\rangle + |01\rangle + |10\rangle + |11\rangle}{2} \tag{9.92}$$

After a small time interval δt, the state will be given by equation (9.26), where in this case Ω represents the induced Wigner angle over that time interval. The probability distributions after δt are shown in figure 9.14.

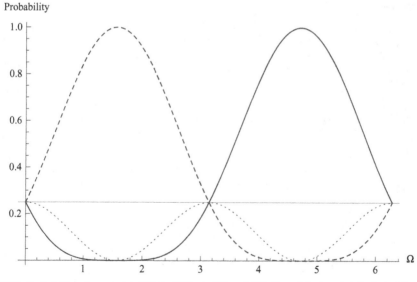

Figure 9.14. Probability of measuring '00' (solid line), '01' and '10' (dotted line) and '11' (dashed line) for a 2-qubit state for a Wigner rotation angle Ω after a short time interval δt.

Clearly, this probability distribution can be used to detect and measure gravitational fields. Indeed, in the absence of gravity the probability distribution is uniform and all the states have equal probability of being measured. So, for example, we could prepare m identical quantum states in the uniform superposition. Then, if the system is immersed in a strong gravitational field that produces $\Omega \approx \pi/2$, measurement of the state '11' will occur with probability very close to 1.

Needless to say, this 2-qubit gravitometer will only be able to detect strong gravitational fields. However, as we will discuss later on, a larger number of qubits can be used to detect weaker gravitational fields. However, for the time being, we can continue exploring the use of this simple gravitometer to detect strong gravitational fields.

The sensitivity $\mathcal{S}_{ij}$ of this 2-qubit system is proportional to the slope of the curves in figure 9.14. That is:

$$\mathcal{S}_{ij} \propto \left| \frac{\partial P_{ij}}{\partial \Omega} \right| \tag{9.93}$$

where P_{ij} is the probability of measuring the state 'ij'. Indeed, if the derivative is large, the slope is large, which means that a small variation in Ω will lead to a large change in P_{ij}. The sensitivity of this system is shown in figure 9.15.

It is possible to increase the sensitivity of this 2-qubit gravitometer by using Grover's algorithm. In this case, we will *use the gravitational field as the oracle of the algorithm*. Recalling that Grover's algorithm is an amplitude amplification algorithm, we expect that the small variations marked by the oracle (e.g. the gravitational field) will be amplified after the application of the Grover operator.

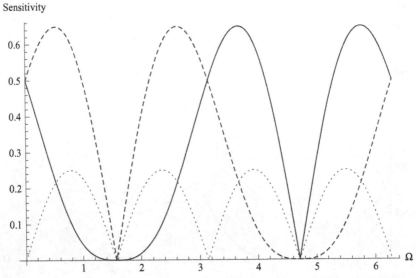

Figure 9.15. Sensitivity of the 2-qubit gravitometer for '00' (solid line), '01' and '10' (dotted line) and '11' (dashed line) for a Wigner rotation angle Ω after a short time interval δt.

Figure 9.16 shows the probability distribution after an iteration of Grover's algorithm using the gravitational field as an oracle. The sensitivity of this new state is shown in figure 9.17. Figure 9.18 shows the superposition of figure 9.16 and 9.17: we can clearly see that even though Grover's algorithm did not amplify the sensitivity of

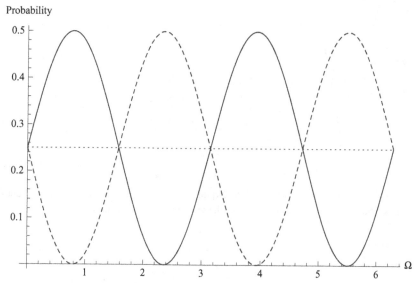

Figure 9.16. Probability of measuring '00' (solid line), '01' and '10' (dotted line), and '11' (dashed line) for a 2-qubit state after one iteration of the Grover operator using the gravitational field as an oracle.

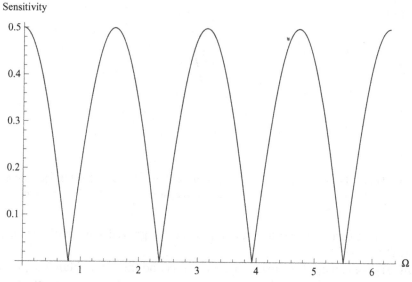

Figure 9.17. Sensitivity of the 2-qubit gravitometer after one iteration of the Grover operator using the gravitational field as an oracle.

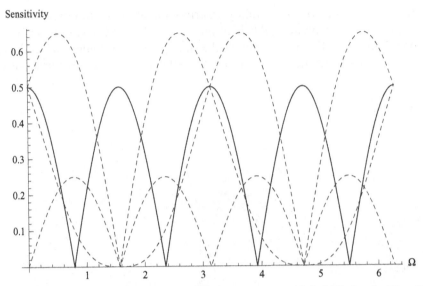

Figure 9.18. Comparison of the sensitivity of the 2-qubit gravitometer with (solid line) and without (dashed lines) an application of the Grover operator using the gravitational field as an oracle.

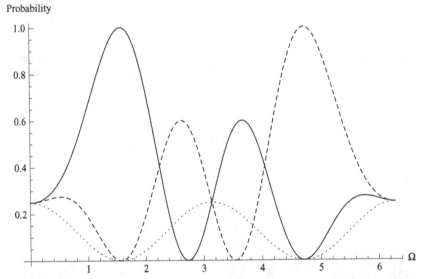

Figure 9.19. Probability of measuring '00' (solid line), '01' and '10' (dotted line) and '11' (dashed line) for a 2-qubit state after two iterations of the Grover operator using the gravitational field as an oracle.

the detector, it certainly created new regions of improved sensitivity (for instance, around $\Omega \approx \pi/2$).

Similarly, we can apply Grover's operator a second time and obtain the probability distributions and sensitivities that are shown in figure 9.19 and 9.20, respectively. The comparison between the sensitivities for 0 and 2 iterations of Grover's algorithm is

Sensitivity

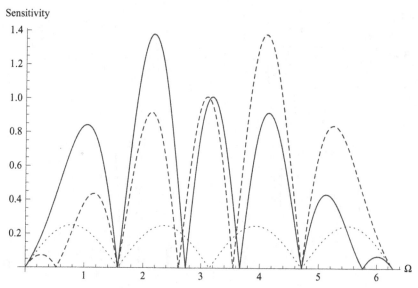

Figure 9.20. Sensitivity of the 2-qubit gravitometer using the states '00' (solid line), '01' and '10' (dotted line) and '11' (dashed line) after two iterations of the Grover operator using the gravitational field as an oracle.

Sensitivity

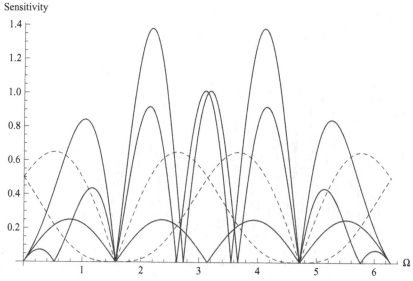

Figure 9.21. Comparison of the sensitivity of the 2-qubit gravitometer with two (solid line) and without (dashed lines) iterations of the Grover operator using the gravitational field as an oracle.

shown in figure 9.21. We can see that, after two iterations of Grover's algorithm, the sensitivity was amplified almost everywhere. Finally, figure 9.22 and 9.23 show the performance of the Grover gradiometer for 9 and 10 iterations, respectively. It can be observed that there are many regions in which the sensitivity of the device increases with the number of iterations.

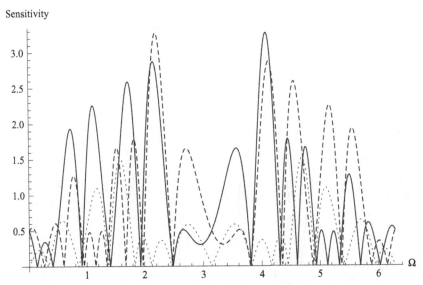

Figure 9.22. Sensitivity of the 2-qubit gravitometer using the states '00' (solid line), '01' and '10' (dotted line) and '11' (dashed line) after nine iterations of the Grover operator using the gravitational field as an oracle.

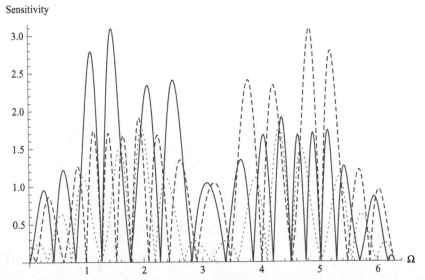

Figure 9.23. Sensitivity of the 2-qubit gravitometer using the states '00' (solid line), '01' and '10' (dotted line) and '11' (dashed line) after 10 iterations of the Grover operator using the gravitational field as an oracle.

N-qubit gravitometer

As we have just shown, Grover's algorithm using the gravitational field as an oracle leads to an amplification of the sensitivity of this simple 2-qubit gravitometer. This is an amazing result, but only if we are interested in measuring strong gravitational fields. However, it appears to be possible to design quantum gravitometers based on Grover's algorithm to measure weak gravitational fields. Of course, to do so we need to increase the number of qubits in the system and/or the number of iterations.

For example, let us assume an n-qubit quantum computer with $N = 2^n$ possible states running at 1 GHz. The system starts in the uniform superposition:

$$|\psi\rangle = \frac{1}{\sqrt{N}} \sum_{i=0}^{N-1} |i\rangle \qquad (9.94)$$

Following the previous analysis for the 2-qubit case, we can expect that, after one iteration, the probability distribution of a quantum computer with $\log N$ qubits in a gravitational field will have N peaks across the Ω axis. We are interested in measuring small angles $\delta\Omega$. Therefore we require the number of peaks N to be at least of the same order as $1/\delta\Omega$:

$$N \gtrsim \frac{1}{\delta\Omega} \qquad (9.95)$$

In addition, we need to apply Grover's algorithm at least $\sqrt{N}$ times to amplify the probabilities.

To summarize, given a quantum computer with $n = \log N$ qubits running Grover's algorithm for $\mathcal{O}(\sqrt{N})$ iterations using the gravitational field as an oracle, in principle it should be possible to measure deviations to the Wigner angle of order $\delta\Omega \approx \mathcal{O}(N^{-1})$.

Sensor performance
Let us assume that we require a highly sensitive gravitometer to be able to measure a deviation of the gravitational constant g of about:

$$\frac{\delta g}{g} \approx 10^{-8} \qquad (9.96)$$

where:

$$g = \frac{GM}{r^2} \qquad (9.97)$$

This means that we would like to measure a variation of the ratio between the mass of an object m and the square distance to the detector d^2:

$$\frac{m}{d^2} \approx 10^3 \frac{\text{kg}}{\text{m}^2} \qquad (9.98)$$

In other words, this is equivalent to being able to measure the gravitational field of an average person (≈ 100 kg) seated 30 cm away from the detector. Alternatively, a surface ship could detect the gravitational field of a 10 000 ton underwater vehicle travelling at a depth of 100 metres[5].

The required task could be accomplished with a 1 MHz quantum computer with:

$$n = \log N \approx \log(10^{27}) \approx 90 \qquad (9.99)$$

qubits running:

$$\sqrt{N} \approx 10^{13} \qquad (9.100)$$

iterations of Grover's algorithm under a gravitational field such that:

$$\delta\Omega \approx 10^{-27} \qquad (9.101)$$

for a total estimated computing time of about 100 days.

Needless to say, larger number of qubits and a faster quantum computer will resolve the gravitational field at a faster rate. Indeed, it can be shown that a quantum computer that operates at T Hz will solve the problem in time τ (in seconds) given by:

$$\tau \approx \frac{10^{10}}{\sqrt{T}} \qquad (9.102)$$

Therefore, a 1 EHz quantum computer with 90 qubits will resolve it in about 10 seconds. In addition, there are potential tradeoffs between the sensitivity, the number of iterations and the number of qubits that remain to be explored.

9.9 Summary

In this chapter we discussed a few instances of quantum information processing in the presence of a gravitational field. Most of the analysis offered in this chapter is similar to that presented in chapter 4 for relativistic quantum information in inertial frames. The main change is, of course, the origin of the Wigner rotation.

First, we analysed the performance of a simple steganographic quantum channel. We determined that the effect of the Wigner rotation due to the presence of the gravitational field will cause the protocol to have zero capacity when the satellite has completed 10^9 orbits. Then, we discussed how quantum teleportation and EPR experiments are affected by curved spacetime. Most interesting is the case of quantum computations, as we showed that, if the error induced by the gravitational field is not corrected, the complexity of most quantum algorithms will be affected in a detrimental manner. In particular, we showed how Grover's algorithm running on a 1 MHz quantum computer is bounded by a database of about 10^{36} elements before the complexity grows from quadratic to linear (assuming that all the running time

---

[5] Even more important, although underwater vehicles can be designed to reduce to a bare minimum their acoustic and magnetic signatures, their gravitational mass *cannot* be shielded. This is true even if the underwater vehicle has neutral buoyancy, in which the average density of the ship is equal to the density of the water. Indeed, because of hydrodynamic considerations, underwater vehicles need to be bottom heavy in order to be stable. In such a case, even though the gravitational mass of the underwater vehicle is masked by virtue of Archimedes' principle, there is a non-zero gravitational field dipole produced by the bottom-heavy underwater vehicle that at least in theory could be detected by a highly sensitive gravimeter.

constants that characterize the quantum computer are equal to one). We also discussed how gravitational effects on states that are not momentum eigenvectors will generate entanglement on the momentum and spin degrees of freedom. Finally, we discussed two approaches to quantum gravimetry.

We need to stress the fact that all the quantum effects produced by the gravitational field are represented by a unitary operation that describes a Wigner rotation. As such, this operation can be inverted if we know the original transformation. However, in general, this may not be a realistic situation.

Bibliography

[1] Bruschi D E *et al* 2013 Testing the effects of gravity and motion on quantum entanglement in space-based experiments arXiv:1306.1933 [quant-ph]

[2] Bruschi D E, Fuentes I and Louko J 2012 Voyage to alpha centauri: entanglement degradation of cavity modes due to motion *Phys. Rev.* D **85** 061701

[3] Terashima H and Ueda M 2004 Einstein–Rosen correlation in gravitational field *Phys. Rev.* A **69** 032113

[4] Lanzagorta M and Uhlmann J 2013 Quantum computation in gravitational fields forthcoming, available upon request

[5] Lanzagorta M and Uhlmann J 2012 Error scaling in fault tolerant quantum computation *Appl. Math. Comput.* **219** 24–30

[6] Lanzagorta M and Uhlmann J 2008 *Quantum Computer Science* (San Rafael, CA: Morgan & Claypool)

[7] Nielsen M A and Chuang I L 2000 *Quantum Computation and Quantum Information* (Cambridge: Cambridge University Press)

[8] Dugas R 2011 *A History of Mechanics* (New York: Dover)

[9] von Borzerszkowski H and Mensky M B 2001 Gravitational effects on entangled states and interferometer with entangled atoms *Phys. Lett.* A **286** 102–6

[10] Kasevich M and Chu S 1991 Atomic interferometry using stimulated Raman transitions *Phys. Rev. Lett.* **67** 181

[11] McGuirk J M *et al* 2002 Sensitive absolute-gravity gradiometry using atom interferometry *Phys. Rev.* A **65** 033608

[12] Petelski T 2005 Atom Interferometers for Precision Gravity Measurements *PhD Thesis* Universita Degli Studi Di Firenze

[13] Snadden M J *et al* 1998 Measurement of the Earth's gravity gradient with an atom interferometer-based gravity gradiometer *Phys. Rev. Lett.* **81** 971

[14] Yurtsever U, Strekalov D and Dowling J P 2003 Interferometry with entangled atoms *Eur. Phys. J.* D **22** 365–71

[15] Parker L and Pimentel L O 1982 Gravitational perturbation of the hydrogen spectrum *Phys. Rev.* D **25** 3180

[16] Didomenico L D, Lee H, Kok P and Dowling J P 2004 Quantum interferometric sensors *Proceedings of SPIE Quantum Sensing and Nanophotonic Devices*

[17] Dowling J P 2008 Quantum optical metrology - the lowdown on high-NOON states, *Contemp. Phys.* **49** 125–43

[18] Lee H, Kok P and Dowling J P 2002 A quantum rosetta stone for interferometry *J. Mod. Opts.* **49** 2325

Chapter 10

Conclusions

In this book we presented a brief discussion of quantum information in classical gravitational fields described by Einstein's general theory of relativity. Our emphasis was to study the effect of the gravitational field on the spin state of qubits. In particular, we analysed in detail the structure of the Wigner rotation for Dirac spinors in Schwarzschild and Kerr spacetimes. In addition we explored the effect of classical gravitational fields on quantum information processing systems. Indeed, we studied the performance of a simple steganographic quantum communications protocol, quantum teleportation, Einstein–Podolsky–Rosen (EPR) experiments, quantum sensors and quantum computation in the presence of gravity. However, perhaps more important, these results suggest an intimate connection between quantum information and gravitational fields.

10.1 Classical information in classical gravitational fields

Other than the trivial fact that if we drop a laptop it will crash into the ground, classical information does not appear to be directly coupled to gravity. That is, rather than the information itself, it is the technological implementation of the classical information processing machine that may actually be coupled to gravity. For example, the internal clocks of two processors in a large supercomputer may be affected by slightly different gravitational fields, which will produce tiny variations as time goes by. In this case, the gravitational field could lead to a synchronization problem between the processors. However, the information itself does not have a direct coupling to gravity.

Indeed, if a noiseless classical computer register is found in a certain binary state, we expect that it will remain in exactly this same state as the time goes by. In a sense, the concept of classical information is abstract enough to be decoupled from physical interactions. In fact, it is very easy to think about classical information without making reference to a specific physical implementation.

Furthermore, a bit may be represented by a voltage difference in a circuit, and the electric fields that make the voltage difference may interact with the environment through electromagnetic and gravitational fields. Then again, the bit as an abstract concept related to a voltage difference will remain unaffected by such interactions. Similarly, classical information may be encoded through modulation on a carrier wave. Once again, all physical interactions with the environment are exclusively mediated through the carrier wave. As such, within the context of classical information theory, the question of how a bit gravitates is completely meaningless.

10.2 Quantum information in classical gravitational fields

On the other hand, the qubit is intrinsically related to the physical properties of electrons, photons and other elementary particles. In this case, the properties of quantum information are a direct consequence of the physical laws that govern the microscopic world. Consequently, *quantum information has a direct coupling to gravity* and other interaction fields.

For instance, the spin of a particle is a quintessential element in quantum information science. Also, the concept of spin only makes sense in a relativistic context. Indeed, from a formal perspective, spin are those extra degrees of freedom of a quantum state that transform in a non-trivial manner under a Lorentz transformation. As such, relativity and gravity are intimately connected to quantum information.

In this book we discussed the explicit coupling between spin-based quantum information and classical gravitational fields described by Einstein's general theory of relativity. The effect of gravitation on quantum information is described by the laws of physics that govern the behaviour of scalar, vector, tensor and spinor fields in curved spacetimes. This is completely independent of the physical implementation necessary to build a quantum computer or a quantum communications system.

Furthermore, in strong contrast to classical registers, if a quantum register is in some quantum state in the presence of a gravitational field, then this state will most likely vary as time goes by, even if the quantum computer is in the idle state. Therefore, a substantial difference between classical and quantum information is the fact that spin-based quantum information has a direct coupling to classical gravitational fields.

10.3 Quantum information in quantum gravitational fields

So far our discussion has been limited to the coupling between quantum information and classical gravitational fields. Even though the results presented in this book can be directly applied to realistic situations of quantum information-processing devices in gravitational fields, our analysis is incomplete. Indeed, a formal theoretical analysis should include the coupling between quantum information and *quantum gravitational fields* [1]. Furthermore, the coupling between matter fields, photons and gravitons could lead to new informatic phenomena [2–4]: for example, the analysis of the propagation of quantum information through gravitons (i.e. a *gravitonic channel*) [1].

Unfortunately, so far, there is not a single consistent and coherent theory of quantum gravity [5–14]. Then again, it may be possible to use the theory of quantum information in classical gravitational fields to shed some light on the structure of a yet-to-be-discovered theory of quantum gravity.

10.4 Summary

Quantum information is intrinsically relativistic and has a *direct coupling* to classical gravitational fields described by Einstein's general theory of relativity. In contrast to the much more abstract concept of a classical bit of information, *quantum information invariably gravitates.*

Bibliography

[1] Lanzagorta M 2013 The gravitonic channel forthcoming, available upon request
[2] Kempf A 2012 Quantum gravity on a quantum computer? Talk presented at the *International Workshop on Horizons of Quantum Physics*, Taiwan
[3] Terno D R 2006 Quantum information in loop quantum gravity *J. Phys.: Conf. Ser.* **33** 469–74
[4] Zhang B, Cai Q and Zhan M 2013 Transfer of gravitational information through a quantum channel *Eur. Phys. J. D* **67** 184
[5] Booß-Bavnbek B, Esposito G and Lesch M (ed) 2010 *New Paths Towards Quantum Gravity*
[6] Ibanez L E and Uranga A M 2012 *String Theory and Particle Physics: An Introduction to String Phenomenology* (Cambridge: Cambridge University Press)
[7] Johnson C V 2006 *D-Branes* (Cambridge: Cambridge University Press)
[8] Kaku M 1998 *Introduction to Superstrings and M-Theory* (Berlin: Springer)
[9] Kiefer C 2012 *Quantum Gravity* 3rd edn (Oxford: Oxford University Press)
[10] Murugan J, Weltman A and Ellis G F R (ed) 2012 *Foundations of Space and Time: Reflections on Quantum Gravity* (Cambridge: Cambridge University Press)
[11] Oriti D (ed) 2009 *Approaches to Quantum Gravity: Toward a New Understanding of Space, Time and Matter* (Cambridge: Cambridge University Press)
[12] Polchinski J 1998 *String Theory* 2 (Cambridge: Cambridge University Press)
[13] Rovelli C 2007 *Quantum Gravity* (Cambridge: Cambridge University Press)
[14] Thiemann T 2008 *Modern Canonical Quantum General Relativity* (Cambridge: Cambridge University Press)